The American Synthetic

Organic Chemicals Industry

The American Synthetic Organic Chemicals Industry

War and Politics, 1910–1930

###

KATHRYN STEEN

The University of North Carolina Press / Chapel Hill

This book was published with the assistance of funds from Drexel University and the Authors Fund of the University of North Carolina Press.

© 2014 The University of North Carolina Press
All rights reserved. Set in Utopia by codeMantra, Inc.
Manufactured in the United States of America
The paper in this book meets the guidelines for permanence and durability of the Committee on Production Guidelines for Book Longevity of the Council on Library Resources. The University of North Carolina Press has been a member of the Green Press Initiative since 2003.

Cover illustration: U.S. Chemical Warfare Association cartoon, "David and Goliath," c. 1925, Charles H. Herty Papers, box 98, folder 21; Manuscript, Archives, and Rare Book Library, Emory University. The original is black and white; Peter Groesbeck (Drexel University) created the color graphic.

Library of Congress Cataloging-in-Publication Data
Steen, Kathryn.
The American synthetic organic chemicals industry : war and politics, 1910–1930 / Kathryn Steen.
    pages cm
Includes bibliographical references and index.
ISBN 978-1-4696-1290-4 (paperback) — ISBN 978-1-4696-1291-1 (ebook)
1. Chemical industry—United States—History—20th century. 2. Industrial policy—United States—History—20th century. 3. World War, 1914–1918—Social aspects—United States. I. Title.
HD9651.5.S73 2014
338.4'766180097309041—dc23
2014003460

Portions of this book are drawn from previously published articles:

"Technical Expertise and U.S. Mobilization, 1917–1918: High Explosives and War Gases." In *Frontline and Factory: Comparative Perspectives on the Chemical Industry at War, 1914–1924*. Edited by Roy M. MacLeod and Jeffrey A. Johnson, 103–122. Dordrecht: Springer, 2006. Used with the permission of Springer Science and Business Media B.V.

"Patents, Patriotism, and 'Skilled in the Art': *USA v. The Chemical Foundation, Inc.*, 1923–1926." *Isis* 92 (March 2001): 91–122. © 2001 History of Science Society. http://www.jstor.org/stable/237328.

"German Chemicals and American Politics, 1919–1921." In *The German Chemical Industry in the Twentieth Century*. Edited by John E. Lesch, 323–46. Dordrecht: Kluwer Academic Publishers, 2000. Used with the permission of Springer Science and Business Media B.V.

"Confiscated Commerce: American Importers of German Synthetic Organic Chemicals, 1914–1929." *History and Technology* 12 (1995): 261–84. Used with the permission of Taylor and Francis. http://www.tandfonline.com/doi/abs/10.1080/07341519508581887.

18 17 16 15 14   5 4 3 2 1

This book has been digitally printed

For My Family

Contents

Acknowledgments, xi

Introduction, 1

1 / Before the War, 19

    German-U.S. Trade in Chemicals, 19

    German Dyes Industry, 23

    U.S. Dyes Market and Manufacturers, 30

2 / American Manufacturers, German Chemicals, 41

    Dyes and Pharmaceuticals, 1914–1918

    American Importers of German Chemicals, 43

    Domestic Manufacturers, 53

    Trade Associations: ADMA and ADI, 71

3 / Mobilization, 78

    Synthetic Organic Chemicals in War, 1914–1918

    The Explosives Industry Mobilizes, 79

    Mobilizing War Gases, 95

4 / Ideology and Institutions, 113

    American Chemists Respond, 1914–1918

    Nationalism and Internationalism, 115

    Nationalism and the American Chemical Society, 119

    Nationalism and the Universities, 126

5 / Xenophobia, Tariffs, and Confiscation, 1914–1918, 138
   The Tariff Fight, 139
   The Trading with the Enemy Act, 149
   The Federal Trade Commission and the Salvarsan Patents, 152
   Alien Property Custodian, 156

6 / Surviving the Peace, 172
   Economic War, 1919–1922
   The Chemical Foundation, Inc., 173
   The Treaty of Versailles and Chemical Reparations, 178
   Tariff and Monopoly, 190

7 / Customs, Courts, and Claims, 204
   The Industry and the Law, 1922–1930
   Tariff Administration and the Department of Commerce, 205
   USA v. The Chemical Foundation, Inc., 214
   War Claims, 230

8 / An "American" Industry, 1919–1930, 237
   German Industry, 239
   National Aniline & Chemical Company/Allied Dye & Chemical Corporation, 250
   E. I. du Pont de Nemours & Company, 255
   Dow Chemical Company, 267
   Union Carbide & Chemical Company, 274
   Bakelite Corporation, 282

Conclusion, 287
Notes, 295
Bibliography, 365
Index, 391

Figures and Tables

Figures

I.1. *Deutschland* submarine in the Baltimore harbor, July 1916, 2

2.1. Bayer Company subsidiary plant in Albany, N.Y., 1918[?], 49

2.2. Federal Dyestuff and Chemical Corporation, c. 1917, 63

2.3. Du Pont's Deepwater dyes plant, 1919, 69

3.1. Edgewood Arsenal mustard gas unit, c. 1918, 100

5.1. The Alien Property Custodian sells Bayer Company, 1918, 164

6.1. U.S. Chemical Warfare Association, "Disarming Germany," 180

6.2. "Why the Dye Industry Needs Protection," 191

6.3. National Aniline & Chemical Company advertisement, 1922, 197

7.1. U.S. Chemical Warfare Association, "National Safety," 222

8.1. Selected aromatic synthetic organic chemicals (ring molecules), 239

8.2. Selected aliphatic synthetic organic chemicals (chain molecules), 240

Tables

1.1. "Big Six" German Dyes Firms, 24

1.2. U.S. Tariffs on Coal Tar Dyes, 1864–1913, 32

1.3. Production of By-Product and Beehive Coke, 1893–1913, 35

2.1. Bayer Company Dyes Sales, 1914–1918, 50

2.2. Selected Dyestuffs Used by 23 Important Cotton Manufacturers, 1913 and 1916, 54

3.1. Production of TNT (in Pounds) on Ordnance Department Contracts, 88

3.2. U.S. Ordnance Department TNT Output (in Pounds) by Firm (July 1917–December 1918), 91

3.3. U.S. Ordnance Department Picric Acid Output (in Pounds) by Firm (November 1917–December 1918), 92

3.4. Output of Toluene (in Gallons), May 1917 to November 1918, in the United States and Canada, 94

3.5. American War Gas Output (in Tons), 1918, 110

3.6. Gas Output of Outside Plants and Edgewood Arsenal, 1917–1918 (in Tons), 111

4.1. Chemistry Ph.D.s in Germany and the United States, 1910–1930, 137

5.1. APC Actions toward Selected German-affiliated U.S. Firms, 169

6.1. Reparations Dyes Received through the Textile Alliance, 188

6.2. U.S. Tariff Rates, 1913–1922, 203

7.1. Competitive and Noncompetitive Dyes Imports, 1928, 209

8.1. U.S. Synthetic Organic Pharmaceuticals Output, 1922–1929, 247

8.2. U.S. Synthetic Organic Pharmaceuticals, Selected Output (in Pounds), 1922–1930, 248

8.3. U.S. Synthetic Dyes Output in Millions of Pounds and Millions of Dollars (Rounded), 1914–1930, 259

8.4. Synthetic Resins, 286

Acknowledgments

Perhaps readers expect the least amount of originality in a historian's acknowledgments because the debts incurred and categories of creditors remain suspiciously similar across scholars, varying only in the particulars. Nonetheless, I want to thank my particulars.

This project required funding for travel to archives and for time to research and write, and I would like to thank several institutions and programs for financial support over the years. First, the Hagley Program at the University of Delaware provided the original support for my studies, and the Hagley Museum and Library additionally contributed a research grant. Through the philanthropy of Sidney M. Edelstein, the Chemical Heritage Foundation supported two important academic years of research. National Science Foundation grants (nos. 9212816, 9411742, and 0080496) were crucial to conducting research in far-flung places. I received travel and short-term research grants from the Herbert Hoover Presidential Library and Smithsonian Institution. At Drexel University, I would particularly like to acknowledge funding from Dean Donna Murasko and the College of Arts and Sciences to help bring the book to final publication. Thanks as well to the contributors to the Authors Fund of the University of North Carolina Press, which also helped to cover publication expenses. Peter Groesbeck, graphic artist at Drexel University, put his talents to work on this project with great generosity.

I visited several archives, some for an extended period, and many archivists have helped me along the way. I particularly want to thank those at the American Heritage Center, the Chemical Heritage Foundation, the Deutsches Museum, the Sidney M. Edelstein Center at the Hebrew University, Emory University, the Hagley Museum and Library, Harvard University, the Herbert Hoover Presidential Library, the National Archives and Records Administration, and the Post Street Archives. Several corporate archives allowed me access to their records: BASF, Bayer, E. Merck, and Merck & Company.

From the Hagley Program at the University of Delaware, I would first like to thank David Hounshell, my primary adviser and mentor. Thanks as well to John Beer, Anne Boylan, Reed Geiger, Peter Kolchin, and David Shearer, all of whom demonstrated that professors could metamorphose into friends. I am grateful to Chuck Grench, Paul Betz, and others at the University of North Carolina Press for assisting in the publication of this book.

Colleagues in the field have provided words of encouragement and support at important times. The cluster of scholars who study the history of the chemical industry and the like is not large, but it is an especially friendly and supportive one: Ernst Homberg, Jeffrey Johnson, J. Peter Murmann, Carsten Reinhardt, David Rhees, John Servos, John K. Smith Jr., Raymond Stokes, and Anthony Travis. I would also like to thank Steven Usselman for thoughtful and helpful readings of the manuscript; thanks as well to Colleen Dunlavy, Louis Galambos, Kelly Joyce, Robert Seidel, and anonymous reviewers for their suggestions. And my appreciation goes to several other people, academic and otherwise, who offered support and encouragement (or kindly distraction), including Regina Blaszczyk, Eric Brose, Kristine Bruland, Jessica Elfenbein, Nora Engel, Robert Feinstein, Carolyn Goldstein, Art Greenberg, Richard John, Gustavo Klurfan, Angela Lakwete, Michelle Marrese, Julie Mostov, Mary O'Sullivan, Nunzio Pernicone, Erik Rau, Pat Squire, Betsy and Ken Selinger, Joanne Thorvaldsen, Laura Tuennerman, and Jean Walker. A big thank you to Eleanor, Duane, Michelle, Carla, Jeff, Peter, and Billie Jo. And a salute to the regulars at the neighborhood coffeehouse, including but not limited to Jeanne, Atul, Nan, Tom, Darrell, and Rich; it's not a Parisian literary cafe on the left bank of the Seine, but scholars, friends, and philosophers of all kinds abound, providing good company at the right decibel level.

The American Synthetic
Organic Chemicals Industry

Introduction

Late in the evening of July 9, 1916, the German U-boat *Deutschland* successfully executed its mission and docked at a warehouse designed especially for the submarine. Submarine warfare in World War I posed a new and troubling menace to military and commercial shipping, and the Germans became adept at exploiting the strategic advantages of the underwater boat. In May 1915, a German U-boat infamously sank the British passenger liner *Lusitania*, which killed 1,200 people and made Americans—still neutral in the European war—more decidedly hostile to Germany. The *Deutschland*'s voyage stunned Americans because its destination was Baltimore, and it carried no torpedoes and very few weapons of any kind. Despite Americans' growing mistrust of Germany, the wartime oddity provoked widespread admiration and curiosity in the United States, creating heroes of the captain and crew. German Americans in Baltimore, noted one news report, celebrated by singing *"Deutschland Unter Alles."* But what kind of cargo could lead the Germans to expend precious resources during war, not only to design and build this unique submarine but also to dispatch it through the British blockade and across the entire Atlantic Ocean? The *New York Times*, one of the saner commentators amid the wild speculating, suggested the *Deutschland* delivered financial securities, a message from Kaiser Wilhelm II to President Woodrow Wilson, and maybe even important diplomatic passengers. Other rumors said the submarine's return cargo included gold and silver. As it turned out, the cargo became one of the most hotly contested material assets during World War I and the 1920s—fought over in the war, during the peace and reparations negotiations, in the U.S. Congress and European parliaments, and on the international markets. Political and strategic concerns over these goods shaped American policymaking in all three branches of government, generating policies as extraordinary as confiscation by the U.S. government. The cargo? Synthetic dyes.[1]

Synthetic dyes were the most prominent set of products within a larger and growing synthetic organic chemicals industry, and the *Deutschland*

FIGURE I.1 *Deutschland* submarine in the Baltimore harbor, July 1916. (König, *Voyage of the Deutschland*, opposite p. 65.)

voyage formed only a curious footnote in a larger story full of colorful characters and international intrigue. That Germany excelled at making these complicated chemicals, supplying some 90 percent of the world's synthetic dyes by World War I, provided the central dramatic tension once Germany became the enemy. The war prompted intense shortages and skyrocketing prices for German dyes in the United States, but the industry took on heightened political and strategic significance because of the chemical relationship to TNT and other high explosives, as well as to war gases, including mustard gas. The industry also included aspirin and other pharmaceuticals, some early plastics, and photochemicals. Dyes and pharmaceuticals, the most complicated of the products, represented the "high tech" industry of the late 1800s and early 1900s, an industry that required advanced scientific research to develop new products. In the midst of war, Americans set their sights on mastering the industry, but it would take enormous political, financial, and technological investments, with no guarantee the Americans could overcome the decades of German experience. Because the industry was so closely identified with Germany, however, building a domestic industry acquired a symbolic significance: it became a patriotic mission.

While Americans greeted the *Deutschland* with fascination, Europeans faced off that summer of 1916 in the trenches at Somme and Verdun, large battles in France that cost more than two million casualties and fully demonstrated the grinding deadliness of the world's first industrial war. Trench warfare, machine guns, artillery bombardments, war gases, and submarines brought widespread death and destruction to Europe and beyond from August 1914 to November 1918. The war of attrition on land and the British command of the seas repeatedly tempted German leadership to pursue submarine warfare more aggressively. After testing unrestricted submarine warfare periodically through the war, the German leadership rolled the dice in early 1917, gambling that submarine attacks on merchant shipping could end the war before an American entry into the war, which came in April 1917, could have an impact.[2] The United States had already been enmeshed with the war economically, not only because of the British blockade but also because Americans supplied the British, French, and Russians with a great range of munitions and raw materials. American manufacturers of high explosives and smokeless powder were among those who rapidly expanded in response to European demand. While the United States was mostly unprepared for war, several of its industries had mobilized before April 1917 by way of European contracts.

Mustard gas, phosgene, and the other war gases left a lasting scar on Western memory, an impact greater than their tactical efficacy on the battlefield, and people quickly dubbed World War I the "chemists' war." Although most notorious, war gases accounted for a relatively small proportion of World War I casualties, and historian David Edgerton, among others, makes the point that traditional weapons, particularly artillery, inflicted the most damage. It's worth noting, however, that modern artillery carried a particularly devastating punch in part because of the relatively new high explosives that each side increasingly used inside the shells: ammonium nitrate, TNT, and picric acid, the latter two synthetic organic chemicals.[3] But quite apart from questions of the war chemicals' military impact, the primary shortcoming with "the chemists' war" moniker is that it underestimates the broader significance of chemicals in World War I and its aftermath, a significance that went well beyond the chemists and the war. Synthetic organic chemicals became deeply entwined with the American political economy during World War I and the following decade, and government policymaking is central to this story. The war brought synthetic organic chemicals to Americans' attention and made the establishment of a domestic industry a patriotic cause. Because of the war and the Germans' dominance of the industry, Americans wanted—even expected—the federal government to establish policies to create an American synthetic organic chemicals industry. Historian Louis Galambos has painted a broad overview of an American nation that became a "creative society" through incremental innovations in government, firms, universities, and other organizations. Between 1890 and 1930, Americans built a "promotional state" in piecemeal fashion, adopting policies case by case and designed to address the particular challenges Americans felt in the swirl of political, economic, and social changes at home and abroad.[4] The case of synthetic organic chemicals demonstrates the complex ways Americans responded to a perceived crisis—the threat to the American economy and national security. Three broad sets of determinants shaped the creation and implementation of policies.

First, some policies grew specifically out of the anti-German emotions stoked by World War I, which created a crisis environment that led politicians and other government officials to push the limits on the kinds of policies they would consider. Most striking was the American confiscation of German property, which included German chemical subsidiary plants and German-owned U.S. patents, an action unlikely to have occurred outside of the context of war. Americans keenly felt their dependence on German chemicals, partly because of the shortages but primarily because of what it represented to Americans—the seemingly inferior industrial and scientific abilities relative to the enemy and rival. Simple profit and loss calculations are entirely insufficient to explain the development of the American synthetic organic chemicals industry.

Second, Americans crafted policies in an international context. Beyond the international and disruptive nature of a world war, the settlement of the peace, war claims, and trade policy provided another dimension to promotional policies. Americans made decisions about the postwar world not only in the aftermath of war but also following upon several decades of globalization. Trade policy, like war, is inherently international, and tariffs played an important role in the development of this infant industry. While the United States emerged from World War I with new international political stature and leadership, Americans attempted to build their domestic industry in the particularly unsettled international context of the early twentieth century. The more autarkic, isolationist turn represented in the attempt to build a U.S. synthetic organic chemicals industry also had ramifications for the international political economy.

Third, knowledge and expertise—who had it and how to build it—shaped American policies toward the synthetic organic chemicals industry. The German pioneers were deeply experienced in the industry's flagship products—the synthetic dyes and pharmaceuticals on which Americans set their sights—and, particularly in these lines, the U.S. government developed policies and administrative mechanisms to redistribute some of the accumulated knowledge and information. Mobilizing for war brought the public and private sectors together to produce war chemicals, a collaboration that had longer-term consequences for the industry, including in the production and distribution of expertise. For peacetime chemical markets, much of the burden of building technical expertise fell directly on the private sector, but a firm's ability to integrate new knowledge, including from government agencies and universities, affected their competitiveness. While the government toolkit did not yet include the massive investments in science and technology that followed World War II, the collective (if not always coordinated) policies created a larger political context conducive to building expertise in the private sector. As other scholars of the era note, however, an innovative economy also relied on creativity in professional fields, not only in science and technology, but in others such as law, and a range of experts in private, public, and non-profit sectors came together in different combinations to promote a domestic U.S. synthetic organic chemicals industry.[5]

###

In this story, most policies were promotional and focused on building the American industry, but because of the wartime context, federal government actions also included measures designed to inflict hardship on the competition. The most hostile policy challenge to the German chemical industry in the United States came from the Office of Alien Property, a new executive agency that Congress established in October 1917 to oversee enemy-owned private property in the United States. Patterned after European laws, the gloriously named Trading with the Enemy Act (TWEA) originally ordered the Office of Alien Property to manage the property to conserve its value, as reflected in the title of the agency's director: the Alien Property Custodian. By the end of the war, the office supervised an estimated $700 million in 30,000 trust accounts that belonged to foreign citizens and firms.[6] But as Americans grew more anti-German, Congress responded to calls to make the law increasingly punitive. A. Mitchell Palmer was among those who lobbied to make the law more aggressive. Perhaps best known historically as the crusading attorney general who presided over the first Red Scare in 1919-1920, Palmer served during the war as the first Alien Property Custodian, and he wanted the power to confiscate German property outright rather than merely "conserve" it for its foreign owners after the war. As Palmer told Congress in the spring of 1918, "You can not strike a heavier blow at the enemy today than to make him understand that he has lost his connection with the industry and commerce of the American Continent."[7] Congress obliged with amendments that permitted the confiscation of both real property and intellectual property, which cleared the way for the Office of Alien Property to strike at the German chemical firms through their American property and personnel.

Shortly after the war, Palmer produced the second annual Alien Property Custodian report, a document that captured the spirit and activities of the office and in which Palmer articulated the deep suspicions he and others harbored about the German chemical industry. For him, German firms with a large U.S. presence constituted "the great industrial and commercial army which Germany had planted here with hostile intent," and his agency went after German shipping, particularly the docks in New York City, breweries, magneto manufacturers, and many others. The chemical industry, however, "was gigantic, perhaps the strongest, and certainly the most remunerative of all Teutonic industries," and it produced military chemicals, all of which made it Palmer's primary target. Decades before the war, he charged, the German government "fully realized . . . that if a world monopoly in the dyestuff industry could be built up," then "the military strength of Germany would be colossally enhanced." The Office of Alien Property served the nation's interest by "securing American industrial independence by dislodging the hostile Hun within our gates," and he attacked the German chemical industry with every power of his office, including full investigative authority, the appointment of directors to firms, and then confiscation of subsidiaries and patents.[8]

Palmer's right-hand man was Francis P. Garvan, who led the APC's Bureau of Investigation and succeeded Palmer as Alien Property Custodian in March 1919. Tenacious, blunt, and aggressive, Garvan led the APC's investigations into German chemical properties and became entirely convinced that the German chemical industry posed a deep threat to American security and economic independence, whether because of their economic and scientific prowess or, as he believed, because their U.S. subsidiaries and importing agencies were centers of spying and espionage. After confiscating German subsidiaries and patents, Palmer and Garvan orchestrated the sale of the property to American owners. They worked with American manufacturers to establish a non-profit corporation, the Chemical Foundation, Inc., to buy the German chemical patents and license them for the benefit of American industry. Garvan became the Chemical Foundation's president, a post he used for the remainder of his life to promote the American industry.[9]

While deeply embedded in the emotions of World War I, the APC's activities on behalf of the American chemical industry also reflected larger debates in the political economy that derived from the emergence of big, science-based industries. First, the rise of big business generated antitrust sentiment in the United States, and the Alien Property Custodian and other allies of a domestic synthetic organic chemicals industry played on those fears. While German industrialization paralleled American developments in many ways, Germany had embraced legalized cartels rather than imposing antitrust laws to ban cartels. Well before the war, the big German chemical companies had formed alliances, and that gave the APC and others a weapon of law and of public relations to attack German interests in the United States. Second, the rise of science-based industry, particularly the chemical and electrical industries, had led to more aggressive patenting strategies by the large firms. They might file for several related patents to bolster their own position or to thwart a rival, and then patents became bargaining tools. Even before the war, some Americans proposed amending the patent law to require patent owners to work the patent to retain it. The APC could use existing fears and debates in the United States against Germany because the German chemical firms were already masters of strategic patenting and regularly filed for U.S. patents. Palmer and Garvan could take American animosities toward big business—animosities generated by big American firms—and redirect them at German firms.[10]

###

Before the war, the German chemical industry sent its dyes and pharmaceuticals around the world—to Manchester, New York City, Rio de Janeiro, Shanghai, and elsewhere: the German chemical industry helped to build globalization in the nineteenth century. The mass migrations of people, the enormous streams of capital, and the widespread exchange of goods all flowed across borders with unprecedented ease in the first wave of globalization. New communication and transportation technologies aided firms, large and small, as they reached around the world to sell goods or acquire raw materials. Economic historians have calculated that the international economy on the eve of World War I reached a level of economic integration not again achieved until the 1960s, by some measures, or the 1990s, by others.[11] The German chemical firms built a sophisticated and effective global marketing and sales network. In the United States, Bayer built a plant near Albany, New York, to produce aspirin and a few dyes, and their subsidiary also handled the goods imported from Germany. The more typical arrangement kept manufacturing in Germany, and each German firm worked exclusively with an American importing agency, many of which were run by German Americans if not German immigrants.[12]

World War I shattered the nineteenth-century globalization, although economic historians Kevin O'Rourke and Jeffrey Williamson argue that the war blew open tensions that had already existed from the period of globalization, particularly hostility to immigration and to cheaper products from abroad. While the Germans' domination of synthetic dyes and pharmaceuticals attracted little animus in the United States before the war, the subsequent American response highlights the deliberate steps and isolationist attitudes that propelled the larger unraveling of the international economy. "Globalization creates vulnerability," as business historian Mira Wilkins puts it,[13] and Americans, suffering through wartime shortages and high prices, resented their dependence on the international networks. The war convinced many Americans that a more self-sufficient, autarkic economy would spare them uncertainties from beyond their borders, and the German chemical industry suffered the consequences.

As the German chemical firms thrived during the globalization, they subsequently suffered when it came apart. The nineteenth-century globalization reflected a "cosmopolitan capitalism," notes historian Geoffrey Jones, in which some firms had ambiguous national identities, not only because they sold products or services globally but also because shareholders and managers and places of registration might reside in more than one site. In the early twentieth century, destruction of the pre-1914 globalization sharpened the national identities of firms as one country after another took steps to insulate themselves from the wider world.[14] As sophisticated and cosmopolitan as they were, however, the German chemical firms retained a strong national identity even during the globalized period, if for no other reason than that they faced such little competition from other nations in synthetic dyes and pharmaceuticals. When World War I came, therefore, little ambiguity existed about their nationality, certainly not in the eyes of Americans, who aggressively and intentionally dismantled the globally integrated economy during and after the war.

As part of their isolationist rejection of the pillars of nineteenth-century globalization, Americans turned more sharply against their immigrant tradition and cast a hostile, mistrusting eye at German immigrants and Americans of German descent. The 1910 census reported that 8.2 million Americans out of a total population of 93 million were born in Germany or were children of at least one German-born parent. Excepting earlier restrictions on the Chinese, World War I marked the beginning of American legal limits on immigration; the open mass immigration of the pre-1914 globalization never returned. A xenophobic "hyperpatriotism" swept through American society during World War I, with consequences as serious as the lynching of a German alien in Illinois and as silly as renaming sauerkraut "Liberty Cabbage." All things related to German culture—and German American culture—became targets, whether opera, the German language, or chemicals. American chemists led the charge to "Americanize" German pharmaceutical names, such as changing the famous anti-syphilitic and first chemotherapy, Salvarsan, into arsphenamine. As historian John Higham described, the fear trampled civil liberties, undermined social cohesion, and aggravated tensions from mass immigration in the decades before the war.[15] The case of the synthetic organic chemicals industry illustrates the economic consequences of the anti-German hysteria.

In setting tariff rates on imported goods, Congress's domestic and foreign policymaking intersected, and the synthetic organic chemicals industry provided a new front for a longstanding and illustrious history of partisan battling. The isolationist impulse guided tariff policy as Congress pushed rates higher, but this fit a larger historical trend that long predated World War I. Great Britain and a few other countries adopted free trade policies during the pre-1914 globalization, but the United States implemented steadily increasing tariff rates between the Civil War and World War I. Synthetic organic chemicals figured prominently in several tariff laws, including the 1916 wartime revenue act, an "emergency" act in 1921, and especially the major overhaul in 1922. Supporters of higher rates on the sector justified them with the same arguments used with other promotional policies, including the importance of the industry to national security and the risk of dependence on "foreign chemicals." The wartime patriotic fervor also exacerbated the anti-European leanings already present in U.S. protectionism. Although the fighting over chemical tariffs fit larger ideological and partisan patterns, the significance the sector acquired in the national conversation during and after the war meant that synthetic organic chemicals were often in the forefront not only of political debates over tariff policy but also in innovations in designing and implementing tariffs, particularly in methods of assessment and in granting the executive branch more authority to raise rates beyond those set by Congress.[16]

The animosity toward Germany—from European allies as well as the United States—carried over to the terms of peace, in which synthetic organic chemicals again figured prominently. First, the Treaty of Versailles (1919), which legally ended the war in Europe, contained specific provisions to address the German chemical industry. The victor nations cited the role of chemicals in Germany's war-making capacity and aimed to disarm Germany, but they also wanted to promote their own domestic industries. The United States, which never ratified the treaty, still participated in many of the clauses related to dyes and pharmaceuticals, including claiming reparations chemicals from Germany. Proponents of the American industry lobbied the Department of State vigorously to implore diplomats not to return any confiscated German chemical property. Second, another postwar task lay in sorting out claims of damages caused during war, which required both a domestic assessment of claims by each country and then an international negotiation, a belabored process. In the United States, the first German claim taken up under the War Claims Act of 1928 centered on the German chemical firms' confiscated American property; the chemical case created precedents and procedures for subsequent claims.[17]

Scholars have explored the role of science and technology in foreign relations after World War II, particularly as the United States and the rest of the world lurched into the Cold War. Physicists lobbied intensively to guide national and international nuclear policy, not trusting that "normal" politicking could adequately manage this supremely dangerous new weaponry. More generally, after World War II, the United States shared its superior science and technology to cement alliances in a welcomed, or "co-produced," hegemony, as historian John Krige writes. After World War I, if to a lesser extent, American chemists and supporters of a domestic industry also sought influence in foreign relations, and chemists notably helped to derail Senate ratification of the 1925 Geneva Protocol that banned chemical weapons. But whereas nuclear technologies led physicists to a more strongly internationalist orientation in which they felt their obligation lay in reaching out across national borders, chemists after World War I pursued policies that fit the mood of the broader American society: a retreat from international engagement.[18]

Like other scientists, chemists saw themselves as part of an international community of scholars, collectively seeking the secrets of nature, but the war brought home the significance of the nation-state in shaping chemists' identities and careers. Until shortly before World War I, Americans journeyed to Germany for advanced chemical education, creating personal ties and professional respect. Professional prestige in the United States rested in part on international recognition. Given chemists' work on frightening weapons for their respective nations, however, the war strained international scientific collegiality, and, along with British and French scientists, Americans remained hostile to the German scientific community for years. Mobilizing for war simultaneously strengthened the national networks among chemists; national security bound science and the nation-state more tightly. While industry advocates promoted an independent American industry, American chemists also embraced a kind of scientific autarky, one that removed bonds to the storied German research universities. Mastering synthetic organic chemicals was a way not only to conquer Germany but also to deprive them of their decades-old identity as the premier home for advanced science.[19]

As American policymakers promoted a domestic industry, they looked abroad for policy ideas, studying not only the German pioneers but also paying close attention to Great Britain, which sought to expand its own synthetic organic chemicals industry after suffering shortages of German chemicals. In the confiscation of property and import restrictions, for example, Americans patterned at least initial proposals after British precedent. When the United States entered the war, the flow of ideas between the two military allies expanded and became institutionalized (if not always efficiently), and the two sides generally took similar positions on the German chemical industry during the peace process. Personal and professional friendships that developed before and during the war only added to Americans' close attention to policies undertaken by the British. After the war, Americans saw the British as a potential trading rival as well as an ally, which created additional incentives to stay abreast of political and industrial developments in Great Britain. Even so, the particularities of the American situation meant that British precedents, as well as German precedents, would be adapted to fit the American political economy, and Americans flatly rejected creating a state-owned dyestuffs firm, such as the British Dyestuffs Corporation.[20]

###

Developing an American synthetic organic chemicals industry in World War I and the 1920s required expanded organizational abilities in government agencies, firms, and universities, and the collective development of a relevant knowledge and information infrastructure. With a mix of fear and admiration, and to plan their own industry, Americans scrutinized the Germans: their enemies in the trenches, their trade rivals, and their role models. By the 1870s, the German chemical industry had caught up to and then pulled away from the British and French competition to achieve a dominating position in synthetic dyes and, in the 1880s, synthetic pharmaceuticals. The German industry's ascendance has captured the interest of historians and other scholars for decades, and a significant body of scholarship explores this first science-based industry.[21] Historians note that the synthetic organic chemicals industry fully embraced advanced science—to a degree rivaled only by the electrical industry—to develop new products. With the aim of establishing a domestic synthetic organic chemicals industry, Americans of the World War I generation studied the German experience and often drew conclusions similar to later historians on the reasons behind the German chemical industry's success, albeit with a more sinister spin. Both the war generation and historians have identified a series of factors that set the German industry apart from its peers: global sales strategies and infrastructure; diversification; and German government policies on patents, cartels, and educational support. But most of all, the German firms institutionalized science within firms by pioneering the development of industrial research laboratories. The depth of German knowledge and experience should have sent "economically rational" American entrepreneurs running to easier fields to till—and that was certainly the case before the war. During the war, in an abruptly altered political economy, entrepreneurs saw new opportunities, but the Germans' enormous head start created a gap in technical and market expertise that Americans labored to close. The level of scientific, technological, and market knowledge, and who possessed it, shaped public policy and the relationship between the public and private sectors.

The early years of the U.S. synthetic organic chemicals industry came after the federal government had expanded and developed greater organizational capabilities, and the industry established ties of varying intensity across the federal government, including ties to build expertise relevant to the industry. Driven by industrialization, the "administrative state" was one response to Americans' demands for "new kinds of services and supports" to cope with the social and economic changes that intensified in the late nineteenth century, argues political scientist Stephen Skowronek in a work that renewed scholarly interest in the role and capacity of the state. The federal government responded, whether in the civil service, military, post office, or Department of Agriculture. Daniel Carpenter discusses how such government agencies achieved a certain autonomy by generating solutions to perceived national problems, and several of the government officials involved in building the synthetic organic chemicals industry fit the mold of the "entrepreneurial bureaucrat" central to his work. The public/private cooperation in World War I mobilization influenced the way many officials imagined the possibilities after the war. As Steven Usselman has pointed out, government officials—like entrepreneurs—had to think carefully about how a competitive economy worked, if only to frame the rules. The drive to build a domestic synthetic organic chemicals industry depended on the expanded abilities of the federal government, and the U.S. synthetic organic chemicals industry was the recipient of a range of initiatives and collaborations during and after the war.[22]

Business historians have asked similar questions about organizational development in the private sector. The classic work of Alfred Chandler focused on the management of large firms—how managers used technological advances in structuring their organization and to engage in mass production and mass distribution. Other scholars have studied the capacity of organizations to build and sustain expertise, particularly in research and development. The primary burden of gaining expertise fell on the private sector, but the American chemical firms' managers shaped their learning strategies in the context of a dramatic and changing political economy.[23] Coining the term "organization synthesis," business historian Louis Galambos has long reflected on the interrelated expansion of bureaucratic organization in the private and public sectors in the early twentieth century. Echoing historians such as William Becker, Galambos notes the shifting coalitions among public and private actors that defy any easy generalizations. Over the years, Galambos has turned his historiographic spotlight on the use of technology in large corporations, on globalization, on the information revolution—and on the expansion of the professions, the people with the specific training to run the bureaucracies of modern America.[24]

Developing the public and private institutional abilities to support a U.S. synthetic organic chemicals industry was a challenge for Americans, but another complicated task lay in building up the human networks that connected the institutions and that became the conduits of knowledge and expertise among them. Over the decades, the German industry had formed relationships with universities, government agencies, trade and professional associations, their customers, and one another. The abruptness with which the Americans launched into synthetic organic chemicals meant that many of those formal and informal relationships had to be built, sometimes from scratch. The professionals that Galambos and others have studied were the key people in the networks. Most obviously, the chemists and chemical engineers were vital to the future of the industry, but other professionals inside and outside firms mattered. Among others, as both Galambos and Skowronek note, lawyers played a crucial role in the political economy, coming up with policy solutions to identified problems in the economy.[25]

Because of the war, Americans destroyed some of the existing networks, shaking loose people from other networks—including the German industry's—and integrated them into new ones that supported the American industry. For example, German and other European employees of the importers or subsidiaries often moved to American manufacturers, bringing along their wealth of knowledge and experience. The industry gathered up expertise from other sources, luring technically trained people from related industries, university faculties, and government laboratories. The Department of Agriculture set up a Color Lab inside its Bureau of Chemistry, and the Bureau of Standards put chemists to work on spectrographic studies of organic compounds. While the industry welcomed the results coming from these small laboratories, perhaps the agencies' largest—and not altogether intentional—contribution lay in training employees that firms regularly hired away. Some of the people drawn to the opportunities in synthetic organic chemicals production, at least initially, were prewar European immigrants. Quite famously, many European scientists fled to the United States in the 1930s, chased out by Adolf Hitler and fascism, but the emigration of European chemists in the 1920s played only a small role in the development of the U.S. synthetic organic chemical industry. Prewar immigration, however, mattered significantly more, and included people such as the British chemist running Kodak's R&D laboratory and George Merck, an American citizen but also a member of the family behind the venerable E. Merck pharmaceutical firm in Darmstadt, Germany, both of whom played notable roles in the American industry during and immediately after the war.

To overcome the German head start, Americans needed to learn the market—which dyes and pharmaceuticals sold in what quantities and to whom—and in this challenge, the manufacturers found the U.S. government a helpful partner. Knowledge of the American market lay primarily in the hands of the German firms, particularly their American importing agencies. Collecting, sorting, and distributing market information about synthetic organic chemicals became a service the U.S. government provided to American producers through the U.S. Tariff Commission and Department of Commerce. The latter started with a "Census of Dyestuffs" (1916), a compilation of import statistics drawn from the U.S. Customs Service records.[26] The U.S. Tariff Commission took over and provided an annual census of synthetic organic chemicals, which included numbers on imports and on the growing domestic production, and the Department of Commerce kept up a steady stream of specialized reports, particularly on foreign industries and markets. Because of the prewar commerce with Germany, the market information existed, and different government agencies were in a position to obtain and share it in ways private firms and associations were not.

In the midst of the 1920s' return to smaller government, the Department of Commerce, under Secretary Herbert Hoover's leadership, promoted and experimented with a different vision of government-business relations, the "associative state." Hoover had been among those whose wartime experiences shaped his vision of public/private relationships. In his well-known scholarly work, historian Ellis Hawley described Hoover's attempts to use the Department of Commerce to facilitate the development of private sector cooperation, especially through trade associations and other mechanisms that involved exchanges of expertise in "enlightened self-interest." Both the Department of Commerce and the U.S. Tariff Commission cooperated with trade associations, particularly the Synthetic Organic Chemicals Manufacturers' Association (founded 1921), to determine the kinds of information to be collected and published. While scholars of associationalism suggest Hoover's vision generally failed, the particular needs of the nascent domestic synthetic organic chemicals industry meant that it benefited more than most from associative efforts. Because much of the industry already existed in Germany, the Department of Commerce (and U.S. Tariff Commission) mechanisms for "remedying the informational failures of capitalism" fit into a broader if not always coordinated agenda to promote the domestic industry at the expense of the Germans.[27]

Whereas federal agencies provided important and respected market information to the industry, their role in expanding production expertise lay mostly in shaping the larger political economy, with one major exception: building the skills and capacity to manufacture military chemicals. The production of high explosives and of war gases provides a particularly explicit example of the way the private sector's level of scientific and technological expertise shaped the public/private relationship. Mobilizing production of military chemicals during the war took on two different models of organization based on the technical abilities in the private sector. In the case of synthetic organic high explosives, particularly TNT and picric acid, American firms had manufacturing experience that predated the war, or at least that they had gained by fulfilling British, French, and Russian contracts in the first years of the war. The military helped to distribute technical knowledge by sharing best practices, sometimes to the chagrin of the firms, but its role in developing new technologies and knowledge was peripheral. However, in war gases, which were more dangerous and technically challenging than explosives, American firms possessed very little expertise even by April 1917. Although the production of gases involved partnerships in the private sector, the military and civilian government agencies dominated the development and manufacture of gases. Leaders in the military, Bureau of Mines, and National Research Council sought out and helped to expand collective knowledge by working with several sources, particularly American universities and British and French scientists and militaries. As Americans went more deeply into the war, research and production of gases increasingly occurred at sites directly controlled by the Department of War. While the organizational models differed, the expanding production of explosives and gases helped the United States generate a larger and more efficient supply of raw materials for a larger range of synthetic organic products. Just as importantly, the war work demanded many more chemists and technical workers at all levels, many of whom gained skills and expertise useful in a growing civilian industry.[28]

Americans' ability to master the industry competitively depended on the construction of a larger knowledge infrastructure. As business historians have noted, to become and remain innovative, firms needed to reach out beyond the borders of their organization to stay abreast of new knowledge emerging from universities, government agencies, competitors, and elsewhere, domestically and internationally.[29] Even if Americans had somehow been able to capture fully the existing German chemical expertise in the United States, the task of becoming competitive in synthetic organic chemicals required the ability to innovate in addition to imitate.

Americans needed to learn what the Germans already knew, but they also needed to position themselves to compete in the future; industrial research and American universities were central to that effort. Particularly in the chemical and electrical industries, the larger firms had begun to invest in industrial research in the years before World War I, but the number of laboratories and the complexity of research continued to expand in the interwar years. As economic historians David Mowery and Nathan Rosenberg have noted, firms with their own in-house research laboratories were best able to interpret and benefit from research conducted elsewhere as well as to generate their own. Increasingly, American chemical manufacturers gained help from American universities in the form of trained students, faculty consultants, and new research. Before the war, a growing number of universities offered rigorous graduate training in chemistry and chemical engineering. With growing opportunities in the chemical industry and with chemistry's higher profile following the war, university programs grew at both undergraduate and graduate levels. The number of chemistry doctorates awarded by American universities quadrupled.[30] Up against German competition, Americans expected they would need to follow the German model of systematic industrial research and of close ties between industry and universities to build a competitive, innovative synthetic organic chemicals industry.

###

By 1930, innovation and discovery in synthetic organic chemicals production—both American and German—brought dramatic change to the industry if not to its contentious political economy. While Americans' abilities to make synthetic dyes still lagged behind Germany, the effort to match and beat the Germans led instead to entirely new product lines—to an American synthetic organic chemicals industry, American proponents claimed. Indeed, the industry took on "American" traits, most notably the development of mass produced, high-volume synthetic organic chemicals. Synthetic dyes and pharmaceuticals had long been an industry of relatively small runs of a diverse set of closely interrelated products, the attribute of the German industry that Americans viewed as the largest challenge to developing a competitive American industry. In volume chemicals, however, Americans could build on emblematic characteristics of American production and consumption.

The German industry's dyes and pharmaceuticals grew from the aromatic branch of organic chemistry, that is, the branch centered on ring molecules, represented in its simplest form by the hexagon symbol of benzene's six carbon and six hydrogen atoms (C_6H_6). During and after the war, chemists and chemical firms increasingly turned their attention to synthesizing chemicals from molecules in the aliphatic branch, such as ethylene (C_2H_4), which form in chains rather than rings. In the 1920s, and even more in the 1930s, the industry brought out new lines of synthetic products like solvents, alcohols, nylon and other fibers, and plastics. The new product lines meant that the boundaries of the industry grew—from dyes to coal tar chemicals to synthetic organic chemicals. The mutable and expanding boundaries of the industry permitted its supporters to claim a larger national political and economic significance for the industry, including national security and health.[31]

As Americans exploited the newer product lines, the German industry remained experienced, powerful, and innovative; in trying to catch and surpass the German industry, Americans aimed at a moving target. The Germans had introduced several new categories of dyes, including the vat dyes, and placed excellent products on the market. But even before the war, the German manufacturers had increasingly redirected their firms to other product lines. The Germans recognized the diminishing revenues they received from dyes sales and suspected their research had left relatively little room for dyes to improve further in any dramatic way. BASF had its new Haber-Bosch process for obtaining nitrogen from air, Hoechst continued its diversification into pharmaceuticals and other products, and the German industry in general expected more dynamic growth from products other than dyes. Among their firsts in the aliphatics was synthetic methanol, which nearly wiped out the American distilleries when it arrived on the American market in 1924. For products like synthetic rubber, Americans and Germans both undertook significant and complementary research throughout the 1930s. The German and American relationship in synthetic organic chemicals remained deeply political throughout the interwar period, a relationship of nationalistic competition and wary cooperation.[32]

For many advocates of the domestic industry, the new aliphatic products "Americanized" the synthetic organic chemicals industry. In the 1920s, the explosive growth in the automobile industry generated demand for solvents, lubricants, lacquers, and antifreeze, among other products. Union Carbide & Chemical Corporation, significant in the expansion of the synthetic aliphatic organic chemicals, gained wide recognition for selling ethylene glycol as its Prestone antifreeze. The volume production played to American strengths in chemical engineering, which helped to scale up operations to plant size. In addition, chemical firms began to abandon coal as the raw material for synthetic organic chemicals, turning instead to petroleum and natural gas, which Americans continued to find in abundance and exploit. Because of the mass production, the value of the newer products exceeded American production of "German" chemicals already in the late 1920s.[33]

Scientific, technological, and industrial achievement was one ruler by which Americans measured themselves against other nations, and Americans of the World War I generation believed they fell short. Despite the industrialized slaughter of the war, Americans ended the war believing they needed more, not less, industrial prowess—for national security, economic security, and national pride. Born in the crisis of war, the mission to establish the U.S. synthetic organic chemicals industry brought together advocates from the public and private sectors to establish promotional policies in complex and often creative ways. Although the United States emerged from the war even stronger relative to western Europe, the war scared Americans in other ways, and they made policy having retreated to a more isolationist worldview in a tense international political economy.

1 / Before the War

German-U.S. Trade in Chemicals

To understand the prewar economy in synthetic organic chemicals, consider a typical, if imaginary, journey of a synthetic dye and a synthetic pharmaceutical from Germany to the United States in 1914. On the eve of World War I, indanthrene blue GCD and Salvarsan represented the most advanced synthetic organic dye and pharmaceutical yet produced by the German industry. Americans had embraced both upon their introductions to the world markets, and the two products illustrate the complex global networks—technological, commercial, and political—underlying this high technology industry of the late 1800s and early 1900s.

More than earlier classes of dyes, the indanthrene dyes of Badische Anilin & Soda Fabrik (BASF) held fast to the fabrics, whether subjected to light, repeated washings, or other threats to a dye's integrity. Chemist René Bohn had discovered the first of indanthrene dyes in 1901 while searching for new compounds that would mimic the properties of indigo, the most important natural dye over the ages and one recently manufactured synthetically. The trade name, indanthrene, was a combination of indigo and anthracene. Bohn and his colleagues expanded the indanthrene class with research on related compounds, cementing BASF's reputation as the most innovative of the large German dye firms at the time. In 1903, Bohn added indanthrene blue GCD to BASF's product line.[1]

The journey of indanthrene blue GCD began at BASF's plant in Ludwigshafen, Germany, a city across the Rhine River from Mannheim. The plant was enormous; it had grown to more than 500 acres and hundreds of buildings since BASF's founding in the 1860s. In 1914, more than 10,000 people worked there. Many of the buildings would have had the equipment of dyemaking, including kettles of various sizes and materials, many with mechanical stirring apparatus, autoclaves for higher pressure processes, pipes overhead, crushing mills, and gauges. The German chemical industry used the country's rich coal deposits; the organic raw materials of the dyes were hydrocarbon distillates of coal tar, such as benzene, naphthalene, anthracene, and other aromatic organic chemicals. Then the manufacturers used inorganic acids, often hydrochloric (muriatic), nitric, and sulphuric, to make intermediates. Further processing, which could include additional reactions, washing and drying, grinding, and filtering, led to the dyes. Each stage could take several steps, and the challenge to manufacturers lay in minimizing the waste in every step and coming as close as possible to the maximum theoretical yield. In indanthrene blue dyes, the key intermediate was 2-amidoanthraquinone, derived from anthracene and cooked in an autoclave with ammonia. Melted with caustic potash (potassium hydroxide), the intermediate yielded blue indanthrene dye. By changing the temperature of the reaction or adding new ingredients, such as zinc or potassium nitrate or halogens like bromine or chlorine, the manufacturers could generate different indanthrene colors. In the case of indanthrene blue GCD, treating a basic indanthrene blue with nitric and hydrochloric acid added chlorine atoms to the molecule and shaded the blue slightly to green.[2] These reactions created new chemical compounds from the organic (or carbon) base, which is what made them synthetic organic chemicals. Because manufacturers obtained their crude organic raw materials from coal tar for decades, synthetic organic chemicals were just as often called coal tar chemicals.

When complete, indanthrene blue GCD formed a paste, which BASF packaged and prepared to ship to textile manufacturers around the world. Dyes destined for the United States traveled north by railway to Rotterdam, The Netherlands, the port from which dyes were loaded onto ships bound for New York City. Indigo and many other blue dyes would be on their way to China, where blue was the most-used color. In 1913 alone, Germany exported 73 million pounds of indigo worldwide, of which 47 million went to China. Americans also consumed large quantities of blue dyes, although blacks sold in larger quantities. In 1913, the United States imported 478,980 pounds of indanthrene blue GCD, more than any other indanthrene dye, which more typically sold 2,000 or 12,000 or 20,000 pounds per year.[3]

The Rhine River flows north, and, downriver from Ludwigshafen, the city of Mainz sits at the junction with the Main River, on which another major chemical manufacturer stood. Not far up the Main, near Frankfurt-am-Main, Meister, Lucius & Brüning (called Hoechst, after the plant's hometown) manufactured a full range of dyes, including synthetic indigo, and recently had achieved success with new synthetic organic pharmaceuticals. One was Novocaine (1905), a synthetic variation of cocaine that carried the pain-relieving properties of the natural drug without the addiction, and another was Salvarsan (1910), the first genuine chemotherapy, used to treat syphilis. Over 9,000 people worked at Hoechst.[4]

Hoechst, like most of the firms, maintained close ties to academic scientists, and Salvarsan was the product of Hoechst's cooperation with Paul Ehrlich, the pioneer in chemotherapy and a Nobel Prize winner (1908), who directed the Institute for Experimental Therapeutics in Frankfurt. Throughout his career, Ehrlich used his synthetic dyes research to inform his medical research, borrowing theories and procedures liberally. In developing Salvarsan, Ehrlich deployed the toxic power of arsenic in combination with an aromatic organic base to kill off the spirochetes responsible for syphilis. He and his co-workers famously tried hundreds of combinations, coming up with a workable formula on the 606th arsenic compound, earning Salvarsan an alternative name, 606. In 1912, Ehrlich developed Neosalvarsan, a variation on the original that was easier for doctors to prepare and deliver to their patients. Hoechst made Salvarsan available commercially, and the American medical community eagerly awaited its arrival, which occurred early in 1911.[5]

When the train carrying the indanthrene blue GCD and Salvarsan arrived at Rotterdam for export, they faced international politics in the form of tariffs. Before the dyes and pharmaceuticals could depart on a ship headed to the United States, the agents of the German firms needed to show their invoices to an American consul, either at the place of manufacture and purchase or at the port of departure. The consul compared the purchasing invoice to local prices, aiming to prevent exporting firms and their agents from either dumping goods below cost in the United States or undervaluing goods to reduce U.S. customs duties. With the paperwork complete, the dyes and pharmaceuticals crossed the Atlantic Ocean in about two weeks and typically arrived in New York City, home to nearly all the American importing agencies affiliated with the German firms. The New York customs house was the nation's largest; the customs agents there processed the arriving goods, assessing the amount of duty based on U.S. tariff rates. The customs office also ran a chemical laboratory in which they could conduct tests on incoming goods, if necessary, to help identify illegal, fraudulent, or impure products. Congress had most recently revised tariff rates in 1913, which determined the duties on Salvarsan and indanthrene blue GCD. Most dyes carried a rate of 30 percent ("ad valorem"), but the additional per pound ("specific") rate had been eliminated in 1883. The tariff schedule imposed a duty of 15 percent on most pharmaceuticals of coal tar origin, although aspirin and a few others carried a 25 percent duty. Congress also exempted certain products entirely from tariffs, placing them on the "free list," and in 1913 the free list included indigo and dyes from alizarin, anthracene, and carbazole. As a dye derived from anthracene, BASF's indanthrene blue GCD passed through customs without any duty. Salvarsan carried a duty of 15 percent.[6]

After indanthrene blue GCD cleared customs, the BASF importing agency, which had been run for decades by Adolf Kuttroff and Paul Pickhardt, took charge of the sales in the United States. The importing agencies served as a vital cog in the machinery that delivered dyes and pharmaceuticals from German firms to American consumers. The importers understood the American market and cultivated ties to a range of consumers. Just as importantly, the importers brought important technical skills to the sales. Many operated technical laboratories to blend the chemicals or to help solve specific application problems for consumers.[7]

Importers taught their customers how to use their dyes and pharmaceuticals properly, and both indanthrene blue GCD and Salvarsan demanded knowledgeable users. Indanthrene dyes, like indigo, were known as "vat dyes," which referred to their method of application. Vat dyes became soluble in a chemical solution that removed oxygen (reduction) and allowed the dye to penetrate the textile fibers, typically cotton. Upon exposure to oxygen in the air (oxidation), the dye adhered to the fiber and resisted fading or washing out much better than most other classes of dyes. The process required careful monitoring of the temperature. The importers' technically trained sales staff worked with the dyers in the textile plants to ensure the dyers understood the best way to deploy the vat dyes.[8]

Salvarsan also posed serious challenges to the end user. Because arsenic was the effective toxin in Salvarsan, the consequences were dire if the attending doctor prepared or delivered the drug incorrectly. In the years before the war, several people in Europe and the United States died after receiving Salvarsan, and the medical and popular press discussed the dangers of careless delivery. In the early years, doctors needed to inject the drug intravenously within three minutes of its preparation, which eliminated preparation by pharmacists and required diligence on the part of the doctors. The solution was alkaline and typically painful for the patient to receive. Ehrlich and his coworkers developed Neosalvarsan to make delivery and reception of the drug easier, but even then deaths occurred. In April 1914, for example, eight patients died at Los Angeles County Hospital as the result of receiving decomposed Neosalvarsan. As the importer for Hoechst, Herman Metz helped to demonstrate to doctors the appropriate method of preparation and application. And, despite the deaths, doctors marveled at Salvarsan's ability to cure syphilis, when earlier regimens had primarily treated symptoms.[9]

The German Dyes Industry

Behind such a journey of indanthrene blue GCD and Salvarsan stood a German industry that possessed four decades of impressive research, manufacturing, and global sales. After 1870, the United States and Germany developed industrially with roughly comparable timing and similar characteristics, such as strong steel and electrical industries. In its synthetic organic chemicals industry, however, Germany stood apart not only from the United States but also from Great Britain, where William Perkin synthesized the first dye in 1856, and from France, which also had a synthetic dyes industry in the 1860s. By 1914, 116 German firms had entered the dyes industry; 91 of them had also left the business. Of the thriving survivors, the most famous included the "Big Six": the larger BASF, Bayer, and Hoechst, and the smaller Agfa, Cassella, and Kalle, nearly all founded in the 1860s. Since the 1870s, the German manufacturers had put significant distance between them and any rivals. By 1881, German firms manufactured half of the world's synthetic dye production. As synthetic dyes improved in quality and variety, virtually replacing natural dyes, the German market share increased to between 80 and 90 percent by 1900 and remained steady for more than a decade afterward. In 1913, only Switzerland, which had five small firms that produced specialty dyes and pharmaceuticals, possessed an export market of any size, slightly less than 10 percent.[10]

The German industry dominated the world's production of synthetic organic dyes between 1870 and 1914, capturing the interest of scholars for decades and yielding a rich and significant body of scholarship on this first science-based industry. Historians note that the level of technical expertise and innovation surpassed that of all other industries of the era, except for possibly the electrical industry. From the 1870s onward, German firms generated a stream of new dyes and dyes classes, synthesizing more than one thousand different dyes in a four-decade period. After the initial aniline dyes, the German manufacturers developed three key types of dyes: alizarin, azo, and vat, which rapidly replaced most of the early aniline dyes.[11] A series of factors set the German industry apart from its peers: German government policies on patents, cartels, and educational support; global market and sales strategies; and, most of all, pioneering research laboratories and a full embrace of science in industrial innovation.

Table 1.1. "Big Six" German Dyes Firms

Common Name	Full Name	Location
BASF	Badische Anilin & Soda Fabrik	Ludwigshafen
Bayer	Farbenfabriken vormals Friedrich Bayer & Company	Elberfeld and Leverkusen
Hoechst	Meister, Lucius & Bruning	Hoechst (near Frankfurt)
Agfa	Aktiengesellschaft für Anilinfabrikation	Berlin
Cassella	Leopold Cassella & Company	near Frankfurt
Kalle	Kalle & Company	Wiesbaden-Biebrich (near Frankfurt)

Source: John J. Beer, *The Emergence of the German Dye Industry*, 49–54; Anthony S. Travis, *The Rainbow Makers*, 74–75.

In their anger during World War I, Americans denounced the German political system and attacked policies related to synthetic organic chemicals. The rise of the German dyes industry coincided with the formation in 1871 of a unified Germany from several smaller German provinces, including the dominant Prussia under Hohenzollern monarch Wilhelm I and his chancellor, Otto von Bismarck.[12] The advent of a national government shaped the political and economic context in which the firms operated, especially where government policy on cartels and competition, patents, and education were concerned.

Unlike the United States, where the antitrust legislation restricted cartel behavior, German cartels enjoyed legal sanction, which meant the cartel agreements could be enforced in courts. The German dyes industry first turned to cartels in 1881, and over the next two decades the various dyes firms signed agreements to cooperate on patents and research and to guard profits. Leaders of the dyes firms took further steps to organize their industry when they formed two rival cartel groups in 1904. Carl Duisberg of Bayer, inspired by trusts that had been recently formed in the United States (U.S. Steel and Standard Oil, for example), strove to convince the directors of the other major dyes firms to join in a single alliance. The Hoechst leadership declined Duisberg's offer but formed one of the two groups when it exchanged stock with Cassella and, in 1907, purchased a controlling share of Kalle. In response, Bayer, BASF, and Agfa created the second cartel through a system of pooling profits. The two cartels reached agreement on the price of indigo and on a few other products and avoided price-cutting, but they remained two very distinct entities.[13]

The German chemical firms patented in both Germany and the United States, among many other countries. Prior to 1877 in Germany, innovations received little legal protection, a situation that facilitated German copying of British and French synthetic dyes in the 1850s and 1860s. Once German manufacturers became the originators of new processes in the 1870s, they welcomed the passage of a national patent law and sent representatives from the dye industry and academic chemistry to advise the government in the early administration of the law. Before the war, the strong patent position of the German chemical companies in the United States attracted the attention of members of the American chemical community, typically in the context of a more general fear that corporations' research and development (R&D) and legal resources had increasingly made the U.S. patent system unfairly beneficial to the "trusts." In 1912, a chemist counted American patents on chemicals granted between 1900 and 1910, and he determined that Germans had received 1,754 U.S. patents related to all chemicals, including more than 1,000 patents on organic chemicals. The chemist calculated that 70 percent (1,237 of 1,738) of U.S. patents granted on various synthetic organic compounds belonged to Germans. In 1915, American chemical consultant and former BASF employee Bernhard C. Hesse cited a standard German reference work from 1914 that listed 921 commercially available dyes, 207 of which U.S. patents protected. That left 714 dyes without patent protection and available to any would-be manufacturer as of 1912, according to Hesse.[14]

The German government also contributed to building universities and research institutes, which, coupled with the dyes firms' pioneering industrial research laboratories, created a depth of scientific understanding and training that supported the rise of the chemical industry. German universities gained a reputation for excellence in organic chemistry, and the Kaiser Wilhelm Institutes, funded jointly by industry and the state starting in 1912, attracted top-notch scientists who focused their attention on selected research projects. Important conceptual breakthroughs led to a more precise understanding of the structure of organic chemicals and guided the search for additional syntheses. Justus Liebig's teaching at University of Giessen from 1824 to 1852 played an important role in the development of organic chemistry. In 1860, August Kekulé formulated a structural theory of organic compounds and, in 1865, of the benzene ring, which provided a theoretical foundation for organic chemistry. Other important academic organic chemists of the late nineteenth and early twentieth century included Adolf Baeyer, Robert Bunsen, Emil and Otto Fischer, and Otto Witt.[15]

Like Americans of the World War I generation, historians have viewed the development of in-house research laboratories as the most significant characteristic of the German dyes firms, a large reason behind their climb to such a dominating market position in the most technically demanding industry of the late 1800s and early 1900s. By 1883, the major German firms possessed laboratories in which research chemists strove to devise new chemicals. Instead of relying on outsiders to discover new compounds, manufacturers brought the discovery process into the firm and sought to regularize as much as possible the production of new products. A team of industrial chemists systematically conducted the research necessary to identify and test promising new products; they expanded their ranges of colors, made dyes more "fast," and applied dyes to a variety of materials. Research laboratories strengthened ties to German higher education; firms provided research funding and a means of manufacturing any developments by academic scientists, and the schools provided expert advice and a steady supply of highly trained chemists to work in the industry. BASF's Heinrich Caro and Bayer's Carl Duisberg especially stood out as successful chemists in their own right and as research directors for their respective firms.[16]

The close relationship between academic scientists and the German dyes industry is evident in the synthesis of indigo, the single most popular natural dye. Working at the Gewerbe Institute of Berlin and funded in part by BASF, Adolf Baeyer, a prominent organic chemist, achieved one synthesis of indigo in 1880 and another from a different raw material in 1882. After 1880, Baeyer also received financial backing from Hoechst, and he and the research chemists in the two companies embarked on a project to develop an economically feasible process for synthetic indigo. BASF finally reached a satisfactory process in 1897 and Hoechst soon after, although the latter developed a different and less costly method by 1904.[17]

Besides the work of developing new products, the research laboratories performed an important tactical duty. As early as the 1870s, German firms contended with the ability of their rivals to use analytical chemistry to decipher the chemical formulas of new dyes placed on the market. In the absence of patent protection, competitors quickly produced their own version of a new product, but even if a patent existed, competitors often manufactured similar products. To slow their rivals' imitations, the German dye firms developed elaborate patent-management strategies. For example, they submitted patent applications for many similar kinds of dyes at the same time, forcing rivals to sift through dozens of patents to identify the one that corresponded to the new dye on the market. "Screening" patents became an integral part of the work of a firm's research laboratory.[18]

In the 1880s, the German synthetic dyes industry became the German synthetic organic chemicals industry. That is, the industry diversified into an important new line of chemicals that bore similarities to dyes: synthetic organic pharmaceuticals. The connection between synthetic dyes and pharmaceuticals was deep. Synthetic dyes and pharmaceuticals shared the same intermediates, including salicylic acid, which chemist Herman Kolbe identified as a therapeutic in the 1870s. Several German firms emerged as specialized synthetic organic pharmaceuticals manufacturers, particularly E. Merck of Darmstadt and Chemische Fabrik von Heyden of Radebeul. Late nineteenth- and early twentieth-century synthetic organic pharmaceuticals included drugs to reduce fever or pain, such as Hoechst's Antipyrin and Bayer's Phenacetin. Hoechst had its Novocaine, and Bayer and Merck each sold a line of sedatives, including Veronal. In 1898, Bayer introduced aspirin, which marked the beginning of one of the world's best-known pharmaceuticals and made the name Bayer synonymous with aspirin.[19]

The German synthetic organic chemicals industry was a global industry almost from its beginning, initially and primarily catering to the world's textile manufacturers. As the firms grew in sophistication about production, they also built a successful worldwide sales and marketing network. During the latter part of the 1880s, the firms shifted from complete reliance on local sales agents and began incorporating sales offices in their own corporate structure. They offered customers technical advice, a service that benefited the textile manufacturers, but such close contact also allowed the agents of the German firms to perceive trends and technical problems and communicate the information back to headquarters. The importing agents played a vital role in maintaining a global market. By 1904, about 20 percent (measured in value) of German-made synthetic dyes went to the United States, 15 percent to England, and smaller percentages to Russia, Austria-Hungary, China, and several other countries.[20]

Following their general pattern, the German companies initially established sales houses in the United States to distribute their dyes. Typically, Americans owned the importing houses, and each made exclusive contracts with a single German firm, agreeing to sell no other supplier's product without permission from the German firm. By 1914, three of the "Big Six" German firms had terminated contracts with independent importing houses and established their own branch sales offices in New York; the other three worked through American-owned houses. The contracts with the importing houses frequently involved an exchange of stock shares, profit-sharing arrangements, and an option for the German firms to buy some or all stock in an importing agency, a provision that later created questions about the legal independence of American importing houses. Even when the importers were American citizens, they often had roots in

Germany, either by birth or their parents' birth, and at Bayer and Merck top executives in the United States had family ties to managers in the German firms. Unlike most German firms, Bayer, Heyden, and Merck established manufacturing subsidiaries in the United States. The only one from the "Big Six," Bayer tried a range of relationships over the decades; it worked with independent importing houses and in joint ventures, and, in 1904, became the sole owner of its subsidiary. The American branch, Bayer Company, sold chemicals imported from Germany and manufactured a limited number of synthetic dyes—perhaps as many as eight—but its Rensselaer, New York, plant specialized more in aspirin and other pharmaceuticals of the Bayer line.[21]

American policymakers and manufacturers understood that the German chemicals firms engaged in cartel behavior, and they consequently watched the importers for signs of illegal cooperation; however, most viewed the vehement competition among the importers as evidence that no widespread collusion existed. The importers agreed, however, occasionally and temporarily, to conventions on price that sometimes included Schoellkopf, the leading, if small, American manufacturer. American policymakers and executives were less sympathetic to a more common vice permeating the textile industry: bribing the head dyer in a textile plant to adopt a particular dye. In 1913, textile firms in Philadelphia brought a lawsuit against the importers, charging that the latter, as the representatives of German firms, violated the Sherman Antitrust Act. The suit began after some particularly egregious bribery by Bayer's Philadelphia agent and then escalated into antitrust charges. The importers countered that they engaged in no collusion and that any profit-sharing agreement among the German firms occurred in Germany and not in the United States. In 1914, the court dismissed the suits; meanwhile, many of the importers, including Bayer, reorganized or changed the terms of their contracts to avoid any further charges of antitrust violations.[22]

As war engulfed Europe, the scarcity in synthetic dyes and pharmaceuticals prompted Americans to examine more closely their dependence on German chemicals. Much as a historian might, Americans of the war generation compared themselves to the Germans and analyzed their shortcomings over the previous decades. Those already familiar with synthetic organic chemicals could readily point to the differences, and many thought it perfectly reasonable that Germany specialized in synthetic dyes and pharmaceuticals while Americans devoted economic resources elsewhere. Early in the war, chemical manufacturer Leo Baekeland dismissed the sector's significance to the American economy: "Any man who could make a law to increase the size of eggs by one gram would make himself infinitely more useful to his country than a man who could increase the production of those aniline dyes," given the value of aniline imports.[23] But those who wished to build an American synthetic organic chemicals industry more rigorously reviewed the history of U.S. dyes firms and their context, finding both concrete reasons for underdevelopment and grounds to hope that the United States could sustain such an industry going forward. Many blamed U.S. tariff policy, while others variously considered U.S. patent policy, the influence of the American textile industry, the state of academic chemistry and the chemical profession, and the supply of raw materials and necessary inorganic chemicals.

By the early 1900s, the German synthetic organic dyes and pharmaceuticals industry had garnered enormous respect, but the German manufacturers faced challenges of their own. First, their profits had declined. In 1890, they paid, on average, dividends of more than 20 percent. Manufacturing synthetic dyes became increasingly less profitable as development costs escalated. After the heady days when alizarin and azo dyes proceeded rapidly and lucratively onto the market, the companies faced rising expenses as they invested in growing amounts of research to find new dyes or improved dyes. The quest to achieve marketable synthetic indigo, for example, required more than two decades and $8 million. In the early 1900s, the dyes firms actively looked at other products—primarily synthetic nitrogen—for future investments. In 1913, BASF's sales broke down to 40 percent azo and aniline dyes, 29 percent indigo, 14 percent alizarin dyes, and 15 percent inorganic chemicals.[24]

A second weakness in the German chemical enterprise came from the academic side. By the 1890s, the top organic chemists dominated the world of German chemistry; the sophisticated centers under prominent chemists methodically and thoroughly pursued research. This highly productive, systematic method of institutional organic chemistry suited the German dyes industry well. The competitive nature of the dyes industry, complete with rapid fashion changes, required industrial chemists to tweak known compounds to create new colors; however, this research rarely provided theoretical breakthroughs, and many of the Ph.D. chemists who trained in this system learned to be worker bees rather than creative, independent investigators. The hierarchical nature of institutional chemistry, with as many as sixty assistants and students under one professor's control, led to orthodoxy in method and subject matter.[25] Ever so slightly, the giant German synthetic dyes and pharmaceuticals industry suggested, at best, a mature industry, or, at worst, an ossifying edifice. Perhaps the door stood open just enough to allow an American industry in when the wartime opportunity arose.

U.S. Dyes Market and Manufacturers

By many measures, the United States was the most industrialized country in the world at the outbreak of World War I, although, like other countries, it remained an "underdeveloped" country relative to Germany in the synthetic organic chemicals sector. The scholarship on technology and on industrial development, from the older "stages of growth" theories to more cultural history approaches, has focused on the significance and considerable challenges that lie in the diffusion of technology and know-how from one place to another. In each case, similarities or differences in culture, political economy, and infrastructure shaped the desire and capacity of the recipient to learn and adapt the technology. Scholars have also noted the extent to which knowledge flows back and forth even during the initial diffusion. The American and German industrial economies bore many similarities, not least of which was their explosive growth and the rise of big business after 1870. But the differences in immigration, political systems, and the fact the two were enemies in World War I made the flow of knowledge and development of skills a matter of national security and national pride, or so believed America's World War I generation. Because of the broad industrial development in the United States, when Americans determined to build a domestic synthetic organic chemicals industry, they therefore had substantial resources and capacities that created the potential for the successful establishment of the industry. That gaining the technological know-how and operating with economic viability still posed major hurdles for the United States suggests just how complex the process of diffusion could be, even among nations with relatively comparable levels of economic development.[26]

Although the United States had a small dyes industry as early as the 1860s, it remained small throughout the ensuing fifty years, and it barely received notice by the Europeans or, indeed, by the giant American textile industry. In 1913, the American dyes industry consisted of seven firms that employed a total of 528 people. Capitalized at $3 million, it produced 6.6 million pounds of synthetic dyes valued at $2.4 million. By contrast, the eight largest German dyes firms produced 140,000 tons of dyes valued at $65 million and employed 16,000 people in dyes production alone and nearly 40,000 in all phases of production. In addition, nearly all the intermediate products that went into the "American-made" dyes came from abroad, primarily from Germany and Britain. Small U.S. firms extant on the eve of World War I included Heller & Merz (Newark, New Jersey, founded 1869); the Central Dyestuff & Chemical Company (Newark, New Jersey, 1898); and W. Beckers Aniline & Chemical Works (Brooklyn, 1912).[27]

The most important American dyes firm, Schoellkopf Aniline & Chemical Works, dated to 1879, when brothers Jacob F. Schoellkopf Jr. and C. P. Hugo Schoellkopf set up shop in Buffalo. Sons of a Buffalo industrial magnate, they had received training in chemistry in Germany before establishing the firm. Often cited as the most "German" firm among American manufacturers because of its greater reliance on chemists and research, as well as on secrecy, the company struggled through nearly two decades before turning a small profit beginning in 1896. Although lacking a formal research laboratory, the firm consistently employed a least a few chemists, mostly German or German-trained, who achieved moderate success in developing dyes. The firm gained seventeen patents, including a line of nigrosine dyes and a direct black dye that brought a healthy export business. In 1908, Schoellkopf Aniline & Chemical Works employed 200 people, including 8 chemists and 3 dyers, and manufactured between 60 and 70 dyes. Six years later, the number of employees had increased to 250, and the company had expanded its line of dyes to 130.[28]

From the hindsight of World War I, advocates of a domestic synthetic dyes industry believed that U.S. federal policy, particularly tariff policy, had damaged the industry's early efforts. Despite increasing average tariff rates after 1870, Congress dropped rates on dyes in the 1883 tariff, which the World War I generation, still largely committed to protectionist policies to help domestic manufacturers, saw as a crucial turning point that undermined a promising start to an American industry. Each time Congress reconsidered the tariff, the American dyes manufacturers seemed insignificant compared to the American textile industry. As a Department of Treasury tariff expert stated in 1882: the American dyes firms "employ so few people and produce such limited quantities of dyes as hardly to amount to an industry worth protecting at the expense, at least, of burthening and crippling the vast industries devoted to the manufacture of textile fabrics" and their consumers. Because dyes were raw materials for textile manufacturers, they preferred to acquire the least expensive dyes, which meant, in tariff policy, that the industrialists supported the lowest possible rates on imported synthetic dyes. With a full complement of active trade associations, the textile industry proved a powerful interest group, lobbying against substantial tariffs on dyes.[29]

Table 1.2. U.S. Tariffs on Coal Tar Dyes, 1864–1913

Year	Ad Valorem	Specific Duty	Free List (Exempted from Tariffs)
1864	35%	$1 per lb.	
1870	35%	50¢ per lb.	
1883	35%	0	alizarin, indigo
1890	35%	0	alizarin, alizarin dyes, indigo
1894	25%	0	alizarin, alizarin dyes, indigo
1897	30%	0	alizarin, alizarin dyes, indigo
1909	30%	0	alizarin, alizarin dyes, indigo
1913	30%	0	alizarin, alizarin dyes, indigo

Source: Bernhard C. Hesse, "Lest We Forget! Who Killed Cock Robin? The U.S. Tariff-History of Coal-Tar Dyes," *Journal of Industrial and Engineering Chemistry* 7 (August 1915): 709.

If tariff legislation consistently remained a priority among American manufacturers before and after the war, their views on the significance of patent law varied. While chemical manufacturers had complaints about patent law, before the war they only occasionally cited patents or patent law as a potential obstacle to development of synthetic dyes and pharmaceuticals, in contrast to assessments once the war began. Critics faulted American patent law for other reasons. American laws allowed an inventor to patent both a product ("composition of matter") and a process. Once a chemical product is patented, little incentive remained to discover new and superior processes to make the product, the critics argued. The laws of Germany, for example, contained nothing equivalent to a product patent. A second major complaint about the patent laws emerged over the absence of a compulsory working clause, which required patentees to use or license the patent or to lose it. Britain had installed such a clause in 1907 specifically to aid its chemical industry, although the provision achieved only mixed results. Hoechst established a plant on British soil to comply with the law, but some commentators remained skeptical that the working clause encouraged domestic industries. Third, partly as the result of corporate power and partly because of the increasing complexity of chemical products, chemists wondered whether the people operating the U.S. patent system, including the examiners and judges overseeing patent litigation, could understand the science well enough to make wise decisions. Also, courts in different jurisdictions might render opposing verdicts on patent matters, and others believed verdicts barely deterred infringement. Finally, trademarks, as well as patents, on pharmaceuticals attracted attention before World War I. The rapidly professionalizing American medical community frowned on patenting medicines, which physicians believed put a monopoly on substances that ethically belonged in the public domain.

The American Medical Association, in particular, waged a running battle with Bayer over the use of "aspirin" and "acetyl salicylic acid," a conflict that would peak in 1917 when the patent expired in the midst of the war.[30]

The American pharmaceuticals industry fared no better than the dyes industry in developing a line of synthetic organic products. Only during World War I did drug makers seriously begin to conduct the kind of research necessary to produce more than a handful of synthetic organic pharmaceuticals. In the decade or so prior to the war, a few manufacturers employed research chemists to ensure standardized goods but rarely in the field of organic pharmaceuticals. "Ethical drug" firms like Parke-Davis and Company, E. R. Squibb and Sons, H. K. Mulford Company, and Eli Lilly and Company, which distinguished themselves from patent medicine makers partly through greater reliance on science, concentrated on other products of pharmaceutical chemistry, particularly biologically based products. A few manufacturers established a small line of synthetic organic pharmaceuticals. Monsanto Chemical Company opened its doors with the production of saccharin (an artificial sweetener) in 1901, a product followed by phenacetin (an antipyretic, which reduces fever) in 1907 and phenolphthalein (a laxative) in 1909. Dow Chemical Company, although specializing in bromine, also produced chloroform, a common anesthetic of the day. Branch plants of German firms—Heyden, Merck, and Bayer—made a limited line of synthetic organic pharmaceuticals in the United States, but they primarily imported products from Germany.[31]

Although the prewar American synthetic dyes and pharmaceuticals industry was almost negligible, other parts of the American chemical industry made the establishment of a U.S. synthetic organic chemicals industry competitive with Germany's seem possible, at least on paper. The production of synthetic organic chemicals on an industrial scale required considerable support from other branches of the chemical industries, many of which existed quite profitably in the United States. Alkalis such as caustic soda and soda ash, as well as sulphuric acid, nitric acid, hydrochloric acid, and chlorine, were among the chemicals necessary to carry out the reactions in the manufacture of dyes and pharmaceuticals. In addition, synthetic organic chemicals producers consumed considerable quantities of alcohols as solvents. In general, the American production of these chemicals kept pace and in some cases surpassed developments in Europe. After the Belgian Solvay & Cie. established a subsidiary, the Solvay Process Company in Syracuse, New York, in the 1880s, Americans gradually acquired a domestic alkali industry. By the mid-nineteenth century, several domestic producers of acids, especially sulphuric acid, had established a bustling and often viciously competitive industry. Eugene Grasselli had founded at midcentury a firm that became almost synonymous with heavy chemicals in the United States. In 1870 William H. Nichols also began his long career by producing sulphuric acid and copper sulphate. In 1899, he engineered the merger of twelve sulphuric acid manufacturers into the General Chemical Company, which accounted for between 40 and 70 percent of the American production at its formation. Until about 1900, the sulphuric acid manufacturers primarily served the petroleum industry, which used the acid in refining. By 1914, the United States produced 4 million tons of sulphuric acid; Britain, France, and Germany combined produced 5.5 million tons. The American inorganic chemicals industry produced in volumes that dwarfed dyes and pharmaceuticals, and it gave birth before World War I to the new discipline of chemical engineering that focused on scaling up to mass production, expertise that was also potentially transferable to other parts of the industry.[32]

The production, or rather, the collection, of the organic crudes that formed the raw materials for dyes and pharmaceuticals rose more slowly than in inorganic chemicals production but had also developed noticeably by 1914. Coal was the starting point: one contemporary estimated that an average ton of coal yielded 72 percent coke, 22 percent gas, and 6 percent tar; chemical crudes essential for synthetic organic chemicals made up only about 10 percent of the coal tar. Those crudes included benzene, naphthalene, toluene, phenol, anthracene, and xylene. By the early 1880s, the German steel industry used by-product coke ovens, but American steel firms continued using beehive coking ovens, which vented these hydrocarbons into the atmosphere. Not until the mid-1890s did the U.S. steel industry began to install by-product ovens (usually made by the German firm, H. Koppers of Essen, or its American subsidiary). The shift created a domestic source of raw materials for the tiny synthetic organic chemicals industry.[33]

The United States even manufactured a few organic intermediate products in the years prior to World War I, although international rivals challenged American makers, especially the Benzol Products Company. A series of mergers culminating in 1896 gave the Barrett Company of Chicago the leading position in coal tar products, which it sought to maintain through a research program designed to expand the uses of coal tar. In 1910, Barrett, Semet-Solvay, and General Chemical formed the Benzol Products Company to manufacture aniline, nitrobenzene, and other synthetic organic intermediates. Although an important intermediate in the manufacture of synthetic organic chemicals, aniline had its biggest market in the United States in the realm of solvents and was increasingly used as an "accelerator" in the rubber vulcanization process. The tariff law of 1909 set

Table 1.3. Production of By-Product and Beehive Coke, 1893–1913

	BY-PRODUCT COKE		BEEHIVE COKE		
Year	Tons	%	Tons	%	Total (in Tons)
1893	13,000	<1	9,465,000	99	9,478,000
1901	1,180,000	5	20,616,000	95	21,796,000
1907	5,608,000	14	35,172,000	86	40,780,000
1910	7,139,000	17	34,570,000	83	41,709,000
1913	12,715,000	27	33,585,000	73	46,300,000

Source: The table appears in *By-Product Coke and Gas Oven Plants*, 67. Published by the Koppers Company, the book used statistics from the U.S. Geological Survey reports on the "Mineral Resources of the U.S."

a 10 percent rate on aniline, which had long been on the free list, and this change may have contributed to the creation of Benzol Products. From its by-product coking operations, Semet-Solvay provided a source of coal tar and General Chemical supplied the inorganic acids necessary in the syntheses of aniline and the other products. Barrett contributed its Frankford, Pennsylvania, plant—the former H. W. Jayne Chemical Company factory and site of Barrett's aniline production—to the new joint venture. Despite Barrett's previous experience with aniline, Benzol Products encountered considerable difficulties in generating satisfactory products during the next several years. Its officers claimed Benzol Products Company fell victim to ruthless and unscrupulous German competition. One executive reported that a representative of Britain's largest aniline firm, which maintained a price and market-sharing convention with German producers, had traveled to the United States in late 1912 to threaten the young firm. The emissary offered to supply the American firm with aniline at advantageous prices if Benzol Products would cease manufacturing it, explaining that otherwise the British and German manufacturers would slash prices and drive the American firm out of business. Benzol Products executives trimmed their firm's sails, produced aniline only when they had contracts in hand, and avoided competition with Europeans on the open market. The European manufacturers nonetheless dropped prices to absorb the American tariff, and the Benzol Products Company lost money until the war.[34]

In addition to the chemical industry, Americans had built industries that were not considered part of the chemical industry before World War I but would later become integrally tied to the drive to build a domestic synthetic organic chemicals industry. These were high explosives, Bakelite plastic, and acetylene and related chemicals. In the nineteenth and early twentieth centuries, the manufacturers of high explosives and the manufacturers of chemicals were almost wholly different sets of companies, but as explosives increasingly demanded the skills of trained chemists, the two industries gradually became entwined. In contrast to dyes and pharmaceuticals, U.S. firms had developed greater capacity in high explosives, partly because the chemical processes for the explosives were much simpler than for dyes and pharmaceuticals, and manufacturing explosives required less skill. American manufacturers recognized that European makers, particularly Germany and Great Britain, still possessed considerably more technical expertise in explosives, but the gap was not so large as to prevent Americans from competing, and explosives therefore possessed a distinctly different global market structure. The United States offered a growing market to explosives manufacturers in the late nineteenth and early twentieth century, as mining, railroad and other construction projects, and agricultural development (blasting stumps and ditches) created a large commercial demand. In 1912, Americans purchased 250 million pounds of high explosives, primarily dynamite, which supplanted black powder as the favored blasting explosive. Besides the civilian market, military leaders found two kinds of synthetic organic explosives particularly attractive. Of these, picric acid (trinitrophenol) initially attracted the attention of Europeans, although its chemical instability made it a risky option, and TNT (trinitrotoluene) especially promised to satisfy military demand for a shell explosive that wreaked havoc upon impact but remained relatively safe during handling and firing. By 1902, Germany had adopted TNT as its official explosive for shells. Besides the synthetic organic high explosives, the American explosives industry gained familiarity with cellulose chemistry, particularly the nitrocellulose used in smokeless powders. While cellulose products, sometimes called semi-synthetics, are not covered in this book, Americans' experience with cellulose developed a related expertise in organic chemicals that built a base, potentially, for a wider industry.[35]

The political economy of explosives created far more controversy before the war than did that of dyes or pharmaceuticals, primarily because the explosives industry ran afoul of American antitrust laws. Founded in 1802 as a gunpowder manufacturer, Du Pont had become the largest American explosives firm in an American industry that became increasingly concentrated. Companies such as Laflin & Rand of New York and California Powder Company competed with the Delaware-based Du Pont through the late 1800s. But by 1907, Du Pont had gradually increased its market share; it purchased outright or acquired a controlling interest in a majority of the American explosives companies and provided about three-quarters of the industry's dynamite. Led by a younger generation of du Ponts, the company aggressively expanded its hold on the industry. A few years earlier, in 1902 and 1903, the firm had established research laboratories to maintain Du Pont's leading position in the American industry. Du Pont's dominant position led to charges of monopoly, and the U.S. Department of Justice successfully sued to have Du Pont break up and spin off two independent firms, Hercules and Atlas Powder Companies, in 1912, to create more competition.[36]

The concentration of the explosives industry occurred internationally as well as nationally, and the American industry also joined international agreements that restrained trade in the world market—and that shared technology. After developing dynamite, Alfred Nobel of Sweden successfully established firms in several European countries, and these firms, especially in Great Britain and Germany, formed the core of a dynamite trust established in 1886 to maintain prices and lower costs; not much later, the agreement included profit-sharing. Shortly after the European trust laid plans to enter the American market, the U.S. manufacturers reached agreement with the European dynamite trust to fix prices and divide the world's markets. Although division of markets and profit-sharing remained the primary goals of agreements, after 1907 the explosives manufacturers tied their pooling agreements to the sharing of technology to avoid American antitrust violations. Technically, the antitrust legislation permitted such agreements. Given that the British and German companies still possessed a solid advantage in the science of explosives manufacturing, the "patents and processes" strategy provided both legal cover and a tangible benefit to Americans, particularly Du Pont, as it broadened its base of technical knowledge. As the result of visiting German TNT plants in 1908, Du Pont's Charles Reese set up the research and manufacturing of TNT, giving Du Pont (and the United States) the capacity and skill before World War I to make this new and important synthetic organic high explosive. These "patents and processes" agreements established important precedents for future exchanges of technical knowledge in new product areas. Although the ensuing war excluded the German Nobel firms from participating in the agreements, it bound Du Pont and the British firms more closely together.[37] As Du Pont diversified beyond explosives into a wider range of synthetic organic chemicals, it would continue to look to Britain for mutually beneficial ways to share the fruits of technological innovation.

Like high explosives, two other prewar developments would help to reshape dramatically the course of the synthetic organic chemicals industry in the interwar years and beyond. First, a Belgian immigrant to the United States nurtured the development of a new product: plastics. In 1909, Leo H. Baekeland patented Bakelite, a phenolic resin that could be molded into virtually any shape. Baekeland had already won respect as an industrial entrepreneur because of his development of Velox photographic paper. To synthesize Bakelite, Baekeland mixed phenol and formaldehyde under higher temperatures and pressures and learned to control the reaction to achieve the desired properties of his moldable synthetic phenolic resin. Although the production of Bakelite and its subsequent imitators remained a comparatively small business in 1914, the range of uses for the new material, which included consumer goods such as combs, pipes, and billiard balls, and industrial goods like electrical insulation, continued to expand. After the war began, Baekeland and other industry advocates strongly emphasized the American origins of Bakelite. Second, over in the gas-lighting industry, firms had learned to mass produce acetylene, which the industry used to enrich their gas. Related firms used acetylene for lights on bicycles and automobiles, and a French process for oxy-acetylene welding quickly created another growing U.S. market for acetylene before the war. Unlike the dyes and pharmaceuticals, which came from the aromatic branch of organic chemicals (those with ring-structured molecules), acetylene and a host of related chemicals came from the aliphatic (or chain molecule) branch. As with high explosives, the chemistry of the products drew them into the wartime drive to build a domestic synthetic organic chemicals industry.[38]

Despite the depth and breadth of the American industrial landscape, which seemed like a promising home for a synthetic organic chemicals sector, both supporters and skeptics worried about the relevant human capital. Were American chemists and university science capable of supporting this sector of the chemical industry? Few industrial products required such advanced scientific and technical understanding, from laboratory innovations to plant production, as did synthetic organic chemicals, particularly the dyes and pharmaceuticals. And the degree of tacit knowledge among German chemists, managers, and other employees in all facets of the industry had required decades to build.

As the wartime demand for knowledge and skill in synthetic organic dyes and pharmaceuticals rose, the United States found useful existing expertise primarily in two places: (1) the prewar European immigrants, whether from Germany, Switzerland, Great Britain, or elsewhere, in the employ of the few American firms or of the American branches of German firms, and (2) the academic chemists, many of whom had received at least part of their advanced training at German universities. But the future of the industry would depend on the ability of American universities to train the next generation.

Measured against Germany, academic chemistry in the United States had little to offer an organic chemicals industry and is yet another part of the explanation for German superiority in the industry's development. The strengths and weaknesses of academic chemistry reflected the contours of the industry. In Germany strong graduate programs produced Ph.D. organic chemists practiced in research, but American colleges and universities churned out analytical chemists with undergraduate degrees geared toward an industrial career that required more practical skills than theoretical depth. From 1861 to 1899, the six leading American schools granted 220 doctorates in chemistry. The number increased after the turn of the century, with 78 doctorates awarded from all schools in 1910 and 97 in 1914; however, Ph.D. chemists constituted only a small fraction of the estimated 16,000 chemists in the country in 1910 and, of these, organic chemists constituted a smaller fraction yet. Only the Johns Hopkins University and the University of Chicago had successfully incorporated research into their graduate curriculum by 1900, and both had patterned their techniques after the German universities. At Hopkins, renowned organic chemist Ira Remsen trained many students who eventually taught chemistry, and he explicitly discouraged them from entering industry. Many university chemists, however, found industrial consulting a lucrative sideline to bolster meager research funds and academic salaries, and one, Arthur D. Little, parlayed consulting into a full-time and well-known business.[39] Whether American chemists could master synthetic organic chemicals sufficiently to support a viable domestic industry remained an unanswered question, one that arose again after August 1914 with much greater urgency and tension.

Despite the scarcity of practitioners of synthetic organic chemistry, chemistry as a profession grew considerably in the late nineteenth and early twentieth centuries, and it had an increasingly effective professional society in the American Chemical Society (ACS). Established in 1876, the ACS remained a professional society largely confined to New York City until the 1890s, after which the society grew rapidly and sponsored three journals: the *Journal of the American Chemical Society*, *Chemical Abstracts*, and the *Journal of Industrial and Engineering Chemistry*. In the decade before the war, the rapid expansion of the industry and profession forced the ACS to grapple with several tensions. One centered on a sharper specialization of the profession into subfields, primarily inorganic, organic, analytical, and physical. Another arose as membership among chemists in industry approached and then surpassed the number in academia. The primary solution for these growing pains was to create new subunits within the ACS to accommodate the expanding array of particular interests. In 1908, the ACS formed the Division of Industrial and Engineering Chemistry, the first of seven divisions established before the war, and began the *Journal of*

Industrial and Engineering Chemistry in 1909. Both the new division and its journal were meant to appeal to the industrial chemists in the ACS. In 1890, to solve an earlier crisis between the New York City chemists and everyone else, the ACS had permitted local chapters, which had helped to address the growing number of chemists and their geographic dispersion across the nation. Over the next twenty years, the increase in membership was dramatic: from 238 members in 1890 (only eight more than in 1876, its founding year) to 5,081 in 1910 and then to 7,170 in 1914. By 1918, membership stood at 12,203, with the greatest wartime increase occurring in 1917 and 1918. In 1912, the ACS moved its headquarters to Washington, D.C., marking symbolically the shift over the previous two decades from its historic identification with New York City to a professional society with a broader geographic constituency. The move served the ACS well when the war drew chemists and the chemical industry directly into national politics.[40]

###

The experience of the U.S. synthetic organic chemicals industry invites that enjoyable but entirely speculative, counterfactual question: what if World War I had not occurred? Would Americans have continued to view the industry as small, specialized, and best left to the German experts? Once the war began and Americans resolved to build their own industry, they consistently measured their progress against the historical development of the German industry, as they understood it. Given the decades of German experience, Americans faced a steep challenge trying to imitate the German development and success. Independent of the trade in synthetic organic dyes, however, Americans in the decades before the war had increasingly been creating conditions more conducive to supporting a synthetic organic chemicals industry, although not necessarily one similar to the German. The steel industry found it more profitable to capture by-products, the rubber industry was helped along by the nascent automobile industry, the giant petroleum industry offered an entirely different supply of organic crudes, German and other Europeans' expertise leached out of subsidiaries and importing agencies, and Leo Baekeland invented Bakelite plastic. American universities increasingly adopted research, though remained notably less prominent than German universities, particularly in chemistry. On the eve of World War I, conditions existed to develop an industry, but whether and how an industry might develop were entirely contingent on a range of entirely unpredictable decisions by any number of people, including many outside of industry. In 1914, the determination of Europeans to go to war dramatically reshaped the context in which Americans made decisions about synthetic organic chemicals.

2 / American Manufacturers, German Chemicals

Dyes and Pharmaceuticals, 1914–1918

At the outbreak of hostilities, American entrepreneurs embraced the manufacture of the signature German synthetic organic chemicals: dyes and pharmaceuticals. While the war heightened uncertainty and risk, not least because of its unpredictable duration, entrepreneurs seized the opportunities born of war. To many in the industry, wresting chemical markets from the German industry also became a patriotic mission larger than themselves. Chaotically launching into domestic manufacturing, Americans faced the daunting reality of German dominance, but many also believed the German experience could offer them guidance and models on which to pattern their industry. Americans wanted the synthetic dyes and pharmaceuticals the Germans had long provided, and most expected an American industry would succeed or fail based on the extent to which they could supply these core German products in such short supply during the war. Americans' lack of experience and the war economy, however, led entrepreneurs to pursue a range of strategies in attempting to acquire expertise and establish a domestic industry.

The German industry's representatives in the United States, the importing agencies and manufacturing subsidiaries, confronted a difficult predicament. Their experience reflected in microcosm the dramatic, if selective, destruction of the nineteenth-century global economy; through war, the nation-states rejected the expanding economic interdependence of previous decades. For those who had spent their entire careers working with the German industry—some bound by family ties—the disintegration of the international commercial networks carried very personal consequences. The importers and subsidiaries of German firms tried to hold their businesses together in the face of increasing centrifugal forces brought on by war: some turned to manufacturing, some distributed American-made chemicals, and some came apart under the pressure. Ironically, the importers' various strategies to survive the disruptions made many of them important contributors to the domestic synthetic organic chemicals industry.

Comparisons to the German industry pervaded American thinking, and, despite a few Americans' attempts to imagine an industrial structure that differed from the German and despite the domestic manufacturers' multiple strategies, Americans had difficulty conceptualizing a successful industry that varied to any great degree from the German, as they understood it. These concerns played out most explicitly in ongoing discussions about the size of firms in the nascent industry. Small American manufacturers responded most quickly and nimbly to the wartime shortages, sometimes offering just one product to the market. They varied greatly in their experience, whether in technical or managerial skills, and arrived and departed from the market with startling rapidity, often failing in a literal bang of explosive chemicals. Some small firms did well during the war, but many observers and participants believed any durable industry would require large firms that could compete with the German giants in scope and market power. In 1916, one group of investors and entrepreneurs aimed for large-scale, integrated production and put together Federal Dyestuff & Chemical Corporation; by the end of the war, believers in big business placed their hope for the domestic industry in National Aniline & Chemical Company and E. I. du Pont de Nemours & Company, the two firms that most deliberately embarked on strategies designed to compete with Germany by manufacturing a broad range of interrelated chemicals.

To a great extent, success in the American industry depended on how quickly and effectively the firms could accumulate sufficient organizational and technical abilities. Firms of all sizes needed to create responsive and functioning organizations, and the war both helped and hindered managers' quest to establish productive routines, whether in laboratories or factories or sales. On one hand, filled with a sense of exceptional opportunity, entrepreneurs and technical personnel took more risks during the war. On the other hand, the war's unpredictability hindered managers' attempts to provide stability long enough to build routines in manufacturing and organization. Manufacturers could hire experienced personnel from elsewhere, but whether they could cohere into a reliable unit was another question.

Armed with widely varying sets of individual and collective skills, entrepreneurs ventured into the field with several plans to capture market share in the volatile but newly inviting industry. If they remained intact, most firms extant before the war readily responded to the high prices. Firms such as Schoellkopf, William Beckers, and Monsanto aggressively expanded. New start-ups might aim to make only one chemical for the duration of the war;

others more ambitiously wanted to build a thriving niche firm that would survive beyond the war. Medium-sized manufacturers experimented with supplying specialized sets of chemicals, whether intermediates or high-end dyes or pharmaceuticals or the few products sold in bulk. Targeting the global competition, the largest American manufacturers wanted to match the German companies. War added several layers of complexity to planning effective entrepreneurial strategies, and, while the German model remained powerful for Americans, their available skills, experience, and resources led entrepreneurs to navigate the choppy waters along a range of routes.

Trade associations were another strategy the manufacturers pursued to acquire and master the information about the industry that they lacked. Founders of the associations expected to address issues common to their membership, such as standardization of dyes, but one group also experimented with an open price society, a novel mechanism to exchange information on costs and prices among members. Formation of the trade associations provoked broad discussion about the most appropriate structure for the associations but also about the most effective structure of the American industry. While common concerns and fear of the German industry brought Americans together, divergent interests, resources, and strategies also worked against cohesion.

American Importers of German Chemicals

The American importers and subsidiaries for German companies were among the first firms forced to confront the new global chaos in chemical markets, and many of them turned to or expanded manufacturing even as they sought to hold together the transatlantic commercial networks in synthetic dyes and pharmaceuticals. Despite their years of experience in the industry, the importers and subsidiaries faced a calculus as uncertain and daunting as the other entrepreneurs who contemplated their prospects at manufacturing profitably in the United States. They possessed an effective distribution network, but one buffeted by the shortages and speculative prices; some already manufactured dyes and pharmaceuticals before the war but relied heavily on their German partners for technical assistance; and they wanted to alienate neither their American market nor their German partners, an almost unwinnable diplomatic undertaking. Immediately upon the start of the war, the European belligerents restricted the exports of many chemicals, particularly explosives or medicines deemed necessary to fight the war, and the importers and subsidiaries coped as they could with the stingier transatlantic trade.[1]

The initial strategies of the importers and subsidiaries centered on trying to keep their prewar networks intact while increasingly depending on American-made dyes and pharmaceuticals. With the governments of Great Britain and Germany determined to place restrictions on trade to and from the United States, dyes and pharmaceuticals transactions moved from the private sphere into the realm of high diplomacy. The importers relied on the pacifist secretary of state William Jennings Bryan and the Department of State to negotiate with the British government to gain the permission and permits that allowed passage of German dyes and pharmaceuticals through the blockade. They also worked to persuade the German government, through Ambassador Johann H. G. von Bernstorff as well as through their German chemical companies, to release the dyes and certain pharmaceuticals to the American market, emphasizing the high-stakes game of public relations underway in the United States. Small batches of high-value pharmaceuticals could travel via parcel post, particularly through Switzerland, Denmark, or Sweden. Otherwise, however, transatlantic commerce during war required permission from three governments, contracting with U.S.-owned ships, working with American textile manufacturing associations to determine orders, obtaining promises from American textile firms not to re-export, coordinating exports with an association of German chemicals manufacturers, and not a small bit of luck.[2]

As Great Britain and Germany jousted for the sympathy, or at least the neutrality, of the United States, neither could afford to give up strategic advantages in war. With the greatest navy in the world, British leaders only grudgingly allowed transatlantic commerce between the United States and Germany. For commercial as well as strategic reasons, the British preferred that none of the German dyes reach the United States; the shortage of German dyes hit British textile manufacturers more quickly, and they had little interest in watching their American rivals thrive and encroach on British market share. In March 1915, British officials in the Foreign Office pondered a recent U.S. request for another shipment of German dyes; they wanted to deny the request. As one wrote, "What I fear is that if dyestuffs are necessary—every other article made in Germany will be claimed as being equally necessary" in the United States. In addition, if the British made an arrangement, the Americans "will ask for more" because "that is what they call business." The official concluded by suggesting that the British "[hold] up a consignment or two of dyestuffs and see what result it produces." Except for dyes on the *Deutschland* submarine in 1916, the Americans received no more dyes from Germany after March 1915. "As a boon to suffering humanity," as Americans put it, the British allowed pharmaceuticals to cross the Atlantic sporadically into 1916.[3]

The Germans, on the other hand, insisted that the United States demand from the British the right to uninhibited transatlantic commerce in order to live up to proclaimed American neutrality. The Germans felt especially acutely the shortage of American cotton, which the British denied on the basis that it formed an ingredient in nitrocellulose explosives. The Germans also worried that their chemicals, destined for the United States, might end up in the hands of the Allies, either because of Allied confiscation or because Americans resold dyes and pharmaceuticals back across the Atlantic. Herman Metz, the Hoechst importer, traveled to Germany at the end of 1914 to facilitate shipment of chemicals, and he was astonished at the fever of war, which even led to "odes of hate" against England performed in music halls. German officials in the United States encouraged their government to cooperate in transatlantic chemicals shipments, but they also warned their fellow Germans about Metz: "He is the real type of the 'smart' American, who as a matter of course uses every situation for his own business purposes."[4] By 1915, perhaps the only point of agreement between the British and Germans lay in their assessment of American opportunism.

With their business tossed into the European war, the American importers and subsidiaries of German firms tried to steady their fortunes by carefully cultivating existing and new relationships with three critical groups: the American textile industry, the nascent domestic synthetic organic chemicals industry, and their partners in the German industry. Satisfying one constituency could complicate relations with another, however, and the importers faced increasingly difficult choices as the war dragged on and they could see with more certainty that the postwar markets would be significantly different from previous decades.

Importers worked strenuously to keep their customers in the textile industry satisfied and tethered to their firms as the dyes crisis deepened. Like other industries, the giant textile industry haltingly refocused their output to the war, responding to large military orders from Europe and then from the U.S. government while coping with supply shortages. Many manufacturers recognized the war's consequences for the synthetic dyes trade almost immediately in August 1914 and purchased reserve inventories as they could, which exacerbated the general shortages. "On August 1, 1914, the news of the European war staggered us all. So far as dyes were concerned, every consumer seemed to realize at once that he had to have at least three or four times his usual quota and urged delivery," Herman Metz later recalled about the beginning of World War I. For the textile manufacturers, the dyes typically contributed less than 1 percent of the cost of production—although it could be as much as 12 percent for heavily plush fabrics—and even the great increase in dyes prices rarely forced a shutdown.[5] Caught without any dyes, however, a textile manufacturer not only had costly production choices to make but would also be tempted to seek dyes from suppliers other than their regular importers.

For the importers, the textile industry provided their source of revenue and means of survival, and they all recognized the need to stay in the good graces of their long-time customers. One strategy pursued by nearly all agencies was to limit price increases and not to appear to take advantage of their customers in a crisis. "We have won this [textile] industry by our conscientious and decent behavior," wrote Berthold A. Ludwig of the Cassella agency in New York to his German partners in 1916. "I cannot share your fears that we have treated our customers too unselfishly," he responded when the Germans wondered why profits had not been higher, given the runaway prices. "This capital will bring in big interest after the return of normal conditions." Ludwig centralized orders in the New York City office and supervised an analysis of prewar sales to give priority to their best customers when distributing dyes. The agency also carefully maintained a reserve of dyes, even if that meant filling orders more slowly, to keep "the firm as a factor in the business through continuous contact with the customers . . . [and to keep] the organization as intact as possible."[6] To maintain their relationships with the textile industry, the importers increasingly turned to American-produced chemicals.

With the war, the importers and subsidiaries for the German firms became a part of the American synthetic organic chemicals industry, either as distributors or manufacturers or both—a path few of them would have chosen without the crisis. The importers' expertise was their means for surviving the war, which they understood very clearly. Protecting and leveraging it in the volatile market, however, took them in several directions, and nearly all lost talented employees, failed in some of their new ventures, and worked in an atmosphere of increasing hostility toward their German affiliations. The new American manufacturers directly threatened the importers' prewar interests, but in many ways the importers and the manufacturers needed one another, particularly as the war grew longer and more severe.

Most of the major importers became distributors for several of the inexperienced American manufacturers, a task that often required them to enforce quality control. With their own technically trained salespeople, the importers could put American dyes of questionable quality through additional processing to provide a more uniform product—sometimes mixing stocks of German-made dyes with those from American producers before delivery. Also, with their thorough knowledge of the American market, importers could advise manufacturers about which dyes to make, although this knowledge they tried to parcel out carefully to protect their distinctive asset in an industry desperately short of information and experience.

Of the major importers associated with the German firms, the importers for BASF steered the most conservative course. Adolf Kuttroff and Carl Pickhardt continued to sell German chemicals as long as they held any, whether from prewar stocks or the sporadic shipments that made it across the Atlantic early in the war. They purchased from American manufacturers and also obtained limited quantities of British dyes. These were lucrative years, and the Americans remitted $1.5 million back to Ludwigshafen as the German share of profits for 1914 to 1916. By the end of 1915, however, business slowed and the importers substantially reduced the capital stock of the Badische Company and partially mothballed their agency. When the *Deutschland* submarine arrived unexpectedly in July 1916, they needed to borrow funds to purchase their share of dyes. In 1918, the importers dissolved the Badische Company and transferred its assets to a new firm, Kuttroff, Pickhardt & Company, a name they had used from 1899 to 1905. Large numbers among their sales staff left the firm, but they also increased the salaries and bonuses of the employees they wished to keep.[7]

At the Cassella Color Company in New York, importer William J. Matheson and his right-hand man, Robert Shaw, thought carefully about their strategies to stay profitably afloat. First, they also targeted particularly important employees for retention. After other employees had abandoned their Providence office, for example, Matheson and Shaw acquiesced without much complaint to demands for a raise from $2,500 to $4,000 for the last experienced salesman there. More intriguingly, they gave a job to the brother of a highly valued salesman to earn the salesman's goodwill and keep him from succumbing to tempting offers elsewhere.[8]

As much as the Cassella leadership wished to focus on sales and distribution, the deteriorating conditions led them to plan manufacturing by the end of 1915. "The selling organization in our business is the key to success," believed Matheson, but the diminishing stocks of German chemicals and growing American industry led the importers to send pointed telegrams to Germany: "Manufacturing here assuming more dangerous proportions ... regard immediate supplies imperative ... we are about the only large importing concern not already connected with the manufacture here." In November 1915, Matheson and Shaw incorporated the Century

Colors Company to pursue manufacturing, but they capitalized at only $500, suggesting an ongoing desire to remain distributors, a goal boosted by the two *Deutschland* cargoes in 1916. During 1916, while they built a small plant in Nutley, New Jersey, they carried on negotiations with the General Chemical Company, an American inorganic chemicals company that had moved into intermediates production, to develop a relationship in which the importers' expertise and patents could be used to the mutual benefit of both firms. As discussed below, they ultimately became part of a much larger merger led by William Nichols of General Chemical.[9]

On the eve of World War I, the Bayer Company in the United States possessed the most extensive manufacturing operations among the German affiliates, making both the strains of war and the opportunities for growth especially pronounced for the Rensselaer, New York, firm. Unlike many of the other importers, they sold their substantial stock of German dyes and pharmaceuticals in the first few months of the war, but then turned to an aggressive expansion of their plant; their broader line of products included intermediates for their dyes and pharmaceuticals, as well as newer categories such as azo and alizarin dyes. They manufactured large quantities of navy blue for uniforms. Between 1915 and 1918, the Bayer Company constructed ten new buildings used principally for dye-making. Arthur Mothwurf, a Ph.D. chemist about forty years old, ran Bayer's dyes operations; prior to the war, he worked in Germany in Bayer's export department, a post that had brought him to the United States on annual trips since 1908. In 1914, the war left him stranded in the United States, and the Bayer Company executives gave him the critical role in dyes production. The company carried an unusual depth of institutional knowledge from prewar manufacturing, but it also struggled to retain employees.[10]

Bayer rapidly expanded its own plant in Rensselaer, but the company also purchased dyes from other suppliers, in particular developing a close relationship with the Williams & Crowell Color Company. In this case, the expertise of the Bayer Company made a direct contribution to the success of an American start-up, and in exchange, Bayer acquired a relatively reliable supply of dyes for its own operations. In 1915, two Rhode Island entrepreneurs established a factory in Packersville, Connecticut, to make sulphur dyes. The manager of the Bayer Company's Providence sales office, Christian Stamm, cultivated strong ties with the inexperienced firm, advising the entrepreneurs on the market and on which dyes to produce. The Bayer Company was their first customer and became Williams & Crowell's selling agency, in which the "Bayer Company did all the standardizing and adjusting of shading" to effect the sale. The relationship grew profitably

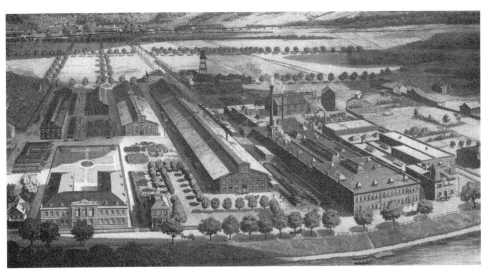

FIGURE 2.1. Bayer Company subsidiary plant in Albany, N.Y., 1918[?] (*Beiträge zur hundertjährigen Firmengeschichte, 1863–1963*, 30. Photograph reproduced with the permission of Bayer Business Services GmbH, Corporate History and Archives.)

for both firms, and demand for Williams & Crowell's sulphur dyes reached new highs with a government contract in 1917. In fact, the founders felt increasingly burdened by their success and wanted out; not for the first time, Stamm pushed his bosses in New York to acquire Williams & Crowell. In January 1918, several of the Bayer Company directors individually advanced money to purchase the firm, while the Bayer Company itself advanced money for working capital. By 1918, the political conditions in the United States lay heavily on the minds of Bayer executives and influenced their decision to buy. Those involved in the transaction subsequently characterized it as an "anchor to windward" for the stockholders, a hedge against the uncertainty of war and the possibility that the Bayer Company would change ownership. As they increasingly worried that the war had irreparably altered the industry and prewar networks, the executives and top employees of the Bayer Company—the firm most readily identified as German—purchased Williams & Crowell as an "American" back-up plan for themselves.[11]

Besides the war crisis, Bayer faced the expiration of their U.S. patent on aspirin in 1917, and the firm placed Hugo Schweitzer, their talented pharmaceuticals chemist, in charge of protecting their aspirin business against the onslaught of competitors. For a few years prior to the expiration, the firm had geared up for intense market and legal struggles over the profitable drug. The Bayer managers advertised extensively, attempting to tie the names

Table 2.1. Bayer Company Dyes Sales, 1914–1918

	DYES MANUFACTURED BY BAYER				DYES PURCHASED (INCLUDING FROM BAYER IN GERMANY)			
	in 1,000 lbs.	%	Value	%	in 1,000 lbs.	%	Value	%
1914	1,219	8.4	$358,000	6.9	13,308	91.6	$4,855,000	93.1
1915	827	16.5	$332,000	9.5	4,177	83.5	$3,161,000	90.5
1916	3,389	77.0	$1,086,000	38.7	1,005	23.0	$1,717,000	61.3
1917	2,667	59.6	$1,540,000	55.0	1,809	40.4	$1,262,000	45.0
1918	3,073	96.8	$1,367,000	89.3	103	3.2	$164,000	10.7

Source: Lybrand, Ross Brothers, & Montgomery, CPA, "The Bayer Co., Inc., Synthetic Patents Co., and Hudson River Aniline Color Works. Report on Examination of Accounts for the Period June 3, 1913 to June 30, 1918," November 1, 1918, p. 13, National Archives and Records Administration, RG 131, entry 155, box 34.

"Bayer" and "aspirin" together in the public mind. They tried to bolster their exclusive claim on the trademark by establishing a corporate entity known as "The Aspirin Corporation" in the belief that a corporate name received more legal protection under the law than a trademark name. But, as in their dyes business, Bayer also developed relationships with other manufacturers and, for example, sold raw materials to Heyden in exchange for a cut of Heyden's aspirin profits. They bought crucial raw materials from American suppliers, including phenol from Thomas Edison, who had begun its manufacture to support his phonograph business in the face of shortages.[12]

Compared to most firms in the nascent industry, Merck & Company—the American branch of E. Merck of Darmstadt, Germany—expanded their manufacturing with relative ease. Prior to the war, their largest-selling products had been morphine, codeine, and cocaine, but the Merck line also included important synthetic organic chemicals such as acetanilid, the leading headache remedy, and chloral hydrate, a sedative they produced in Midland, Michigan, under license from another German manufacturer. After preliminary research in 1914, Merck began production in 1915 of hydroquinone, an important X-ray and photographic chemical, and quickly gained 75 percent of its market; they also made phenol, the base for a number of synthetic pharmaceuticals. Adjacent to their plant in Rahway, New Jersey, Merck helped to establish, under the English chemist T. H. Davis, Rahway Coal Tar Products, a firm that it purchased outright in 1917 and that made many of the necessary coal tar bases and intermediates, including phenol, aniline, and nitrobenzene. Under the direction of George Merck, Merck recapitalized in 1917 from $250,000 to $1 million, and by the end of the war employed more than 700. From sales of $3.9 million in 1913, Merck grew to over $8 million in 1917 and 1918, which included exports to Allies and other markets around the world. Unlike many other German affiliates, Merck's profits and sales kept growing in the war, rather than peaking in 1915 and 1916, indicating the exceptional degree of their success in making products previously received from Germany.[13]

The relationship between the American importers and their German partners suffered enormous strains during the war. While the United States remained out of the war, the two sides carried on intermittent communication, but the decades of shared business interests only partly mitigated the uncertainty and misunderstanding resulting from war. Until 1917, the importers continued to interpret American conditions for their German partners, but the stress of the situation and the infrequency and confiscation of the mails left each side wondering about the intentions and actions of the other. Why, asked the German executives at Cassella, had the importers and textile industry not pressured Congress to demand from the British unimpeded trade in German dyes? The Germans had an unrealistic picture of American politics and anti-German feelings, replied the importers. They warned that hardball tactics would only backfire: "If the German Government refused absolutely to permit the export of products, the real victims will be, not the spinners or consumers in this country, but the producers on your side and the importers on ours.... The consumers will accept undyed products as a necessity of war." Meanwhile, the German manufacturers had agreed among themselves not to expand production in the United States, and each "was placed in an embarrassing position" among their colleagues for the independent behavior of their American partners, which could include unapproved exports as well as new production facilities. Still, as long as transatlantic commerce permitted it, the importers sent over reports on their activities, plus the German share of revenues, as they had long done prior to the war. Nearly every importer undertook a corporate reorganization during the war to place legal control more firmly in American hands and to remove the more politically charged German names. The war pushed the partners apart, and on both sides of the Atlantic they increasingly understood they would not return to a pre-1914 market. The challenge lay in devising strategies to cope with the war and the unpredictable postwar world; the war damaged the commercial bonds between the American and German partners, and the nationalism on both sides threatened to destroy the common transnational interest they shared.[14]

Herman Metz, the Hoechst importer, used his political connections from years in New York City government (1906–1910) and in Congress (1913–1915) to push officials in the Departments of Commerce and State to facilitate the diplomacy necessary to trade through the Atlantic war zone. While Metz was in Germany in late 1914 to extract a shipment, he grasped immediately the diverging interests between the German and American partners, and he returned home determined to undertake manufacturing himself, even if he maintained his doubts throughout the war about the viability of an American industry. His tactics in the war and beyond showed Metz to be a political and entrepreneurial master—the opportunist the German diplomats had described in 1914. Metz wasted little time before enlarging his own Consolidated Color Company, and in 1915 he also bought a half interest in the Central Dyestuff & Chemical Company, another small Newark, New Jersey, dyes firm that predated the war. Initially, Metz continued his dyes and pharmaceuticals sales through Farbwerke Hoechst, the official name of his importing agency since 1912, but like several other importers he created new corporate entities over which he had more control and which sounded less German. First came H. A. Metz & Company, Inc., in 1915 to handle dyes, and then H. A. Metz Laboratories, Inc., in 1917 to carry Novocaine, Salvarsan, and other pharmaceuticals. The agreement with his Hoechst partners contained provisions prohibiting Metz from manufacturing as a rival, and through 1915 the two sides carried on a tense correspondence, captured by the British. "We are pained at the thought that you should . . . be interested . . . in the manufacture of . . . new articles," wrote a Hoechst executive to Metz. Over the course of the year, however, as the prospects for renewed commerce diminished considerably, the dismayed German executives resigned themselves to the new reality and agreed to allow Herman Metz's brother Gustave, a chemist, to spend time in their Salvarsan plant in 1916 to learn their process.[15]

The Bayer Company executives attempted to maintain close working ties to the German firm. First, the American subsidiary took over the markets of South and Central America, which had also suffered shortages as the result of the British blockade, and collected the revenue for the German parent firm. Herman Seebohm, the president of Bayer Company and brother-in-law to Carl Duisberg, could occasionally ship the money to Leverkusen, but he also maintained an account in New York on their behalf. Second, the German firm provided the American subsidiary with explicit technical advice to help expand their production, particularly for making aspirin with the latest techniques used in Germany.[16]

In Germany, executives of the chemical firms tried to cope with the change that war forced on their industry. At times, they became very caught up in the war fever: in April 1915, after early German military successes, an E. Merck correspondent pondered the usefulness to the industry if the Germans completed a canal from the Rhine through newly conquered Belgium to the giant port of Antwerp, as rumors suggested. By the end of 1915, with the war's end no longer in sight, the tenor of the conversation between the

American and German partners of all firms changed, reflecting a certain acquiescence that the war had greatly diminished the control they had over the future of the industry. In 1916, the German executives responded in part by combining into a stronger cartel that brought the six major firms and several others into closer cooperation. With their combination, the Germans acknowledged that they would need to fight after the war for their pre-1914 market position, and they hoped their consolidation would reduce their costs even further below their growing global competition.[17]

Given their familiarity with the experienced German industry and the American market, the importers stood in a unique position to assess the progress and prospects of the domestic industry, assessments their German partners persistently requested. The importers maintained great confidence in the technical and commercial superiority of the German industry, but they believed the nascent industry posed a threat, particularly if Congress stepped in to protect the American manufacturers. Hugo Schweitzer of Bayer believed the United States certainly possessed the resources to build a domestic industry but doubted that it could efficiently produce the more complicated dyes under normal conditions. He expected Americans would find the "greatest lack" in "technically trained chemists." Metz shared Schweitzer's skepticism; in 1916 he publicly declared: "All of these efforts are temporary . . . real relief will not be had for many years to come, no matter how hard we try to bring it about." The price differentials were simply too great. But he understood better than most the opportunities in abnormal conditions. Cassella received mixed messages from their American partners: in 1916, Ludwig wrote that "organisation and the cooperation of isolated businesses is lacking, as well as the scientific and technical staff. The total of capital concerned dazzles the American observer before all things. He overlooks the fact that the most well-intentioned critic cannot find people in the ranks of this industry." Ludwig's boss, William Matheson, largely agreed, but expected that "a few limited specialties which can be made in a large way in the highest perfection" could carry an American industry beyond the war.[18] The irony of the importers' position was that they made significant contributions to the American industry; whether as distributors or manufacturers, the importers voluntarily and sometimes involuntarily transferred their expertise to inexperienced American producers.

Domestic Manufacturers

The nascent American industry embarked on a wild ride with a heady mix of unparalleled opportunity and uncertainty. The existing and new

Table 2.2. Selected Dyestuffs Used by 23 Important Cotton Manufacturers, 1913 and 1916

DYESTUFFS	TOTAL AMOUNT USED				AVERAGE PRICE PER POUND		PERCENTAGE CHANGE FROM 1913 TO 1916	
	1913		1916					
	Pounds	Value ($)	Pounds	Value ($)	1913 ($)	1916 ($)	Pounds (%)	Value (%)
Synthetic Indigo	3,758,195	380,350	1,349,418	1,540,063	.15	1.15	-64.1	165.4
(Natural Indigo)	—	—	963,547	1,905,746	—	1.98	—	—
Sulphur black	501,968	88,268	810,971	483,705	.18	.60	62	448
Sulphur brown	189,943	51,216	463,238	187,274	.27	.40	143	265.6
Indanthrene blue	653,718	175,703	112,450	112,450*	.27	1.00	-82.8	-36
Hydron blue	85,967	25,613	25,781	25,781*	.30	1.00	-70	.2
Diamine black	96,763	31,741	25,032	54,122	.33	2.16	-74.1	70.5
Paranitraniline	11,862	1,752	7,210	9,466	.15	1.31	-39.2	440.3
Benzopurpurine	6,144	1,397	4,629	4,953	.23	1.07	-24.7	254.5
Alizarin blue	4,241	7,845	—	—	1.85	—	—	—
Alizarin	53,840	6,840	2,785	22,280	.13	8.00	-94.8	228.6
Primuline	9,407	2,124	1,175	4,276	.23	3.64	-87.5	101.3
Diazo indigo blue	446	267	—	—	.60	—	—	—
Total selected dyes	5,381,901	804,857	3,766,236	4,350,116	.39	2.03	—	—
Total all dyes	10,289,030	1,840,582	9,072,734	6,258,633	.18	.69	-11.8	241.5

*Estimated

Source: "'Textile Consumption of Dyestuffs," *Drug & Chemical Markets* 4 (February 6, 1918): 5–6, which takes the numbers from the U.S. Tariff Commission.

entrepreneurs drew on a patchwork of expertise where they could find it—certainly the importers and the few small prewar firms, but also the substantial inorganic chemicals industry. They constructed their firms and larger industry in a complex—even trial and error—set of strategies that necessarily had to address domestic resources and business culture but also incorporated their perceptions and analysis of the German industry's success.[19] Americans understood that the German industry had developed networks of interdependent chemicals—by-products from one process could often be used as inputs in another—and that these networks of exchanges within the firms and industry could be the sites of great cost efficiencies or inefficiencies. They wondered how to organize the production of raw materials, intermediates, and the finished dyes and pharmaceuticals—stages that could each have multiple steps. Could a small firm rely on the market for supplies, particularly during a war? Should a middle-sized firm develop its own sales network? Or, could a large vertically integrated firm best control the heightened risks of a market disrupted by war and more realistically compete with the German industry? While most industry observers looked for the emergence of large firms, in practice, American manufacturers tried to make synthetic dyes and pharmaceuticals with a great variety of strategies and organizations.

After forty years of competing with the German imports, the Schoellkopf organization claimed "hard knocks" boasting rights among American manufacturers, but it had largely remained dependent on German supplies of intermediates and suffered from shortages early in the war. The firm stood in the best position to benefit from the crisis. As J. F. Schoellkopf boldly proclaimed in 1916: "This is the first chance we have had in 35 years to make a decent profit, and we are making it." From 250 employees in 1912, his firm grew to 600 in 1916. To cope with the shortages and to take full advantage of the wartime possibilities, Schoellkopf pursued four strategies in the early years of the war. First, given the advantage of a three-month stock of intermediates in August 1914, the firm "hastily improvised [sic] temporary plants" to manufacture three intermediates—benzidine, dinitrobenzene, and H acid (a sulphonated naphthalene compound)—key to the most basic blacks and blues. Second, as its stock of German intermediates dwindled, the firm cut its range of colors from 120 to 15, concentrating on blacks and blues, the best-selling colors. In early autumn 1916, the firm began constructing a new plant with a capacity of 30 million pounds of dyes per year, 10 times its prewar output, and 30 to 40 times its early 1916 output. Third, to reduce their risk, the firm entered into long-term contracts with dyes consumers before it undertook the next stage of plant expansion. The consumers benefited from a nearly guaranteed source of dyes, although they viewed this arrangement impractical for the great variety of dyes purchased in small lots; Schoellkopf secured a market share for the duration of the contract, which was more predictable than the length of the war. Like most entrepreneurs, Schoellkopf also charged prices high enough to amortize quickly the cost of new plants.[20]

Finally, Schoellkopf established its first research laboratory, hiring Professor Clarence G. Derick away from the University of Illinois to become the research director in the summer of 1916. The firm promised Derick both financial resources and freedom from intense managerial oversight. As the company constructed a large new laboratory building, which opened the following April, Derick began hiring his staff. In the interim, he also pored over the available literature, including files of information on dye consumption in the United States collected over the years by Willard Watkins, a Schoellkopf manager. Derick believed that Watkins's files provided the specialized and accurate information that only an experienced firm could possess, and it gave them a marked advantage over competitors. The files helped the R&D staff to learn the dye processes in which the Schoellkopf firm specialized prior to 1914, and they more easily identified which intermediates the firm should manufacture to place it in a strategic and strong place in the market in future years.[21]

William G. Beckers arose as one of the most prominent dyes manufacturers during the war. Beckers originally came to the United States from Germany as an employee of Bayer Company and had established his own dyes firm in Brooklyn in 1912, selling a line of fifteen dyes in 1914. In November 1914, Beckers's Brooklyn plant exploded disastrously, in which "Beckers [was] shot out the window" and "had an ear nearly torn off." A textile manufacturer brought Beckers to the attention of Eugene I. Meyer, a successful Wall Street financier, and Meyer provided half the capital for Beckers to rebuild his plant on a larger scale and exerted influence on the firm for years to come. When Beckers came to Meyer, he had less than $25,000 in equipment; he left with a $200,000 capital investment from Meyer and other investors. By the spring of 1916, Beckers claimed 10 percent of the domestic market. Whereas Schoellkopf provided many dyes for cotton goods, Beckers focused more on dyes for woolens, including some blue dyes sold to the navy, in which the quality of the dye "might not have been as good as it might be," noted Meyers, and Beckers himself acknowledged that he sent dyes out "not in the regular market form, that is powder, as we should, but in paste form," which contained 60 to 70 percent water. "Our customers are willing to take the material in any shape, no matter how weak it is."

Significantly, Beckers manufactured a few dyes from underrepresented categories, such as anthraquinone and alizarin, and he expanded his firm through both internal growth and the acquisition of Standard Aniline Products of Wappingers Falls, New York, a start-up that manufactured sulphur black and intermediates such as aniline and beta naphthol.[22]

The wartime experience of synthetic organic pharmaceuticals manufacturers paralleled that of the dyes manufacturers: shortages of raw materials and the lack of technical know-how hindered the production of synthetic organic pharmaceuticals. In other respects, pharmaceuticals differed from dyes. First, as a newer product line, there were simply many fewer synthetic pharmaceuticals than dyes, and they were produced in much smaller quantities; while individual products could require enormous investments to replicate, the larger endeavor was more manageable. Second, synthetic organic pharmaceuticals composed a relatively small portion of the American drug market, and many of the established pharmaceutical manufacturers, such as Eli-Lilly or Parke Davis, showed little interest in adding synthetic organic products to their traditional biologically based product lines.[23] Even for Merck & Company, which manufactured one of the broadest lines of synthetic organic chemicals, biologicals made up the vast majority of their product line. However, the development of synthetic pharmaceuticals coincided with larger trends in the drug industry, including a growing reliance on science, contentious but changing attitudes toward patents, and increasing federal regulation.

In synthetic organic pharmaceuticals, the small Monsanto Chemical Company found itself in a position parallel to Schoellkopf's in the dyes industry. Dependent on German intermediates at the outbreak of war, Monsanto saw few options other than to produce its own. Founded and led by the colorful John F. Queeny (when angry, he was "Jehovah blowing fire from the heavens"), who had once worked for Merck & Company, the firm grew quickly during the war. By the beginning of 1916, the company had invested between $150,000 and $200,000 in new buildings and equipment for the manufacture of intermediates. Later that year, Monsanto successfully produced phenol, an intermediate the firm sold as a disinfectant. The war boosted Monsanto's earnings dramatically, from $81,000 in 1913 to $905,000 in 1916, providing the capital for the plant expansion. In 1917 and 1918, the St. Louis firm began production of aspirin and phthalic anhydride. Phthalic anhydride was a necessary intermediate in the manufacture of phenolphthalein, a component of a laxative, and a product of Monsanto's since 1909. The firm had many Swiss chemists on staff, but Monsanto was one of the few pharmaceutical firms without direct ties to Germany.[24]

Abbott Laboratories in Chicago had a long tradition of providing patent medicines to the American market, but it made a transition, often difficult, to a more scientific producer during the war. The firm successfully collaborated with the University of Illinois, particularly with organic chemist Roger Adams. Military contracts for several synthetic pharmaceuticals provided Abbott with the incentive and capital to embark on a R&D program. As a research consultant for Abbott, Adams developed processes to manufacture barbital and then Novocaine in 1917 and 1918, conducting research in his university laboratory and also advising Abbott's plant operations. Abbott had also undertaken the manufacture of Atophan (cinchophen), a treatment for gout, by 1918. Despite help from Adams and his student Ernest Volwiler, who became Abbott's research director, the firm struggled with the more complex synthetic products. In 1918, Abbott ended up buying the Novocaine from another manufacturer, the Rector Chemical Company, to fill its contractual obligations. At least one chemist at Rector Chemical privately labeled Abbott a "low-grade patent medicine house," although part of his animus derived from Abbott's price-cutting activities to obtain a share of Rector's Novocaine business.[25]

###

One of the most striking characteristics of the new American industry lay not in the expansion of familiar firms but in the sheer numbers of new manufacturers trying their hand at making synthetic dyes and pharmaceuticals. Some spotted opportunities, but others depended on a particular chemical for their core business. In the latter case, examples included perfume manufacturers such as Florasynth Laboratories (Unionport, New York) and Van Dyk & Company (Jersey City, New Jersey); typewriter ribbon and carbon paper manufacturer, Neidich Processing Company (Burlington, New Jersey), which needed methylene blue and methyl violet; and paint and pigment manufacturers, colorists of another stripe, who found themselves short of dyes used in their products. By the end of the war, both Sherwin Williams (Pullman, Illinois) and Ault & Wiborg (Cincinnati, Ohio) presided over substantial dyestuffs departments; in the case of Ault & Wiborg, the dyes division became larger than the original paint and pigment business.[26]

Thomas A. Edison's firm normally purchased phenolic resins to manufacture his phonograph records, but the war caught his suppliers short, and Edison undertook the production of phenol, one of the important intermediates for dyes and pharmaceuticals. In a crash construction program, Edison built a plant and assembled the equipment to begin operations in 1915. As one of the first new manufacturers of phenol, Edison found a ready market for excess production in the early years of the war, including

Bayer's aspirin production. The firm also needed paraphenylenediamine as a catalyst for the phenolic resins, which led to the production of a variety of important intermediates such as aniline, nitrobenzene, and acetanilide in much smaller quantities. By 1917, however, competition had flooded the market, prompting Edison to close his two phenol plants.[27]

One failed start-up would have remained obscure if not for the subsequent prominence of one of its founders. Early in April 1916, three friends from Harvard University, Stanley B. Pennock, Chauncey C. Loomis, and James B. Conant, formed the Aromatic Chemical Company, to take advantage of the high prices for synthetic organic chemicals. They had developed a process to manufacture benzyl chloride, a product requiring "cautious chlorination" of toluene, as a modern reference book still states it. Conant withdrew from the firm in September 1916, when as a fresh Ph.D. he received an offer from Harvard to become an instructor and began his illustrious career. Two months later, Pennock and Loomis blew themselves up in their Newark, New Jersey, plant. Pennock died immediately, along with two other employees, but Loomis survived. As of early 1917, the firm and its patent were up for sale.[28]

A few firms new to the industry emerged as middle-sized producers with distinct strategies. The Newport Chemical Works (Carrollville, Wisconsin), the 1915 baby of coke and gas interests, hired Swiss chemist Ivan Gubelmann away from Monsanto in 1916; they likewise acquired the sales expertise of Elvin H. Kilheffer, who had previously worked for the Kalle importing house. Newport specifically aimed at the newer, more difficult dyes. Another notable firm was the Calco Chemical Company (Bound Brook, New Jersey), which Robert C. Jeffcott founded in 1915; he was drawn into the production of intermediates and synthetic dyes and pharmaceuticals initially when the dye crisis threatened his "wall-covering fabrics" business. Jeffcott wanted to specialize primarily in providing intermediates to other manufacturers.[29]

The Dow Chemical Company (Midland, Michigan), another of the middle-sized producers, emerged as one of the most intriguing firms in the industry, not least because Herbert Dow embraced a stereotypically "American" strategy in a sector that most contemporaries believed the antithesis of a mass production industry. Although Dow diversified over time, he anchored the firm's production of synthetic organic chemicals on two of the largest-selling products: indigo and aspirin. The scale of production appealed to Dow because, he argued, "Germans have no basic advantage over us where material is made on a large scale, as would be the case with indigo." When the opportunity of war arrived, Dow's experience in synthetic organic chemicals thus far had centered on making chloroform from carbon tetrachloride, which had been a sideshow to his primary business in bromine extraction and chlorine production. Dow's experience as a manufacturer of bulk inorganic chemicals most certainly shaped his strategy when he ventured to make dyes and pharmaceuticals.[30]

Indigo was among the most heavily consumed dyes in the textile industry, largely in blue denim (and, during war, in navy uniforms), and the price of indigo increased from 15¢ per pound before the war to more than $1 per pound in October 1915. Chemists familiar with the dyes industry recognized indigo as one of the most difficult and complex dyes to synthesize, but the potential rewards tantalized Dow. He hired Professor Lee Holt Cone from the University of Michigan during the summer of 1915 to investigate the possible production of organic chemicals. By July, synthetic indigo consumed Cone's attention, and promising early research encouraged the company to lay plans to manufacture the necessary intermediates, although the firm's top management remained uncommitted. Moreover, Dow hoped to sell brominated indigo dyes, known for faster, brighter—more profitable—colors, as an additional outlet for the firm's primary product.[31]

Herbert Dow faced several challenges in the production and promotion of indigo, the first of which centered on convincing the other top executives at his firm. When Dow wanted to press forward by October 1915, others in the company advised caution. One of these was longtime technical consultant Albert W. Smith of Cleveland's Case School of Applied Science, who graduated from Case a year before Herbert Dow and spent most of his career in the chemistry department at Case, funneling advice and students to Midland. Smith opined that Dow Chemical's resources might be more wisely spent elsewhere, perhaps on simpler organics. As a compromise, the firm designed a plant that could produce the 5,000 pounds of indigo per day Herbert Dow desired but began by installing equipment sufficient for only 400 pounds per day. Dow and his associates took comfort in the fact that two-thirds of the equipment purchased for the indigo plant could function in alternative processes should the indigo operation fail.[32]

Once underway with the indigo venture, delays occurred regularly, often the consequence of a firm rapidly extending its reach during the war when chemical and equipment suppliers could be backlogged with orders. Failing to find a supply of the chemical constituents for indigo, the firm decided to synthesize its own aniline and to build a chloracetic acid plant, both of which Dow considered temporary delays rather than serious roadblocks. In addition, Dow had to consider existing brominated indigo patents, primarily owned by the Swiss. His staff first attempted to solve this obstacle by altering an expired BASF patent, but that legally shaky ground led him ultimately to conclude a patent license with the Swiss. In early 1917, Cone departed for Semet-Solvay and the loss of at least one other indigo chemist to military conscription deprived Dow of technical leadership, slowing production during 1917 especially.[33]

Despite delays, in January 1917, Dow became the first American manufacturer of synthetic indigo; his firm ran test batches through the spring, and gradually increased and improved production over the next two years. As his plant expanded, Dow advertised his intentions to produce large amounts of indigo, partly to alert prospective consumers but also to discourage potential entrants. Dow knew other firms would find indigo prices tempting, but he hoped (unsuccessfully, as it turned out) that the complexity of indigo and its attendant expenses might look more daunting to a potential rival who knew Dow had achieved a solid headstart.[34]

Similar to indigo among dyes, aspirin stood out among synthetic organic pharmaceuticals in consumption, accounting for 40 percent of total U.S. sales in 1918 (in part because of the influenza pandemic). When Bayer's patent expired in 1917, several firms rushed to manufacture aspirin, including Dow. Dow diversified into aspirin production as a peacetime outlet for its large phenol production. Despite two fires in the aspirin plant, Dow strengthened its market position as "aspirin" became a generic name.[35]

As a smaller firm new to synthetic organic chemicals, Dow opted to rely on an established dealer to sell his dyes and pharmaceuticals. Dow sold dyes through the Aniline Dyes & Chemicals of New York City (as did Ault & Wiborg). The dealership dated to the 1870s and just prior to the war specialized in Swiss chemical imports. By 1918, the staff of Aniline Dyes & Chemicals included Dow's former sales director, William Van Winckel, and W. H. Fieldhouse, formerly at the Badische importers. As with the German importers, Aniline Dyes & Chemicals' sales experience aided the novice manufacturers.[36]

###

Before the end of the war, three American firms aspired to create large, vertically integrated operations in dyes and related synthetic organic chemicals. Although the three shared the same general goal, their resources and strategies were notably different. The entrepreneurs behind Federal Dyestuff & Chemical Corporation attempted to build their firm from scratch; National Aniline & Chemical Company was a merger of several firms in related lines; and E. I. du Pont de Nemours & Company, an established explosives manufacturer, primarily grew internally through a strategy of diversification. Exceptional profits and investment capital from the war economy shaped the planning of all three, and all hoped the integration would reduce the risks of the market, particularly volatile during the war, but the differing levels of experience and expertise shaped the firms' strategies and their fate and carried consequences for the American industry more generally.

Before the United States entered the war in 1917, just one American firm announced plans to manufacture synthetic organic chemicals on a scale and scope comparable to the German firms. In 1915, as the imports dwindled significantly, a group of Cleveland capitalists staked an entrepreneurial start-up with headquarters in New York City. The people behind Federal Dyestuff thought big: they capitalized for $15 million in October 1915, several times the capitalization of any existing dyes maker, and they expected to build at least eight plants around the country, the first of them at Kingsport, Tennessee.[37] They envisaged a vertically integrated firm from raw materials to distribution, and they aimed to survive beyond the war.

Hired by the stockbrokers, consulting chemist Edward C. Worden visited Federal Dyestuff's Kingsport plant several times in the spring of 1916, inspecting everything from location and raw materials to personnel and equipment. Like other less informed observers, Worden left Federal Dyestuff a full believer in the potential of the firm. If the firm's management continued in its current policies, "I cannot see . . . where this corporation will be anything but a large financial success." He expected Federal Dyestuffs to "have a most appreciable bearing" on the American dyestuffs industry more generally.[38] Worden's enthusiasm for the future of Federal Dyestuffs rested on several points: the Kingsport site offered convenient access to coal, salt, water, zinc, and limestone, all of which contributed to lower costs; railroad lines provided good transportation to markets in New England and the South; a temperate climate; and sizable pools of unskilled labor, black and white.

The single greatest selling point of Federal Dyestuff, however, centered on the educated, skilled workforce the firm's management had gathered in Kingsport. "I never before visited a chemical manufactory in its incipient stage of production, which had such a large number of experienced chemists, physicists, and college graduates," wrote Worden. Nearly forty "scientific men" worked at the plant, among whom "a large number" had been "specifically trained in the manufacturing of products." In addition, Worden noted an unusual degree of "harmony and esprit de corps" among the managers in the plant. Worden was also impressed with John C. Hebden, the firm's vice president and point man for building and operating the plant. Worden noted Hebden's "thoroughness of learning in both the theoretical field and the practical manufacture of dyestuffs."[39]

FIGURE 2.2. Federal Dyestuff and Chemical Corporation, c. 1917. (Courtesy of the Archives of the City of Kingsport, Tennessee, KC Manuscript Collection 181.)

Under Hebden and company president George T. Bishop, the firm grew rapidly. Large-scale construction in Kingsport began almost immediately, starting with a plant designed to manufacture one ton of sulphur black per day by late January 1916. With "800 men a day on the construction work," Federal Dyestuff's leadership aimed to have completed by early July 1916 a second sulphur black plant, ten times larger than the first, an aniline facility, a building for caustic soda and chlorine, and large steel storage tanks. They expected to produce all the necessary intermediates for their expanding line of dyes, and they ambitiously planned to develop azo and alizarin dyes, newer classes of dyes more popular among consumers but also more difficult to manufacture than the relatively simple sulphur dyes. In June 1916, the firm's management announced a shipment of twenty-three tons to a customer in New Jersey, the largest single shipment of dyes "made in America," proclaimed the main trade journal, and "unmistakable evidence that American chemical manufacturers have awakened to the possibilities opened up by the war in Europe." Rumors of an annual profit of $2,000,000, "or 14 per cent" on the common stock, flew around Wall Street.[40]

But the future for Federal Dyestuff fell apart quickly. In January 1917, six of the corporation's officers, including the president George T. Bishop, resigned their positions. Publicly, Bishop stated only that he and the others "had resigned because of pressure of other duties," but in May the shareholders proposed a reorganization, seeking new capital to cope with the increasingly burdensome debt of the firm, particularly the interest payments on $2 million in mortgage notes. Over the next two years, as interest due dates arrived and expired, officers and creditors generated several plans to rescue Federal Dyestuff from bankruptcy, including leasing the plant to two different Canadian explosives firms, seeking (successfully) more government contracts for explosives and khaki dyes, and recapitalizing. Late in 1918, the firm underwent a court-ordered sale, and new ownership and directors purchased the company and formed the smaller, more focused Union Dye and Chemical Company. The government contracts gave the firm attractive assets, despite its troubled history.[41]

Throughout the financial difficulties, rumors swept Wall Street that Du Pont would buy Federal Dyestuff to "be the kingpin of the non-Teutonic dye situation." Indeed, the Du Pont executives sent emissaries to Kingsport to investigate the firm. Reports gathered from at least one of these visitors suggest that by June 1916, at the same time Edward Worden predicted such a positive future for Federal Dyestuff, serious difficulties had faced the ambitious newcomers. One unsigned report found the Federal Dyestuff estimates of future output unrealistic in nearly every product but the simplest sulphur black; in addition, "Bridging . . . the gap between test tube experiments and unit production must of necessity involve quite some time, especially when working with inflammable and explosive combinations." Although the research chemists possessed workable recipes from the laboratory, it was "highly improbable" that the firm's technical staff could scale up to "a state of efficiency" on the factory level as planned. Looking at one process after another, the writer pronounced the ambitious schedule "practically" outside "the realm of manufacturing possibility."[42]

In the end, the company survived through the war largely because of its government contracts for explosives and for dyes for military uniforms. The grand plans to integrate from raw materials to sales, to scale up a full range of dyes, and to operate like a German firm all collapsed, and the doubters who questioned the wisdom of trying to emulate and compete with the Germans added more evidence to their case. After the United States entered the war, however, the vision for a fully integrated large-scale dyes firm reached fruition in the notable merger of existing firms.

In May 1917, a merger created the National Aniline & Chemical Company which instantly became the largest player in the domestic synthetic organic chemicals industry. The General Chemical Company and five other firms integrated raw materials, intermediates, and final dyes; the firms included the nation's two largest dyes firms in 1916, Schoellkopf and Beckers, the intermediates producer Benzol Products Company, and raw materials suppliers Semet-Solvay, Barrett, and General Chemical. The merger negotiations also included a sales agency: the Cassella importers, William J. Matheson and Robert Shaw, but the Cassella discussions required another five months before they could reach common ground on such a politically complicated agreement. By September 1917, National Aniline combined the most experienced firms the U.S. synthetic dyes industry possessed thus far. The corporate cultures of the constituent firms, however, varied dramatically; the first few years saw fiercely fought battles in the top management, and National Aniline developed an ambiguous—even mysterious—relationship with the wider chemical community.[43]

In the merger negotiations, the various partners assessed one another's abilities, conducting their evaluations in the context of an unpredictable market and the expectation of harsh competition from domestic as well as German manufacturers. William Nichols, head of General Chemical and Benzol Products Company, led the negotiations; among other things, he worried that unless Benzol joined forces with Schoellkopf and Beckers, Du Pont might arrange an exclusive contract to provide intermediates to those firms, taking away Benzol's best customers. The Cassella group, on the other hand, believed Benzol to be the most important potential ally and privately scoffed at Schoellkopf for its history of minimal profits and at Beckers for showing little potential to manufacture dyes competitively in a peacetime market.[44] Others thought Schoellkopf's new research lab, its technical and sales knowledge accumulated over forty years, and its aggressive wartime expansion all placed the firm in a powerful bargaining position in the merger negotiations through the spring of 1917.

Cassella's German ties simultaneously attracted and concerned Nichols and the other National Aniline organizers. The importers' expertise had enticed Nichols for months, but the growing anti-German sentiment in the country made Nichols skittish because Matheson and Shaw wanted to retain their ties to the German partners. Uncertain of the outcome of the war and whether the American dye industry would ever amount to anything, Matheson and Shaw refused to promise they would never again work with the German firm or sell German dyes. Instead, they proposed an arrangement that would allow them to be the selling agency for both National

Aniline products and Cassella products after the war. Although the National Aniline constituents sensed a conflict of interest, Matheson felt otherwise. In his estimation, success in dye manufacturing required satisfactory returns that generally came with a monopoly of a particular color. Matheson believed that as selling agent for both firms, he could render service as a go-between, keeping both companies from competing needlessly in the market with the same dye. In September, the two sides reached a compromise position: Matheson and Shaw retained the right to leave National Aniline and to revert to the German Cassella's agency, but they agreed to give up their stock in National Aniline in such a case.[45]

Almost immediately, the top executives scuffled, and the Cassella group soon made its presence felt in the managerial ranks. The May 1917 merger agreement named the original officers and directors: in addition to Nichols, the top leadership of the firm included the officers of the company—Jacob F. Schoellkopf Jr., C. P. Hugo Schoellkopf, I. F. Stone, and William T. Miller (from Schoellkopf Aniline & Chemical Works, Inc.) and William Beckers—and fourteen directors—four from Schoellkopf, three from Beckers, three from General Chemical, two from Barrett, and two from Semet-Solvay. The Schoellkopfs disliked the managerial style of the other executives; contemporary rumors suggest an incompatibility between the Schoellkopf's tradition of family ownership with close control and the more impersonal, decentralized management of the new corporation. Eugene Meyer, the primary investor in Beckers, recalled it differently: the Schoellkopfs "managed it badly" and "were not equal either to the difficulties of the occasion or to the opportunity to develop." "We outvoted the Schoellkopfs and they got mad and resigned." Nonetheless, the Schoellkopfs retained their stock ownership when the firm subsequently made a major shift in the organizational hierarchy. Nichols also stepped down from active management.[46]

The Cassella group's William J. Matheson became president and chairman of the board; Robert Shaw joined Beckers, I. F. Stone, and L. C. Jones (of Semet-Solvay) as vice presidents. Over the next several months, disgruntled National Aniline employees complained of the "Prussianizing" of the corporation as Matheson implemented a stricter hierarchical organization and streamlined the managerial ranks by releasing employees from the other constituent companies while retaining Cassella staff. National Aniline moved its headquarters to the ten-floor Cassella building in New York City, further augmenting Matheson's control of the company.[47]

The new corporation quickly became the most dominant domestic dyestuff manufacturer, producing about 70 intermediates and 110 dyes, being the sole producer of 24 of the latter. Most of its dyes belonged to the azo class (nitrogen compounds), but the corporation also produced a line of sulphur dyes and synthetic indigo.[48] The location of the National Aniline plants reflected the geographic distribution of the larger industry. With the exception of firms in St. Louis, Missouri, and Midland, Michigan, most of the new industry settled in the mid-Atlantic and New England, particularly New Jersey, eastern Pennsylvania, northern Delaware, and upstate New York.

Du Pont became the third large vertically integrated synthetic dyes manufacturer in the United States; as explosives manufacturers, they ballooned in size from Allied purchases long before the United States entered the war. Their enormous war revenues allowed them to accelerate their prewar diversification strategy. By the time they committed to dyes manufacturing in late 1916, Du Pont could build on experience from manufacturing synthetic organic chemicals related to their enormous explosives operations. In 1918, Du Pont manufactured forty different dyes—an amount equaled by three other American firms—and more than sixty intermediates. The number of intermediates, the ability to produce synthetic indigo, and a commitment to expand to a wider range of dyes made Du Pont a recognized second to National Aniline among American dyes manufacturers in 1918. For all of Du Pont's resources, they struggled for years to master the difficult industry, and the effort devoured millions of dollars that the firm could only have afforded with the exceptional war revenues.[49]

With the know-how to manufacture explosives-related intermediates but without a steady market for the products during times of peace, Du Pont explored the possibilities of diversifying into dyes manufacturing. Shortly after the war began in Europe, Du Pont concentrated its efforts on manufacturing diphenylamine and toluene derivatives, synthetic organic chemicals integral to its production of explosives. By mid-1915 the company had begun producing a small range of intermediates derived from benzene, toluene, and phenol, and expected to produce naphthalene intermediates in the near future. Estimating the demand for intermediates by firms such as Schoellkopf and Heller & Merz, Du Pont executives believed they could profitably sell organic intermediates to the growing dyes industry. Another potential customer, the rubber industry, purchased large quantities of dinitrobenzene to make aniline for its "rubber accelerators."[50] Partly as a result of this initial break into synthetic organic chemicals, the company executives contemplated a much larger undertaking, expanding production to include a range of synthetic dyes, even though the firm possessed neither the necessary skills nor fully understood the complexity of their proposed undertaking.

Du Pont's inexperience with coal tar chemicals encouraged its executives to seek sources of technical information from outside the firm. Du Pont first looked across the Atlantic to Great Britain, where Levinstein Ltd. operated a confiscated Hoechst plant, and the two firms sporadically negotiated for technical information on synthetic dyes processes through 1915 and 1916. In late 1916, Du Pont agreed to pay Levinstein £25,000 annually for ten years in exchange for Levinstein's technical knowledge in dye manufacturing. To execute the transfer of knowledge, Du Pont sent a dozen chemists and draftsmen to England in December and January to examine Levinstein's plant in depth. Chased out of England by German threats to torpedo neutral ships, the Du Pont employees rapidly returned with the technical information that they articulated in a series of large reports, which became the basis for the company's dyes manufacture. Useful as this material proved, it still left Du Pont without the working know-how that came only with hands-on experience. "We knew the principles on which it operated," recalled Irénée du Pont, "but we could not steal the personnel; we could not get the chemists and the trained men; we could not steal the process." He described the deal as an arrangement between "two amateurs"; "we thought they knew more about it than they did."[51] Such a complicated industrial technology simply did not transfer easily, and it posed a double challenge to obtain the German know-how via the British.

Through Du Pont's expanding research program, chemists began acquiring technical familiarity with synthetic dyestuffs. In 1915 and early 1916, as executives eyed the domestic dyes crisis, Du Pont research directors instituted a preliminary program to acquaint their increasing number of organic chemists with dyestuff chemistry. When Du Pont formally decided to enter manufacture, the executives initiated construction of a dyes plant—at Deepwater Point in New Jersey—and a new laboratory specifically geared to the research of dyes. The company patterned the plant after the Hoechst plant in England. The new laboratory, Jackson Laboratory, first opened in autumn 1917, and plans already existed for its expansion. At the beginning of 1917, Du Pont employed approximately eighty organic chemists. Jackson Laboratory, like the Deepwater plant, grew in proportion and cost far beyond original expectations, and within two years the total personnel at the lab, chemists included, exceeded 500. In May 1918 Du Pont also acquired technical experience by purchasing the United Piece Dye Works of Lodi, New Jersey.[52]

Du Pont's management addressed the task of selling dyes. From the Badische importing agency, they hired Morris R. Poucher, long-time director of sales, and eight other salesmen, who became the base of Du Pont's dyes

FIGURE 2.3. Du Pont's Deepwater dyes plant, 1919. (Courtesy of the Hagley Museum and Library, Wilmington, Delaware.)

marketing organization. By 1918 Du Pont executives had learned they would need to employ the importers' methods as well as their salesmen. To maintain quality control, investigate the application of dyes to various materials, and solve problems raised by customers, the company added a technical laboratory to the Jackson complex.[53] By employing Poucher and his colleagues, Du Pont essentially followed the same strategy as the National Aniline: engage the group of people most apprised of the art of selling dyes, the importers of German chemical firms. Although Du Pont merely hired some of the Badische employees rather than acquiring the entire agency as National Aniline had done with Cassella, the actions of both large corporations highlight the significance of the importers to the establishment of the domestic industry. They were the most obvious beneficiaries of refugees from the importing houses but not the only ones. Without the war's shock to the industry, experienced personnel would have had little incentive or opportunity to leave the importers in such quantities.

The nascent industry faced another difficulty in the manufacture of synthetic organic chemicals. Quite simply, they often made bad products, and the question of quality dogged manufacturers for years. In the case of pharmaceuticals, regulatory procedures usually prevented manufacturers, if not criminals, from placing inferior products on the market. But American dyes suffered from a public relations problem. Store managers frequently announced that they would not guarantee the dyes in the clothing they sold, and consumers learned to associate faded or runny colors with American dyes, an inferior substitute necessarily foisted on them as a result of the war in Europe. Very obvious and public failures drew comment even in the industry's trade literature. In April 1917, the military had "modified" their criteria for khaki dyes because they could not find enough American-made dyes that met their more stringent prewar standards.[54] Leaders in the textile industry publicly maintained that they found the quality of American dyes satisfactory, but privately expressed disappointment and frustration in many instances: "very inferior to the old imported dyes," "uniformity not quite as good," or satisfactory in sulphur black but no other color. By 1918, textile manufacturers expected some categories of simpler dyes to be made well, but not the more complex dyes.[55]

Privately, and sometimes publicly, the dyes firms acknowledged their difficulties. "It is only fair to say," noted a Du Pont executive, "that . . . most of the azo colors . . . are not entirely pre-war standards and whether they will live in after war competition depends entirely on the attitude of our Government." In 1918 Central Dyestuff and Chemical Company warned chemist Paul Aronson that he would likely be released soon. After working on "patent blue" for four months, Aronson wrote, the "company found it an impossible problem for the present."[56]

Among consumers, however, the domestic manufacturers attempted to counter their negative image. They pointed out that some blame belonged to fraudulent operators and consumers ill informed about the limits and appropriate uses of certain dyes types, applying dyes to fabrics they were never designed to color, but legitimate complaints persisted. I. F. Stone, a seller at Schoellkopf and then National Aniline, attempted to argue his case before textile manufacturers. The "popular impression" that American dyes lacked the same qualities as the German dyes was "not a fact." American manufacturers' costs might be higher, but the end products were "practically" identical to the German: "we are able to supply chrome colors . . . practically equal as to fastness" to German dyes and "to furnish practically every color which is needed."[57] No doubt skeptical textile makers noted the abundance of the qualifier "practically" in Stone's statements.

American inexperience in the synthetic organic chemicals industry emerged most dramatically in plant accidents. Rarely a week went by without the main trade journal, *Drug and Chemical Markets*, reporting at least a destructive fire, if not a fatal explosion, occurring somewhere in the industry. "Fatal Experimental Work, Explosion Follows Efforts to Learn German Dyemakers Secrets," said the headline after the laboratory of William Beckers's plant in Brooklyn exploded in November 1914. Initial reports laid suspicion on an overly pressurized autoclave used for diazo dyes. Beckers's persistence in rebuilding assured his place in the patriotic folklore associated with the World War I dyes industry. Early in 1916, fire led to an explosion in a Bayway, New Jersey, plant when the flames reached a storehouse with aniline dyes and heaved burning aniline through the air, injuring two night watchman. Leaking benzene tanks at Calco created flammable soil, and construction workers digging holes were met with flames. Later that year, Dow's chlorbenzol plant exploded, killing a man and injuring several others. Dow hypothesized that a safety valve gave out, leaking benzol vapor, which ignited near flame, and he suspected that an employee had violated rules and smoked on the premises. Soon after, his friend and adviser Albert Smith suggested constructing the benzol and phenol plants as "open sheds" to dissipate explosive fumes.[58]

John F. Queeny, the head of Monsanto, viewed the accidents and hazards as a consequence of Americans' limited practice in the field. Forty years of experience had taught the Germans how to guard against accidents and ailments common to commercial production. Although American laboratory chemists had acquired years of experience with benzol, for example, "none have dreamed of the danger that lurks in its use in large quantities." To combat "insidious systemic derangements" from the fumes, Queeny directed the company to provide workers with free shoes, "daily changes of fresh underwear," and mandatory baths at the end of their workday.[59] People inside and outside the chemical community could hardly ignore the dangers of the industry; the real and potential harm to people and the environment, however, never became a significant issue in the public debate over whether to support and encourage the domestic industry.

Trade Associations: ADMA and ADI

By late 1917, members of the American dyes industry increasingly believed they needed to form a trade association to advance their interests. The creation of trade associations was a regular occurrence in the early twentieth century, partly born of Progressive ideas that cooperation and coordination, rather than competition, might be more productive under certain circumstances. Members also hoped their trade associations would help them lobby more effectively, negotiate to some degree about output, and impose a kind of planning on their economic sector.[60] Some wanted their association to pursue standardization to get a grip on the daunting array of dyes, but standardizing dyes required sharing knowledge and information, a scarce resource that more successful manufacturers could be reluctant to share. In the discussions leading to the formation of their association,

Americans in the dyes industry had the opportunity to explore and imagine how their industry might take on a structure different from the German. The original organizers hoped for an association that included not only large and small manufacturers, but also dealers and even representation from dyes consumers. But the American dyes industry founded their association in the war's particular and peculiar circumstances; a sense of crisis and urgency attended their deliberations. They feared the complicated chaos of their market, the small manufacturers increasingly feared the large manufacturers, and the manufacturers and dealers mistrusted one another's intentions. Looming over the dissensions internal to the industry was the German industry: still the model to emulate but potentially a devastating rival awaiting only the war's end to be unleashed. A strong sense of national identity infused the proceedings, and as the American dyes industry formed their association, the growing anti-German sentiment led feuding members to question their opponents' patriotism.

Over the course of 1918, American dyes manufacturers sought to build a trade association, partly to lobby on their behalf but also to help compensate for inexperience and a lack of information. At the organizational meeting in January 1918, Benjamin Kaye, the lawyer hired to organize the new association, argued that the structure of the American industry—large numbers of small firms—required the industry to cooperate and coordinate if it were to stand any chance against the German industry. H. Gardner McKerrow from Marden, Orth & Hastings, a dyes importer and manufacturer, was the primary organizer of the meeting, and he particularly wanted the industry to standardize the dyes made in the United States. McKerrow believed standardization would give the sellers and buyers a similar level of expectation about the quality and consistency of the dyes. He saw "dangerous and insidious" motives among those who refused to support standardization. Opponents would keep "the American dyestuff industry in a chaotic and confused condition and thus render it a fertile field for re-exploitation in the interests of the German dyestuffs manufacturer." Frank Hemingway, a manufacturer of intermediates, who was appointed the interim president to preside over the initial meeting, took up McKerrow's call for standardization, pointing out that the government would need standardization to enforce tariff laws, and it would be better for the industry to take the lead on the issue, despite those who thought it unfeasible.[61]

Dissension appeared at the convention very quickly, and influential manufacturers argued that the interests of manufacturers and dealers were too different to combine into a single association. Almost immediately after the opening speeches, Lee A. Ault of the Ault & Wiborg Company rose from the floor and insisted that the organization include only manufacturers. Usually associated with foreign firms, the dealers suffered from insinuations that they did not have Americans' best interests at heart. Although treated better than importers for German firms, even dealers for French or Swiss companies frequently found themselves defending themselves against disloyalty. The organizers discussed making a separate, non-voting dealers' section within the association, but before the group could meet a second time, representatives of the largest manufacturers, many of whom sat on the organizing committee, decided they wanted the dealers to have no membership whatsoever in the new association and recommended that dealers form their own separate organization. *Drug and Chemical Markets* wondered whether the bitterness among dealers would lead them to oppose higher tariffs. Another viewed the exclusion of dealers as "needlessly" putting "weapons . . . in the hands of the German dye manufacturer."[62]

The rift also highlighted the potential gulf between the large and small manufacturers, and, in the midst of the debate about the membership's composition, some observers believed the small manufacturers and dealers might align in an association against the larger firms. The large firms, with their vertically integrated organizations that encompassed sales and distribution, saw the dealers as inefficient middlemen. The small manufacturers of only a few products, however, relied upon the dealers to sell their limited line of dyes. The largest firm by far, the National Aniline & Chemical Company, avoided the association, displaying little interest in the association's formation or proceedings until the end of 1918. As a result, Du Pont was the largest member firm, despite the fact they had not yet placed many dyes on the market, and when in April the association elected their officers, Du Pont's Morris R. Poucher became president. "It is said," reported *Drug and Chemical Markets*, "that the Du Pont interests were strongly represented in the preliminary conferences." In any case, "the officers of the Association are free from all taint of Teutonic affiliation," and no firms with "affiliations" to Germany could join. And, having finally determined the new association's composition, the leadership also chose a name for their association: the American Dyestuff Manufacturers' Association (ADMA). The new association initially appeared poised to represent the manufacturers, whether publicly attesting to the quality of American dyes, testifying before Congress, or working with the U.S. Tariff Commission to collect trade statistics. The dealers formed a separate organization in the middle of 1918, the United States Dyestuff and Chemical Importers' Association, which its founders had to defend against charges of disloyalty.[63]

Despite the manufacturers' success in excluding the dealers, they were also unsatisfied with the new ADMA, and its official founding did not mark the end of the association's contentious situation, nor the quest to understand their industry better through gathering information collectively. Key founders of the ADMA, particularly Frank Hemingway, Morris Poucher, and Robert Jeffcott, pursued the formation of an "open price society," which led them to create yet another trade association, the American Dyes Institute (ADI), almost simultaneously with the founding of the ADMA. By exchanging information on prices, costs, and products, the manufacturers in the ADI sought to eliminate inefficiencies in their production programs. Proponents of the open price society hoped the information sharing might stabilize prices. In theory, each member would submit to the ADI secretary copies of orders received from customers, revealing the customer, product, and price involved in the contract. The secretary would organize the information by dye or intermediate and customer and then inform all manufacturers of a single product about the prices the other manufacturers received. Originally, the ADI members also planned to exchange certain types of cost accounting and production information but never executed those exchanges.[64]

As with the ADMA, Du Pont played a leading role in the formation of the ADI, but National Aniline again chose not to join, and opponents of the open price society saw bullying. A National Aniline representative, writing anonymously but identifiably in a trade journal forum, expressed considerable hostility to the ADI; the representative suspected that the ADI's leadership sought to force National Aniline into the association to "dictate prices or policy" to the Buffalo firm. Perhaps the representative was wary of antitrust legislation (and Du Pont's dubious record with antitrust) or National Aniline's own challenges in warding off a reputation for "German" behavior, but National Aniline never joined the open price society and only joined the ADMA late in 1918. A small manufacturer, also writing anonymously, said he would not join the association, either, because "naturally we little fellows will be absolutely at the mercy of the big manufacturers in the Association when it comes to prices." He hoped the small manufacturers and dealers might form their own association.[65] The dyes community recognized that Du Pont and a few of the other larger firms increasingly asserted their leadership, and the small manufacturers worried about the big American firms as well as the big German firms.

Large and small opponents of the open price society might have been reassured by its mediocre performance. Manufacturers certainly wanted to learn about their competitors' products and future plans but only reluctantly disclosed the facts about their own firms.[66] In fact, through most of 1918, the ADI secretary never distributed price data on more than five products per month. Although a few manufacturers found the exchanges valuable contributions to their decision making, the open price society never became a reliable mechanism for obtaining market information. Other manufacturers, very conscious of the antitrust laws, felt skittish about participation, despite reassurances from Arthur J. Eddy, the lawyer who devised the open price mechanism, and despite the knowledge that the ADI shipped copies of their minutes to the Federal Trade Commission without the FTC disapproving of the ADI's activities. On a more functional level, the secretary was frequently unable to identify the same chemical products with different trade names; standardization would have helped. By the end of 1918, Poucher began an effort to have the ADI absorb the ADMA, which ceased to exist in January 1919. With many of the same leaders and goals, the two associations overlapped considerably. Many ADMA members, which by then included the National Aniline, wanted no part of the open price society, however, and the ADI relegated the mechanism to a sub-section of the organization, where it continued functioning in its limited capacity until discontinued in 1920. Four years later, when the dye manufacturers faced charges of monopoly and price fixing through the ADI, the industry's representatives claimed in their defense that the open price mechanism failed, an argument the evidence supports.[67]

The formation of the trade associations, especially the American Dyes Institute, marked a small milestone in the development of the dyes industry. Except for the open price society, the activities of the ADI in 1918 received only minimal notice in the contemporary records, and participants subsequently admitted that the open price society had accomplished little. While the organizational beginnings would serve the industry well in the two years immediately after the war, the formation of the associations exposed the fault lines in the American dyes industry. The divide between the manufacturers and dealers is surprising perhaps only in the original hope they could have cooperated within the same association while possessing diverging interests. But the tension between small and large manufacturers, as well as among the large firms, appeared almost as starkly. The American firms knew they would face stiff competition from the Germans after the war, but they increasingly understood the domestic competition might be just as fatal.

While the American industry collectively knew less than the German, American firms varied in their level of inexperience, and even the efforts to standardize faltered in part because the effort required the more knowledgeable firms to reveal information about their dyes, a step they viewed as giving away their advantage. The common national identity forged in the face of shortages and anti-German sentiment was tempered by the reality of market competition. Negotiations over the associations' formation provided a forum for articulating competing visions for the structure and strategy of the new industry, and the results left no doubt about the relative power of a big firm like Du Pont to influence the shape of the trade association as well as the industry.

###

This chapter has centered on classic concerns of business history—the strategies of entrepreneurs and managers, the structure of the firms and industry, the scope of the product lines in synthetic dyes and pharmaceuticals—and we see the American manufacturers of all sizes respond to the crisis in many ways. In 1918, according to government records, 78 dyes firms produced over 300 dyes, and 32 synthetic organic pharmaceuticals reached market.[68] By the armistice, the United States possessed a domestic synthetic dyes and pharmaceuticals industry, but it was precariously built on unsure foundations. Two indicators alone—the smaller line of products and the high cost of production relative to the prewar German industry—boded ill for American manufacturers. In addition, the accidents, poor quality of products, and inability to manufacture the most complicated products all indicated that the four years of intensive effort was still insufficient to master the industry's complexities and to match the Germans' experience.

If this history had been the whole story, one can easily imagine the American synthetic dyes and pharmaceuticals firms encountering a quick postwar demise once German chemicals returned to the American market. But the war disrupted far more than trade, and the domestic industry waged its battle on many fronts. The distinctiveness of the firms' strategies lay not so much in what they did internally or on the market, but in their multifaceted relationships with the state. The political and ideological reaction in the United States gave the industry significance well beyond short-term opportunities for profits. Instead of collapsing before the more experienced German industry, the American dyes and pharmaceuticals manufacturers lobbied for and received beneficial federal policies, many of which would have been politically inconceivable without the war. Chapter 5 returns to the dyes and pharmaceuticals manufacturers, but it focuses on the war's ideological and political ramifications for the industry on both sides of the Atlantic.

The military chemicals that belligerents used to wage war, particularly several high explosives and war gases, also affected the industry's development. The chemical relationships among dyes, TNT, aspirin, and phosgene, for example, led industry proponents to emphasize the industry's connection to national security, but the production of military chemicals on an enormous scale also carried direct and indirect consequences, which ranged from government contracts for dyes and pharmaceuticals to competition for the growing number of chemists. Until 1917, the demand for military chemicals came primarily from the Allied governments, but once the United States entered the war, American mobilization brought a closer and more complicated cooperation between the government and industry. Contracts and conscription mattered for the domestic industry, but so too did the organization of mobilization and the unprecedented investment in chemical expertise.

3 / Mobilization

Synthetic Organic Chemicals in War, 1914–1918

If synthetic organic chemicals had included only dyes and pharmaceuticals, proponents of the domestic industry might still have argued convincingly for its survival in the name of national interest. But because many high explosives and war gases were also synthetic organic chemicals, industry backers could make a much more explicit case for national security. Although some proponents exaggerated the speed with which dyes and pharmaceuticals plants could be converted to explosives and gas arsenals, the chemical relationships among the civilian and military products existed. By the armistice, war chemicals began contributing to the long-term establishment of the domestic industry in important ways, both technologically and politically.

Technological contributions of the war chemicals centered on the quantitative accomplishments of wartime production more than any grand new innovation. American industry possessed an enormous capacity to expand and supply the belligerents with the matériel for waging war, whether weapons, fuel, food, or any of thousands of items that crossed the Atlantic. This was also the case for synthetic organic chemicals used in World War I, particularly high explosives and war gases. The demand for expertise in war chemicals affected the nascent industry, both by competing for talent with the dyes and pharmaceuticals makers but also by training thousands in at least some aspect of production, if not all in Ph.D.-level research. The political benefits to the industry accrued from the increased hostility to Germany once the United States entered the war, along with the high profile of the war chemicals. Politically it became easier to promote a domestic industry as necessary to the national defense; and personal and professional alliances forged in war would also facilitate politicking for the industry after the war.

The connections and comparisons to Europe continued through the war. While Germany was the point of comparison for the United States as it sought to make dyes and pharmaceuticals, Great Britain became the primary model for the United States to consider during the war. The close collaboration among military and scientific personnel, as well as cultural and political sympathies, influenced features of American mobilization, from the movement to place chemists in chemically related war work instead of in the trenches to sharing new research and processes to the vast movement of matériel across the ocean.

This chapter compares the production of synthetic organic high explosives to the production of synthetic organic war gases. The nature of the two groups of chemicals, and the varying levels of American expertise in each, shaped the mobilization.

The Explosives Industry Mobilizes

The American economy was drawn into the European conflict thirty months before the United States officially entered the war. Although the British blockade hurt certain sectors of U.S. transatlantic trade, it also created new opportunities for entrepreneurs as a result of the shortages. Moreover, in the first year of the war, unprecedented Allied purchasing produced a chaotic bliss for American suppliers. Allied governments sent armies of representatives to the United States in search of everything from mules, food, and leather to steel, artillery, and explosives. France, Britain, Russia, and Italy established offices to supervise purchasing and inspections. The American explosives industry rapidly grew enormous and sent millions of pounds of high explosives to Europe. When President Woodrow Wilson and Congress brought the country into the war in April 1917, the scope of the federal government increased dramatically, but the experience of the American explosives manufacturers since 1914 meant that most of the organizational and technical expertise in the United States lay in private hands. The armistice arrived long before the government could supplant private industry as the principal source of technical expertise. The government, however, played an important role not in generating expertise in high explosives, but in diffusing expertise across the industrial landscape. By 1917, the military organization responsible for procuring explosives for most of American consumption—the Department of War's Ordnance Department—became an important agent in the dissemination of industrial expertise in synthetic organic explosives production.[1] For a country attempting not only to win the war but also to establish a domestic organic chemicals industry, the enormous production of high explosives would expand the raw materials available and give many chemists experience in the field.

The Allies sought explosives of all kinds for the war effort. Military high explosives, first used extensively in World War I, included three main types: trinitrotoluene (TNT), trinitrophenol (picric acid), and ammonium nitrate, the first two of which were synthetic organics. Nitroglycerin, also a high explosive, found fewer military applications. Whereas vast quantities of smokeless powder (nitrocellulose) functioned as propellants to launch or discharge shells from the guns, the high explosives, contained within the shells, exploded on contact with the target. Ideally, from the perspective of military strategists, a shell would yield a powerful explosion when it reached its destination but possess enough resistance to the shocks of handling and discharge to prevent a premature and accidental blast. TNT's "high power and great stability" made it an especially desirable explosive, particularly to navies because it could be stowed safely on board for long periods of time. By 1915, the British favored TNT, while the French preference throughout the war remained picric acid, which yielded a more potent blast, but which also carried a greater risk of instability and inopportune detonations. Neither explosive was new to American industry, but there was little reason before the war to produce either on a large scale. At the outbreak of war, manufacturers, whether experienced or not, signed contracts with the Allies to produce them in unprecedented quantities.[2]

Different sets of firms made TNT and picric acid, and after April 1917 the government-industry relationship developed variations based on the two groupings. A list of the largest manufacturers of TNT included well-known names, most having corporate connections with Du Pont. Before August 1914, Du Pont was the only company in the United States producing TNT; although it manufactured only relatively small quantities, it had the skills to scale up production. Entering the war, the firm possessed facilities to produce annually 660,000 pounds of crude TNT—nearly the entire American capacity—in plants in Repauno, New Jersey, and Barksdale, Wisconsin. By April 1917, when the United States declared war, Du Pont had delivered about 50 million pounds of the explosive to Europe. In the meantime, an explosion in the Repauno plant in 1916 led to its temporary abandonment, and Du Pont expanded the Barksdale plant and built a new plant at Deepwater, New Jersey, to accommodate the tremendous demands. Du Pont investigations of tetryl, which served as a "booster" explosive to set off the main charge of TNT or amatol in a shell, culminated in the construction of a tetryl plant, also at Deepwater, at the end of 1915.[3] With its enormous production of smokeless powder in addition to the synthetic organic high explosives, Du Pont grew into one of the nation's largest firms during the war.

Three other firms related to Du Pont produced high explosives for the war. Atlas Powder Company and the Hercules Powder Company, the two explosives manufacturers created in 1912 from Du Pont as a result of the antitrust suit, both expanded their output dramatically. Atlas became the largest producer of ammonium nitrate (the most important high explosive that was not a synthetic organic chemical), and Hercules became a large producer of TNT as well as ammonium nitrate and cordite, a type of smokeless powder. In May 1915, Russia gave Hercules its first war contract for TNT, which it manufactured in California and later sold to the British as well. In 1913, a former Du Pont executive set up Aetna Explosives Company, which manufactured TNT, picric acid, and other military explosives in its plants in Emporium, Oakdale, and Nobleston, Pennsylvania. However, while Du Pont and Hercules charged customers for the cost of plant construction, Aetna borrowed heavily to expand and collapsed into receivership during the last two years of the war.[4]

Americans had even less prewar experience with picric acid than with TNT; it had been made only experimentally in the United States. Burgeoning French demand lured some American makers into picric acid contracts, but these firms tended to be smaller than those that made TNT. The ranks of picric acid manufacturers comprised such unfamiliar names as Butterworth-Judson, Goodrich-Lockhart, Everly M. Davis, and the Nitro Chemical Company. Du Pont made negligible amounts, and these were primarily used as raw material for ammonium picrate; Hercules and Atlas made none. The Butterworth-Judson Company, one of the two largest producers of picric acid, began to manufacture the explosive in the spring of 1916. Organized in December 1915, the new holding company combined the sulphuric acid production of the Butterworth-Judson Company and the organic chemicals of the recently organized American Synthetic Dyes, both of Newark, New Jersey. The Semet-Solvay Company, a builder of by-product coking ovens since the 1890s, moved into the production of synthetic organic chemicals, including high explosives, during the war. It became the other large producer of picric acid with its plants in Split Rock and Syracuse, New York. Only Aetna and Semet-Solvay held contracts to make significant amounts of both TNT and picric acid.[5]

In addition to the prewar and new explosives manufacturers, other producers equipped themselves to take advantage of the opportunity provided by the war. In Midland, Michigan, the Dow Chemical Company seized the opportunity to produce phenol, the organic base for picric acid. Although nitrated into the explosive elsewhere, Dow's phenol eventually made its way to France and, through Japan, to Russia. The research staff of Dow developed a workable synthetic process, and construction was underway in the fall of 1915. Dow faced delays, however, because Du Pont had engaged around the clock the single maker of a necessary piece of equipment, and

Dow simply had to wait. By November, however, the Dow staff had placed the phenol plant in operating condition and planned immediate expansion, believing it their most profitable line of manufacturing.[6]

Given Americans' relative inexperience with high explosives production, it is no surprise that American firms occasionally received help from the Allies, desperate to increase the flow of explosives across the Atlantic. Du Pont built on its prewar relationships with European firms. The war disrupted Du Pont's market- and profit-sharing agreements with the British and German explosives trust, and the trust itself was an early casualty of war between Britain and Germany. The cordial prewar relations between Du Pont and Nobel in Britain, however, most certainly influenced wartime cooperation between the Allied explosives makers. Although Du Pont imported toluene prior to the war, the international agreements had kept the Delaware firm abreast of the latest innovations in military explosives. At the outbreak of war, Du Pont still lacked the requisite organic chemicals manufacturing experience, but its chemists and engineers could more readily learn the processes based on the technical knowledge acquired from abroad, particularly after the latest market-sharing agreement in 1907 with Nobel. In another case, the French offered necessary help. While the chemistry of picric acid posed few scientific difficulties for Americans, quickly developing a safe and practical manufacturing process proved more challenging. In 1915, for example, Aetna delivered its first French contracts for TNT and picric acid five months late; such delays obliged the French government to send technical advice as well as capital. By the end of 1917, Aetna was supplying two-thirds of the TNT and one-half of the dry picric acid that the French government purchased in the United States.[7]

The impact of the Allied purchasing on certain sectors of the U.S. economy came quickly and, notably, the chemicals trade press expressed astonishment at the sheer size of the contracts. The amounts of money that changed hands often remained private, but occasionally the public might learn that the French had signed a contract for picric acid in May 1915 worth nearly $1 million, rumored to produce profits close to 20 percent. By the end of the war, $1 million would not seem like a large contract in explosives. In addition, high explosives shared with dyes and pharmaceuticals the same organic raw materials (benzene, toluene, and phenol) and similar processes (nitration and sulphonation), although explosives consumed dramatically greater quantities. Prices for toluene and benzene and other chemicals raced upward as the British and French governments, and private speculators, competed for scarce raw materials. Toluene went from 30¢ to $5 per gallon, and benzene soared from 25¢ to $1.75.[8]

When tumultuous market conditions overwhelmed Allied purchasers, they turned to J. P. Morgan & Company, a company that had made its fortune in previous decades partly by directing European investment into American railroads and other assets. In January 1915, J. P. Morgan & Company agreed to coordinate and supervise purchases for the British, and the venerable banking house concluded a similar agreement with France four months later. To oversee these enormous contracts, the bankers hired Edward R. Stettinius Sr., a respected executive of Diamond Match, to run a new "Export Department." For the next two years, Stettinius was the mastermind of a complex organization that handled more than $3 billion in Allied contracts. Contracts to explosives firms accounted for at least a fifth of this amount.[9]

At his Export Department, Stettinius devised strategies to obtain reliable deliveries of explosives and other products at better prices. He not only guided J. P. Morgan & Company but also significantly influenced U.S. military purchasing guidelines after April 1917. First, he stopped publicizing the products the Europeans demanded, a process that had only alerted the market to raise prices. Instead, he sought contracts privately and quietly, settling on generous but not outrageous prices. Second, as he valued reliability and dependability, he approached large, established firms, which in high explosives meant Du Pont, Hercules, Aetna, and Semet-Solvay. As other historians have noted, this policy encouraged further concentration in the economy. Finally, in fields in which too few or no experienced suppliers existed, he identified possible candidates. For example, he encouraged the steel industry to accelerate its conversion from beehive to by-product coke ovens, the primary source of raw materials for high explosives. Meanwhile, French demand for picric acid led J. P. Morgan & Company to cultivate new manufacturers, such as Butterworth-Judson, which became one of the largest picric acid contractors. To run his department, Stettinius chose knowledgeable people to oversee industries they already knew, and his office grew to a staff of 175.[10]

While the supply of matériel brought unprecedented prosperity to many makers of high explosives, the demands of the war also brought about a limited, private form of mobilization. Because of President Wilson's official commitment to neutrality and public resistance against joining a foreign war, the United States in April 1917 was militarily unprepared to participate in a world war. The war-related economy, however, was much better prepared, and the asymmetry soon became very apparent.[11]

For high explosives and every other war industry, the declaration of war made the U.S. federal government immediately relevant to production. As in the other belligerent countries, the government harnessed the economy to wage total war. The relationship between the explosives industry and the federal government reflected many of the larger challenges to mobilizing the nation's resources for war. Historians have noted the speed of U.S. mobilization relative to that of the Allies, from whom the Americans borrowed many policies and procedures. However, even a relatively rapid transition could not transform the American economy quickly enough to integrate smoothly into an ongoing war. Knowing the problems and difficulties of Great Britain and the other belligerents could not alone prepare Americans for massive redirections of resources or for government management of the economy. The U.S. Army, U.S. Navy, and rapidly mutating civilian mobilization boards struggled among themselves, and with their enormous tasks, showed little of the efficiency so valued by the "Progressives" of the early twentieth century. Generous estimates give April or May 1918, a full year after the U.S. entry into the war, as the time when American mobilization achieved effective coordination, although one could delay that date by many months in some sectors. The havoc in Washington, D.C., tested Allied patience. In American mobilization, Stettinius noted to a colleague, "we are repeating all of the mistakes that were made by England during the first months of the war." The "autocratic control" developed in Britain to coordinate production appeared to be as "offensive to our democratic ideals" as it had initially been to the British.[12]

Even under government direction, mobilization relied heavily upon the private sector; that private firms possessed the necessary technical know-how was the underlying assumption. No one in the United States seriously contemplated a military takeover of the economy, which was not only politically impossible but logistically irrational. Still, the federal government alone possessed authority to coordinate national efforts. Legislation in 1916 gave the federal government the power to commandeer plants, but this typically compelled cooperation rather than takeovers. The logic of mobilization depended upon creating partnerships—between firms with know-how and the federal government with legal and financial resources.[13]

The task of coordinating high explosives production fell to the military and the mobilization boards—first the Council of National Defense (CND), set up in 1916, and then the War Industries Board (WIB) in 1917. Initially a sub-agency within the CND, the WIB became the primary mobilization board, particularly after its reorganization in 1918. These industrial boards had one thing in common: they brought people from business and government together. While it pained Wilson and other Progressives to give more power to representatives of business, in the year before April 1917 most political, military, and business leaders came to see this as necessary to coordinate supply and demand and establish priorities. Wilson aimed, however, to use private expertise without losing overall control.[14]

Both the CND and WIB attempted to coordinate the economy through planning, but they lacked authority to enforce decisions. After President Wilson strengthened the WIB in March 1918, coordination occurred more smoothly, although the WIB still coddled as much as coerced business into cooperation. The WIB created "cooperative committees" to work with individual industries. In most cases, concentrated industries with few firms, such as the explosives industry, proved much easier for the WIB than the management of more competitive industries. Within the WIB, the chemicals division comprised about twenty sections. One section dealt specifically with explosives; others addressed supplies of raw materials (such as toluene and nitric acid), synthetic dyes, and intermediates. The chemicals division worked with firms through an association of manufacturers and also maintained a file of statistical information. The WIB established priorities, evaluating government contracts and their claims on economic resources, whether toluene, building materials, or petroleum.[15]

Scientists and engineers contributed to "preparedness" and then mobilization through two quasi-governmental organizations. In July 1915, the U.S. Navy called upon Thomas A. Edison to head its new Naval Consulting Board, which drew on the major engineering societies for technical advice. More importantly, the National Academy of Science created the National Research Council (NRC) in 1916 to mobilize research for national defense. The NRC identified technical problems, which they assigned to scientists and engineers in universities and government agencies. Marston Bogert, a Columbia University chemist, chaired the NRC's chemistry committee. In July 1918, the NRC established an explosives committee, which did not formally convene until late August. While the explosives committee garnered praise for collecting and organizing information, the NRC's impact in the field of synthetic organic explosives was relatively slight.[16]

For the high explosives industry, the military as well as the mobilization boards influenced production. When the explosives firms came into contact with the government, they typically dealt with the Department of War. Historian Paul Koistinen blames the Department of War for many of the failures in mobilization, noting that it never underwent the reorganization in procurement such as the Department of the Navy had implemented at the turn of the twentieth century and that wartime demands overwhelmed its bureaucracy. Secretary of War Newton D. Baker and his military leadership vigorously resisted attempts to subordinate army purchasing to civilian-run mobilization boards, and only after a year of disarray, culminating in a serious crisis in the winter of 1917–1918, did the Department of War begin to cooperate with the War Industries Board. Both military branches made their own purchases throughout the war, however.[17] Charged with purchasing munitions for the Department of War, the Ordnance Department inside the Department of War worked closely with the explosives manufacturers.

The performance of the Ordnance Department mirrored the larger pattern of American mobilization.[18] Unprepared for war in April 1917, Ordnance officials faced a steep learning curve and attained a respected level of competence only a year later. When the United States entered the war, the department contained ninety-seven officers and grew over the next nineteen months to 6,000 or 7,000 officers overseeing 8,000 factories with 5 million employees. In high explosives, the department inherited a relatively effective munitions industry. Du Pont and Hercules, among others, had expanded rapidly, adjusted quickly, and by 1917 had little need for outside technological input. In the quest to expand American technical expertise in synthetic organic explosives, the Ordnance officials made their primary contribution in the diffusion of existing knowledge. The officers built on the experience of the private firms and made a notable contribution in collecting "best practice" techniques and distributing them more widely across the expanding industry. Had the war continued into 1919, the Ordnance Department's contribution to this transfer of technology would have been even greater.

In the spring and summer of 1917, the Ordnance Department gathered information and advice, especially from Edward Stettinius. If Stettinius was annoyed to receive duplicate inquiries from officers of different ranks and offices, he hid it gracefully. Brigadier General William Crozier, Chief of Ordnance, tried to recruit Stettinius into his department, and although Stettinius turned down the offer, many of his Export Department staff found positions in Ordnance and other agencies. Identifying manufacturers of explosives and explosives' raw materials was a surprisingly difficult task, not least because firms entered and exited the market with regularity, and Stettinius shared J. P. Morgan & Company's list of manufacturers. Stettinius also advised the Ordnance Department on contracts, prices, and cancellation clauses. The officers adopted almost entirely his advice to stay with large, established firms whenever possible. After the war, Ordnance officers justified this policy, arguing that it would have been "suicidal" to grant contracts to "companies who were not entirely familiar in every respect" with high explosives. The department was unable to avoid inexperienced firms completely, but the officers preferred and encouraged the "companies with the best technical knowledge and organization."[19]

The Department of War also approached the Allies for information. When the United States entered the war, production of explosives and other munitions became a coordinated transatlantic effort. Through 1917, representatives from the Allied countries considered their respective strengths and deficiencies in ordnance production and agreed on manufacturing programs to employ their resources most effectively. Because Britain and France had been fighting since 1914, they were already fully mobilized and abreast of the latest weapons development. In fact, the two countries manufactured a quantity of artillery sufficient to satisfy not only their own needs but also the needs of newly arriving American troops. Consequently, in devising a production strategy, the British and French armies would outfit American troops with guns in exchange for the matériel the United States could supply most rapidly and plentifully: explosives.[20]

Ordnance officers acknowledged the existing relationships between Allied purchasers and American suppliers, and they strove to avoid interrupting a working system. Unlike J. P. Morgan & Company, the government could bring the force of law to negotiations with American manufacturers and therefore compelled more favorable prices than Allied governments had paid. What struck the trade press about picric acid contracts in October 1917 was the disparity in prices paid by the Allies, ranging from $1 to $1.25 per pound, as contrasted with the prices of 60¢ and 63¢ per pound paid by the U.S. government. Consequently, when existing contracts expired, the Allies placed subsequent orders through the U.S. government to obtain the better prices. This was not always simple. When, for example, the French realized that the U.S. government had contracted for Aetna's output of TNT and picric acid, which they had expected for the duration of the war, they sent harsh letters to the department. In this case, the plant's output continued to go to France, but through American contracts.[21]

After April 1917, Allied technical advice typically flowed through the U.S. military, rather than directly to firms. Attachés stationed in London and Paris stayed abreast of new technical developments and became conduits for information passing in each direction. For example, in the spring of 1918, the British offered the United States technical information on a continuous process for the manufacture of TNT. Ordnance officials shared the British information with several firms, soliciting manufacturers' assessments of the method's applicability to the United States. In this case, the American manufacturers rejected the process as more dangerous and no more efficient than their current practice. In early 1918, the National Research Council posted scientific attachés to London, Paris, and Rome, but the NRC believed that the Department of War had undercut them by insisting on using military attachés.[22]

Table 3.1. Production of TNT (in Pounds) on Ordnance Department Contracts

July 1917	400,000
Aug. 1917	1,300,000
Sept. 1917	1,750,000
Oct. 1917	2,550,000
Nov. 1917	4,200,000
Dec. 1917	3,950,000
Jan. 1918	6,041,000
Feb. 1918	5,262,000
Mar. 1918	6,689,000
Apr. 1918	6,145,000
May 1918	6,890,000
June 1918	6,864,000
July 1918	10,821,000
Aug. 1918	9,250,000
Sept. 1918	10,140,000
Oct. 1918	12,926,000
Nov. 1918	11,390,000
TOTAL	111,944,000

Note: The figures are rounded to the nearest 1,000.
Source: "History of Explosives Section: Explosives, Chemicals, and Loading Division," part II, vol. 6, "Manufacture of TNT," December 16, 1918, p. 2, National Archives and Records Administration, RG 156, entry 528, box 3.

The most significant transfer of technology from the Allies was the British process for making amatol, a mixture of TNT and ammonium nitrate. When, in 1917, the demand for TNT outstripped supply, the British extended their store of TNT by mixing it with the more plentiful ammonium nitrate—first in proportions of 50–50, and then 20–80. Ordnance officers expected the United States would encounter similar shortages and made inquiries of their own arsenals, of Du Pont, and of Britain. In October 1917, two British military experts traveled to the United States, bearing stacks of confidential reports that encapsulated the British state of the art. In inorganic explosives as well, the Americans also borrowed British technology: a huge Atlas plant at Perryville, Maryland, which reproduced a Brunner Mond facility to make ammonium nitrate, began manufacturing in July 1918.[23]

However much Ordnance officers learned from J. P. Morgan & Company and the Allies, American firms were the primary suppliers of technical knowledge and advice to the Ordnance Department. The voracious market for high explosives stretched the American pool of expertise, as when employees from established firms struck out on their own, aiming to land the rich Allied or Ordnance contract that would give their "start-up" a share of the market. An Ordnance officer later argued that the expansion of the industry had left the "experienced talent . . . consequently very thinly distributed over the existing production." For example, two former Butterworth-Judson employees established a new firm and obtained an Ordnance contract to make picric acid, but when the technically competent partner died, the firm was left scrambling to find a replacement. Before prohibitive legislation in August 1918, firms also poached employees from rivals (or from the Bureau of Mines and other government agencies).[24]

By April 1917, the explosives manufacturers had already become a major employer of chemistry graduates fresh from the universities, and that continued for the duration of the war. Correspondence from the Harvard University chemistry program reflects the great demand for employees of any and all skill levels. Competition became so stiff that Edward Mallinckrodt, seeking to hire a new chemist for his pharmaceuticals firm in St. Louis, wondered if the Harvard professors might know of "any young men who, on account of some unimportant physical defect, may have been rejected in the draft." Word was out quickly at the university about the great expansion at Du Pont, which hired many Harvard graduates. One newly employed graduate wrote back that "the opportunities for chemists here are quite varied as one may choose between research on dyes and explosives, to analytical work and supervisor work in plant operation." Another hired as an explosives inspector found the culture in New Jersey sufficiently different from New England to welcome his seventy-hour work weeks, but he was grateful for the type of work. He was also grateful that, while he had had "several experiences" with the explosives plant that "serve to break the monotony and furnish a little excitement," nothing had been too damaging. The older hands at Du Pont's explosives plant were not above hazing, and one Harvard alumnus wrote back that his co-workers had "kindly assured me that if the TNT was hit [by lightning] that I would never know anything about it," and then they staged a fake explosion that sent him running.[25] Pranks aside, the reports of the young chemists at Du Pont suggested they received crash courses in the practical aspects of making synthetic organic chemicals.

With the demand for expertise, Ordnance officers believed that increased production depended upon their ability to expand and distribute technical knowledge. The Ordnance officials took measures to improve the level of knowledge within their own ranks and in the industry more generally. Aware of the limited number of experienced personnel in the department, Ordnance officials consciously worked to expand the number of specialists trained to supervise the manufacture of explosives. The department "distribute[d] this limited force of experts as equitably as possible" among various plants and began training programs at established firms to create quickly "a vastly enlarged force of competent operators and supervisors." In such cases, the government agency essentially commandeered or bought the technical knowledge and know-how of some private firms and transferred them to other manufacturers. In a similar manner, if one by-product coking plant discovered a more efficient method to recover raw chemicals, the Ordnance Department would communicate the improvement to other plants.[26]

In August 1917, for example, as part of the military's quest for alternative high explosives, the Ordnance Department investigated TNA (tetranitroaniline) and decided to encourage production. When Du Pont declined to undertake the project, the Ordnance officers "pushed" the Calco Chemical Company to build and operate a TNA plant. As part of the government's obligations in the contract, the officers contracted with the original patentee, Bernhard J. Flurscheim, a Belgian, to provide to Calco "a detailed description of the process of manufacture" and "all technical advice as may be obtainable from the patentee." The transfer of knowledge did not occur in the manner planned, partly because Flurscheim returned to England for four months during construction in early 1918. In his stead, the Ordnance Department facilitated Calco's employment of two other technically trained men from the Aetna plant, where Flurscheim started working in 1915. Calco made little use of the two Aetna men, perhaps afraid Aetna would learn too much about the plant and process modifications that Calco implemented as a result of its own experiments. In reviewing the arrangement after the war, the Ordnance Department concluded that Flurscheim's knowledge had been indispensable to Calco's manufacturing program but that Calco had also brought significant in-house knowledge to bear on the project.[27]

To manage production, statistics were vital. To compare firms, the department gathered data on raw materials, yields, and rate of output. To a degree that made some executives unhappy, Ordnance officers shared their information. When manufacturers signed contracts with the government, they agreed to maintain accounting records and to make these available on demand to Ordnance officials. By mid-1918, Ordnance officials were visiting under-performing plants and suggesting changes in equipment and procedure based on their observations of better performing firms.[28]

Throughout the war, Du Pont not only dominated the production of explosives but also set the standard by which Ordnance officials judged the performance of other firms. After the department invited explosives makers

Table 3.2. U.S. Ordnance Department TNT Output (in Pounds) by Firm (July 1917–December 1918)

Du Pont	53,310,578
Canadian Explosives Company	31,925,000
Hercules	22,354,400
Nitro Powder Company	3,475,000
Aetna Explosives Company	613,880
Bartlett Howard Company	213,800
Federal/Union Dye & Chemical	51,242
TOTAL	111,944,000

Source: Explosives and Loading Division, Explosives Section, "TNT Production," December 20, 1918, National Archives and Records Administration, RG 156, entry 764, 471.86/880, box 135.

to Washington to learn each firm's capacity and potential, Ordnance officials tried to set prices for TNT based on Du Pont's cost structure. If Du Pont could sell TNT for 50¢ per pound, then the others should, too. However, when the other firms hesitated, the department had to increase its offer, not least because other firms' managers knew that Du Pont already had most of the necessary raw materials locked up in contracts. To obtain as much TNT as it wanted, the department had to offer 52¢ or 55¢ per pound to a few of the other manufacturers.[29]

Du Pont rarely needed (or heeded) advice from Ordnance officials. Officials recommended practices they most likely learned from Du Pont, and for the most part the department seemed satisfied with Du Pont's cooperation. On at least one occasion, however, the Delaware firm protested. In the summer of 1918, Ordnance officials asked for plans showing plant layouts, which most firms supplied. Du Pont's executive committee, however, refused to turn over the plans of its highly productive TNT plant in Barksdale, Wisconsin. In the end, Du Pont's executives relented because their engineers had inspected the plans that were submitted by other firms, and Ordnance officials used the visits as leverage to demand a fair return. No doubt other firms gained more than Du Pont from such exchanges.[30]

The Ordnance Department personnel encouraged the sharing of technical knowledge through several other means. First, the Explosives Section of the department accumulated an extensive technical file that contained detailed written instructions and reports about manufacturing processes used throughout the United States. Second, when informal sharing of experienced personnel could not meet the demand for technical experts, the Ordnance Department established more formal instructional programs. In one case, when the district Ordnance officers struggled with insufficient numbers of competent technical inspectors, they ran a two-week seminar

Table 3.3. U.S. Ordnance Department Picric Acid Output (in Pounds) by Firm (November 1917–December 1918)

Aetna Explosives Company	16,118,576
Semet-Solvay Company	10,650,000
Nitro Chemical Company	4,593,518
Goodrich Lockhart Company	4,443,791
O'Brien Synthetic Dye Company	2,708,699
Lansing Chemical Company	2,546,333
Hooker Electro-Chemical Company	797,300
TOTAL	41,858,217

Source: Explosives and Loading Division, Explosives Section, "Picric Acid Production," December 12, 1918, National Archives and Records Administration, RG 156, entry 764, 471.86/880, box 135.

to train a group of inspectors in techniques specific to supervising the construction and operation of explosives plants. In a more general case, to encourage continued expansion in the number of qualified technical experts, the Ordnance Department established the School of Explosives Manufacture at Columbia University. Designed as part of officers' training, the school convened its first class of twenty in the fall of 1918, although the armistice arrived before the class completed the program in mid-December.[31]

To expand the output of explosives, particularly TNT, the Ordnance Department also needed to encourage increased output of raw materials. The basic raw materials for TNT are sulphuric acid, nitric acid, and toluene, but military administrators focused on toluene as the crucial input, and by 1918 the officers no longer left firms scrambling to purchase their own raw materials. The Ordnance Department pursued three strategies to increase the supply of toluene to meet the demand for TNT. First, it encouraged the trend toward more by-product coke ovens, which steel firms had already accelerated with the incentive of European purchases. The Ordnance officials attempted to improve the average amount of toluene recovery at coke ovens by monitoring the yields at plants and then advocating the most successful methods. As in the prewar years, however, the steel manufacturers' first priority was the quality of coke, and they would only consider improvements in toluene recovery if the resulting coke proved adequate to smelt iron, and the department strove to identify improvements that satisfied both needs. By-product coke ovens provided 80 percent of the toluene supply in 1917, but a newer, second source had emerged during the early war years when prices of toluene multiplied several times over. Gas companies that supplied heating and lighting to cities began to strip their manufactured gases of toluene before distribution to customers. Removing toluene reduced the lighting power of manufactured gas by at least 30 percent and the heating power by 5 percent. Consequently, one of the Ordnance Department's tasks was negotiating with local regulatory authorities to accept a lower quality gas. Finally, in the quest for greater supplies of toluene, the Ordnance Department evaluated the investigations of new processes for cracking petroleum, one of which yielded about 35,000 gallons per month starting in July 1918.[32]

Smooth delivery and increased output drew Ordnance officers into issues of safety and health. Manufacturing high explosives was dangerous, even more so in the hands of inexperienced producers, and disastrous explosions interfered with output. Two debilitating explosions leveled TNT plants in the United States and another two factories working under U.S. contracts in Canada. Picric acid made in the traditional English "pot process" released so many dangerous fumes that Ordnance officials pushed firms to adopt newer, safer nitrating units. In choosing contractors, officials evaluated the risks associated with each plant. Sometimes, in periods when demand neared total plant capacity, officials felt obliged to give contracts to manufacturers whose plants frightened even the inspectors. Although officials fought to keep a dangerous plant operating if they needed the output, they also pushed firms to make their plants safer. This produced a course of action that neither endorsed dangerous practices nor effectively eliminated them. In general, Ordnance officials believed, usually correctly, that the more experienced producers like Du Pont and Hercules were the safest. To reduce the risks, the more established manufacturers evolved standard procedures in everything from plant layout and equipment selection to daily sanitary and cleaning habits.[33]

Explosives firms occasionally asked the department for help with other branches of government. Especially troublesome was the general shortage of labor, aggravated by the loss of skilled men to the draft. Several firms submitted to the department lists of essential employees, only to be told that Ordnance officers could not interfere with the rationing of labor, which was handled by a different agency. After August 1, government orders forbid companies from hiring away another's employees. Instead, they needed to appeal for additional laborers to government agencies, which also struggled to meet the shortages. Goodrich-Lockhart executives in Perth Amboy, New Jersey, complained about their allotment of workers for their picric acid plant. When they needed one hundred, they received twenty-seven, and of the twenty-seven, seven lasted at the plant for three days, and two stayed as long as two weeks. The company disparaged the quality of men they received, but the turnover surely reflected the unpleasant nature of the work. In some cases, the officers believed firms underpaid their labor. By

Table 3.4. Output of Toluene (in Gallons), May 1917 to November 1918, in the United States and Canada

Month	Coke Ovens	Gas Plants	Other	Total
May 1917	690,000	80,000	30,000	800,000
June 1917	700,000	90,000	30,000	820,000
July 1917	700,000	100,000	30,000	830,000
Aug. 1917	710,000	100,000	30,000	840,000
Sept. 1917	720,000	110,000	30,000	860,000
Oct. 1917	740,000	110,000	30,000	880,000
Nov. 1917	750,000	110,000	30,000	890,000
Dec. 1917	690,000	100,000	30,000	820,000
Jan. 1918	530,000	150,000	30,000	710,000
Feb. 1918	510,000	140,000	60,000	710,000
Mar. 1918	770,000	200,000	60,000	1,030,000
Apr. 1918	820,000	210,000	50,000	1,080,000
May 1918	850,000	240,000	60,000	1,150,000
June 1918	810,000	320,000	80,000	1,210,000
July 1918	840,000	430,000	70,000	1,340,000
Aug. 1918	930,000	470,000	80,000	1,480,000
Sept. 1918	980,000	400,000	70,000	1,450,000
Oct. 1918	1,040,000	600,000	50,000	1,690,000
Nov. 1918	1,050,000	520,000	70,000	1,640,000

Source: Horace C. Porter to Myron S. Falk, January 16, 1919, "History of Service in Ordnance Department," pp. 1-6, National Archives and Records Administration, RG 156, entry 528, box 9. The numbers are drawn from "Production of Toluol, United States and Canada, 1917-1918," a graph in exhibit B of Porter's report, and could have inaccuracies of up to 10,000 gallons due to the difficulties of reading the graph.

the summer of 1918, the going labor rate at picric acid plants hovered around 40 to 44¢ per hour, plus various bonuses. The managers who paid less ran into labor shortages, and the Ordnance officers might pressure a firm to raise wages if the labor shortages hindered output. Firms also faced restrictions from state governments, which regulated labor safety and health, and they were also subject to local town and neighbor complaints. When the Calco Chemical Company expanded its plant for TNA, the local residents protested the intensifying fumes wafting over their community. They appealed to their U.S. senator, who demanded and received an explanation, although the Chief of Ordnance opined that "inconvenience and annoyance to the individual must be made subject to the national interests." In those cases, Ordnance officers vouched for the importance of a firm, but a state government could and sometimes did close down an exceptionally dangerous plant unless modifications were made.[34]

In the end, the department managerial superstructure contributed significantly to the transfer of technological knowledge across the U.S.

explosives industry. Statistics indicated where to find best practice, and the department used their information to improve the performance of firms with less expertise. Information gathered from the plants of Du Pont and Hercules ended up in the hands of smaller firms in South Bend, Indiana, or Lansing, Michigan. When the war drew to a close, the Ordnance officers sought to codify and preserve the technical information they had acquired from firms and the Allies. They systematically gathered process descriptions and wrote technical reports in the hope that the information would be useful in any future conflict and cut time required for any future mobilization.[35]

Mobilizing War Gases

In the winter of 1917 and 1918, Col. Amos Fries, the Chief of Gas Service in the American Expeditionary Force, sent reports to his military superiors explaining the types of war gases and their strategic uses as they had evolved in the European theater. Mustard gas, first used by the Germans in the summer of 1917, was "very dangerous and insidious as its action is delayed from 4 to 12 hours and . . . one does not know he is being gassed." The gas "burns worst in damp weather and against soft parts of skin." Tactically, a commander might deploy mustard gas several days before a planned attack because it lingered in the target area for at least 48 hours, longer in damp or low-lying areas. A second major gas, phosgene, made a deadlier strike, but soldiers more immediately detected its presence, and it worked best when catching enemy troops by surprise, especially "at night as confusion and sleepiness conduce to casualties." It dissipated rapidly enough to contemplate sending one's own troops into the target area within three to five hours (or even twenty minutes in an open, breezy area). The third major gas was chlorpicrin, which the combatants sometimes mixed with phosgene. It was "considerably more persistent though less deadly than phosgene." As a "lachrymator," it also irritated eyes and forced enemy troops to wear gas masks. Even the best gas masks made breathing difficult and limited the physical exertions of soldiers, making it tactically valuable at times to keep the enemy's masks on. The Germans began gas warfare in 1915 at Ypres, Belgium, with the release of chlorine from canisters, relying on the wind to carry the gas to its destination (a climatic condition that favored the Allies far more often than the Germans). Both sides continued to use these "cloud attacks," despite "the utter lack of enthusiasm on the part of the infantry" for them, because they "produce more casualties per weight of material used than any other form of attack known." But both sides also learned to deliver the gases more accurately by shooting them over in shells. To engage successfully in gas warfare, the United States would need to manufacture the three major gases and various others, learn to modify gas masks quickly for defense, and develop new gases to outmaneuver the enemy. They started almost from scratch in April 1917.[36]

The mobilization of gas warfare in the United States took on a significantly different organization than that of high explosives. Despite the gas warfare underway on European battlefields since 1915, the American military had been slow to recognize "the great importance of gas warfare."[37] Nor had American industries supplied the Allies with gases as they had explosives or a thousand other products. American gas mobilization involved not only the coordination of resources and existing expertise, as in explosives, but also the generation of an enormous amount of new expertise. In addition to the lack of expertise in military and industrial circles, other characteristics of gases contributed to a fundamentally different relationship between government and industry in the mobilization of gases. The extraordinarily toxic nature of gases made them more dangerous to manufacture and transport than high explosives. Secrecy mattered more, as military advantage in gases derived from surprises, including catching the enemy unprepared for a new substance. Plus, few manufacturers expected to be able to sell such gases on the market once the war ended.

These disincentives for firms meant that the loci of gas expertise lay outside U.S. chemical firms. Much more than in explosives mobilization, Americans relied on government agencies and on European allies to make the synthetic organic chemicals for gas warfare. As the war progressed, the growing gas program became increasingly centralized under military control.

Initially, the primary impetus behind the American gas program came not from the military or from the chemical manufacturers, but from a nonmilitary government agency and from the academic chemistry community. The Bureau of Mines, familiar with gases and gas masks from their work on American mining, led the development of the American gas program under director Van H. Manning. As Manning offered his bureau's help to the military in February 1918, the National Research Council (NRC) almost simultaneously asked the bureau's personnel to direct gas research. This began a fruitful collaboration between the bureau and the NRC. Acting with extraordinary bureaucratic entrepreneurship, Manning orchestrated a research program that peaked at 1,900 people, including 1,200 chemists and other technical staff, concentrated in Washington, D.C., in new laboratories. Whereas the NRC's Committee on Explosives was almost an afterthought, its Committee on Noxious Gases played a vital role in identifying significant research problems, coordinating academic chemists, and communicating with the Allies. It performed as the "clearinghouse" of scientific information that the NRC's founders had envisaged and helped to guide the bureau's research program. The bureau also received advice and guidance through its advisory committee, comprising military officers, chemical manufacturers, and government and academic chemists. In addition, Manning's agency held a census of chemists taken in February 1917 to facilitate directing chemists into laboratories and not into the trenches. Conscious that many good European chemists had died at the front early in the war, Americans placed 4,003 of the 5,404 drafted chemists in chemistry-related positions in the war effort.[38]

Advice also came from abroad. The relationship with the Allies developed differently in war gases, because the American gas program needed Allied expertise in a way the U.S. high explosives industry simply did not. When Americans contemplated the organization of their gas program, they took the British model into consideration. Americans also knew that an "understanding" existed with the British government, which was willing to "exchange experts whenever necessary." In fact, the director of gas research, George Burrell, later recalled that the bureau designed its research strategy only after "literature . . . from France and England had been digested." Although the French and British gas programs communicated rather badly with one another, the Americans developed regular channels with the Allies, particularly the British. Through the NRC and the military, American emissaries were in close touch with the British gas efforts, learning from British successes and failures. As research progressed, the British shared their reports in meetings or even via cable, and Americans developed a system of bi-weekly reports that communicated research results not only internally but also to the Allies.[39]

In early April 1917, the NRC committee convened a series of meetings that proved instrumental in structuring the American gas effort. Manning marched the committee through a series of basic questions: Who should do the U.S. research? Should it be under the military? Where should it be located? Should the military (in addition to the NRC) send chemists to Europe? What organizational structure should the gas research take?[40]

The early months required planning and rapid expansion, which came with an element of organizational trial and error. Among other things, the bureau made the development of an adequate gas mask its first priority; gas "defense" research centered on charcoal and other materials. "Offensive" research, which started more slowly—getting underway in July 1917—by early 1918 attracted some of the best chemists in the country. In May, the bureau sent representatives around the eastern United States to recruit chemists to gas research, and several universities became "branch laboratories" in the bureau's gas program. Increasingly, however, the program's leadership attempted to consolidate the bulk of the research in Washington, D.C. Already in April 1917, the American University had offered its facilities, and in September the bureau centered its research on the campus, which grew from two buildings to "over sixty" by the armistice. Important work still occurred in the branch laboratories, but at its American University Experiment Station, the bureau pioneered focused interdisciplinary research involving groups of scientists.[41]

The challenges posed by mustard gas, for example, illustrated the research organization's strategies (and sometimes lack thereof) for developing gases from the laboratory to full-scale production methods. First, the Americans met with their British and French counterparts in September 1917 to learn the results of the Allies' preliminary investigations. In the following months, cablegrams and visitors flowed back and forth across the Atlantic, sharing results and suggestions. The process ultimately chosen for the primary method of production (sulphur monochloride) was virtually a simultaneous discovery by American and British scientists, not least because they shared the research leading up to it. Second, the Americans placed some of their best organic chemists on mustard gas development, both those chemists in Washington and outside. In particular, Moses Gomberg at the University of Michigan, fending off persistent importunities to join his colleagues at the American University Experiment Station, labored in early 1918 in his university laboratory to devise a workable method for ethylene chlorhydrin, an intermediate in the original method of manufacturing mustard gas. Finally, despite the scientific literature, German samples collected on the front, and the accumulation of top scientific minds, the American chemists required several months to develop technical familiarity and competence with mustard gas. Lacking any experience with the gas themselves, the Americans had to devise workable and affordable processes in the laboratory before they could guide full-scale production.[42]

While the Bureau of Mines and the NRC concentrated on replicating existing war gases and designing new ones in the laboratory, the full-scale production of gases fell under the military's jurisdiction. As with the Department of War more broadly, the gas program underwent several reorganizations during 1917 and 1918. Initially lodged in the Trench Warfare Section of the Ordnance Department, the gas program ultimately achieved the status of an independent military unit within the Department of War with the creation of the Chemical Warfare Service (CWS) at the end of June 1918.

The CWS consolidated gas-related programs scattered elsewhere in the government, including that of the Bureau of Mines. These changes partly reflected the increasingly central and complex role of the military in gas production.[43]

In October 1917, the Trench Warfare Section of the Ordnance Department established the Gunpowder Neck Reservation, a piece of the Aberdeen Proving Ground in rural Maryland, as its center for gas development. Initially, when it still expected private firms to make the gases, the military planned to build at Gunpowder Neck a large plant in which to fill shells with the gases. The scope of the reservation expanded over the next several months as the military added manufacturing and laboratory facilities. From the British and French, the designers of the complex "sought . . . advice," which became "the inception of what later developed," although ultimately most technical advice came from American chemists and manufacturers. Given the immense scale of the planning, the American designer recalled that the Allies "thought we were nothing short of crazy," although he believed the size reasonable, if one "extrapolated" from the increased use of gases.[44] In March 1918, the army removed the Gunpowder Neck Reservation from the jurisdiction of the Trench Warfare Section and gave it autonomous standing within the Ordnance Department, renaming it the Edgewood Arsenal in early May. The arsenal would become a highly centralized, government-owned, and government-operated complex responsible for the production of poison gases.

The Edgewood Arsenal contained the two primary functions in the preparation of war gases—manufacturing the gases and filling the shells with gases, an equally hazardous endeavor. Plants for chlorpicrin, phosgene, chlorine, and mustard gas were underway in the spring of 1918 with production ready in June, July, August, and September, respectively. In most cases, chlorine—and not the organic chemicals—formed the key ingredient in creating the desired toxic effect and was the "bottleneck" in the manufacturing process, although enough chlorine plants existed prior to the war that manufacturing know-how was sufficiently abundant in the United States. At the armistice, Edgewood Arsenal was immense, containing 558 buildings for the manufacture of war gases and gas-filled shells and for housing 7,400 workers.[45]

The most ambitious organizational change in war gas mobilization came in June 1918, when all the gas-related programs in the federal government were consolidated into the new Chemical Warfare Service. The formation of the CWS, and the May arrival of General William Sibert to take command of the military's war gas efforts, brought significant transition to the research

FIGURE 3.1. Edgewood Arsenal mustard gas unit, c. 1918. (*The Edgewood Arsenal*, published as an issue of *Chemical Warfare* 1, no. 5 (March 1919), 38. Courtesy Othmer Library of Chemical History, Chemical Heritage Foundation, Philadelphia.)

organization of the Bureau of Mines as well as the military parts of the gas program. Known even by the secretary of war as one of the "most stubborn" of military men, Sibert was convinced that the war gas projects needed consolidation and should be placed entirely under his command. Even before Sibert's arrival, William Walker used his position as the commander of the Edgewood Arsenal to expand the range of military investigations into chemical warfare research, which already had the Bureau of Mines leadership wary of competing jurisdictions. Dismayed at the possibility that he would lose his project, Van Manning protested against military control and attempted to retain autonomy for the bureau's research program. His external advisory committee came to his defense as the military attempted to take over gas research. Manning's operation, they argued, was "complex and delicate but well articulated and working with an efficiency and enthusiasm which has greatly impressed us." Many of them had the model of corporate research in mind and believed that "best practice . . . place[d] the research work under separate control," protected and away from the immediate demands of the producers. To Sibert, however, the analogy fell short. In war, the only demands were immediate demands, and he believed that the chemists were too isolated and autonomous. He wanted them to respond more directly to the military's needs. His view prevailed, and President Wilson signed the executive order transferring the Bureau of Mines research program to the new Chemical Warfare Service on June 26, 1918.[46]

The protests by the Bureau of Mines obscured the reality that a research chemist would often follow his project from the laboratory stage through development and production. Several personnel on the military side began the war with the research program, including such high-ranking officers as William Walker and William McPherson.[47] In most respects, however, the Bureau of Mines and the military retained a division of labor until the creation of the Chemical Warfare Service.

Gas manufacturing evolved over the course of the war. From semi-scale production at small firms cooperating with the Bureau of Mines, manufacturing increasingly became more militarized. Initially, the Ordnance Department contracted with several firms to make war gases in 1918. For strategic, legal, and technical reasons, however, the military began to consolidate production into government-owned and -operated facilities, and also onto the single, giant compound in rural Maryland. Among other things, the military officers wanted "to have it under entire military control ... without spying eyes," and they feared wide dispersion of such toxic material. The quantities demanded in Europe grew from a "few hundred pounds per month" in April 1917 to "a few thousand pounds per month" at the armistice. In May 1918, the military determined to expand gas production by five times, and in the summer of 1918, the American forces went from six companies of soldiers devoted to gas warfare to fifty-four.[48] While the relationship between the military and industry was problematic in both high explosives and war gases, the balance of power was very different because of the chemical characteristics of the war gases and because of less experience with them.

Through the fall of 1917, the military officials still expected to purchase war gases from private manufacturers, but they ran into a host of problems that forced them to consider alternatives. Above all, manufacturers were reluctant to make war gases, and the military simply could not line up enough companies to meet their demands. They pursued larger contracts with the firms operating semi-scale works in cooperation with the Bureau of Mines, and they attempted to lure established chemical firms into gas production. In particular, they asked Du Pont to undertake gas manufacture, but the firm's managers declined, justifiably claiming their works were fully occupied with other war work. And Semet-Solvay "flatly refused to make either phosgene or chloropicrin."[49]

Manufacturers hesitated to make gases for several reasons. They knew the opportunity for profits existed only for the duration of the war, and this figured into their calculations for investment and returns. Phosgene could be sold in a peacetime economy to a domestic dyes industry, but most war gases were understandably rather limited in their potential for widespread commercial applications. Manufacturers therefore needed to master a new technology quickly, knowing their production run would end with an armistice, whenever that might come. Still, most government contracts guaranteed a profit, and the evidence suggests that the safety hazards played a significant part in manufacturers' decisions to avoid making gases. They knew the gases placed their plants, employees, and surrounding areas at risk if not properly managed, and few had the confidence to experiment under such conditions. Chemists spoke of batches of mustard gas "go[ing] wild," putting out hydrochloric acid and "unknown" toxic gases.[50] Among the manufacturers who ventured into gas production, every single one demanded that the government assume the liability for any accidents, which it did.

Even as the government and military officials in charge of the gas program became increasingly committed to the idea of centralized production, they also perceived distinct strategic benefits if even a few firms would manufacture gases. First, the large size and scope of the Edgewood Arsenal meant that smaller plants could be constructed more quickly and generate gases sooner. Second, the satellite plants served as a back-up supply in the event of an accident at Edgewood. Finally, in many cases, administrators in charge of chemical warfare were unsatisfied with the known processes for manufacture and hoped that research and development at other firms might yield improved methods. In the case of mustard gas, Edgewood Arsenal's managers encouraged experimental plants at the Dow Chemical Company, Zinsser & Company of Hastings-on-Hudson, New York, and at a small government plant in Cleveland largely because the government officials did not yet have a process for which it wanted to commit the resources and space of Edgewood Arsenal. If the experimental plants proved workable, the arsenal could not only learn new techniques but also expand the satellite plants. At the armistice, two large plants for mustard gas neared completion at Zinsser & Company and the National Aniline & Chemical Company at Buffalo.[51]

Nine manufacturers contracted with the U.S. government to make war gases, and their collective history reveals a series of problems, from safety to technical difficulties. Tension permeated the relationship between the military and manufacturers as they sought a workable division between public and private. The two sides struggled over managerial autonomy, safety, supplies, liability, and costs, among other things. The balance of power clearly lay on the military side, however, as its officers asserted themselves more and more, and the manufacturers' limited technical expertise offered no countering leverage. The experiences of four companies are discussed below; each reflected the difficulties of developing technical expertise under stressful wartime conditions and with extremely hazardous materials.

As early as 1916, the Oldbury Electrochemical Company in Niagara Falls, New York, began working on an experimental plant to produce phosgene. The firm's managers sought to use a by-product efficiently: in the firm's production of phosphorus, it generated significant amounts of carbon monoxide, one of the ingredients in phosgene. When the United States went to war, the firm's general manager, who was British, reached out to the Bureau of Mines and the Ordnance Department to offer his firm's expertise to the gas program. The bureau cooperated with Oldbury, but the firm developed enough of the work to retain proprietary rights to the process; later reports confirmed the process was similar to the one eventually adopted in the phosgene plant at the Edgewood Arsenal, however.[52]

From the military's first discussions with Oldbury in summer 1917, the firm's managers explicitly and consistently agreed to build a large-scale phosgene plant only if the government assumed the liability. The military agreed, but Oldbury's managers remained deeply skeptical and skittish about the liability, and they required five months of negotiations before they were satisfied they would be protected legally in case of an accident. A designer of the Edgewood Arsenal later recalled that Oldbury clearly wished to avoid the project but went forward as the result of "moral ... compulsion." In December, Oldbury finally signed a contract (ten tons of phosgene per day at cost plus 10 percent) and began construction in January. Hampered by winter in Niagara Falls, construction took another six months, and the plant first ran in August. At the armistice, the total output of the plant was 435 tons, which meant the firm ran at about half of capacity once it began to produce. The military blamed the low output on the supply of shells (for Livens projectors) rather than Oldbury's technical competence.[53] Despite the slow pace of negotiation through 1917, the relationship between Oldbury and the military would appear exceptionally smooth compared to the experience of other firms.

In the fall of 1917, the military officers learned of a second manufacturer of phosgene, Frank Hemingway, Inc., in Bound Brook, New Jersey. The officers' initial contact with the firm was "very unsatisfactory" from the military's perspective, and they dismissed the firm as a "paper company." In December, when Hemingway mastered an experimental plant, he resumed negotiations with the government for full-scale production of phosgene. The military issued a contract in January for a plant to produce five tons per day, although provisions were quickly made to expand the plant to ten tons. Hemingway's crew finished construction and first operated at the end of April. As with most other firms making gas, Hemingway's firm produced in fits and starts and ultimately well below expectation.[54]

Hemingway blamed the military for undue red tape, and the officers thought Hemingway slow to recognize they were the ultimate authority in the plant, but their collective problems centered on the difficulties in assessing and working out a new process. Hemingway and his chief technical advisor, Charles Baskerville, a chemistry professor at the College of the City of New York, developed a different method for making phosgene, and Hemingway was determined to retain the proprietary rights to the process. The officers accommodated that demand; to assess the technical potential of Hemingway's proposal, the officers assembled an independent panel of chemists who would respect the proprietary secrets. Neither the panel nor Hemingway's staff anticipated the level of difficulty, however. For three months following the start-up of the plant, it was riddled with technical failures: clogged pipes, leaks in pipes and generators, vacuums appearing where they ought not to have appeared, constant repairs to equipment, and "several explosions." The firm's construction costs were $32,000 over the initial $120,000 budget. The actual operating costs of 20¢ to 22¢ per pound exceeded by a third the anticipated cost of 14¢ per pound. The officers acknowledged that many such problems should be expected in a new process, but they also believed the Hemingway crew had made several preventable mistakes in their planning. And it drove them to distraction that the firm's managers seemed to move too slowly to make repairs. In the late summer, when the plant operated better, shortages of oxygen for the phosgene also hindered output. Continuing shortages led the military to mothball the Hemingway plant in October, and it remained shut down for the duration of the war. By the armistice the plant had produced 410,000 pounds of phosgene, a fraction of its potential.[55]

The problems between the military and the phosgene manufacturers, while clearly frustrating to both sides, were not so different from the kinds of production problems that might occur when undertaking the manufacture of any new difficult chemical process. The wartime pressure to produce quickly only highlighted the technical failures. Given the experience the military would have with the manufacturers of chlorpicrin and mustard gas, the problems in the phosgene plants appeared almost routine.

The American Synthetic Color Company of Stamford, Connecticut, began operations early in the war as one of the small new dyes firms attempting to exploit the war shortages. To obtain technical advice, the firm's managers contracted with two MIT chemical engineers, Leslie T. Sutherland and E. S. Wallace. The dyes firm had already diversified into the manufacture of picric acid when a Bureau of Mines representative contacted Sutherland about undertaking semi-scale manufacturing of chlorpicrin, made by chlorinating picric acid. The bureau had access to a workable British process that they would share with the firm. Sutherland and his bosses accepted readily and told the military in September their experimental plant could be scaled up in two weeks to produce 4,000 pounds of chlorpicrin per day. The military officers knew the firm's finances were shaky, but the American Synthetic Color Company's managers were the only private manufacturers experienced with the gas who were willing and seemingly able to make chlorpicrin, and the military contracted with them in December. The procurement order specified immediate delivery of 5,000 pounds per day, but also expansion to 50,000 pounds per day until the firm reached 9 million pounds total.[56]

The American Synthetic Color Company never lived up to expectations, and the disarray that prevailed in the firm's plant and front offices had the military supervisors pulling their hair—and fearing a major accident. First, the firm failed to produce gas in quantities anywhere near the contracted amount; by March, they had made just one tenth of the specified quantity. Second, it was an inefficient producer, consuming more raw materials than the reactions ought to have required. Third, the multiple owners provided incoherent leadership. By the end of February, the Ordnance Department prepared to commandeer the plant. In March, the military sent in an investigator, Captain Edward E. Free, to make a thorough report on the failures of the company.[57] His reports abandoned staid and bureaucratic military vocabulary to express his growing fear that the American Synthetic Color Company posed an insufferable risk.

Free roundly criticized the management of the plant from nearly every angle: "The plant is inexcusably dirty, materials are carelessly handled and wasted, the labor is discontented, dissensions are rife amongst the foremen and higher employees, the laboratory controls are inadequate, the business system is loose, and the books are neither up to date nor complete." The workers also faced excessive risks from the twelve-hour shifts and shortage of gas masks, and the plant lacked necessary security measures against sabotage. The problems lay primarily with "incompetent and vacillating management," and Free believed that only the excessive humidity in prior weeks had prevented the mishandled picric acid from exploding. The military's first priority, wrote Free, must be to secure the plant "against explosion and fire.... There is a real danger of an explosion which would practically wipe out the city of Stamford."[58]

On March 27, the military moved in and shut down the plant because it had become too dangerous and because the managers had made only mediocre attempts to clean up the plant and the finances in previous weeks. The Ordnance Department commandeered the plant and installed an officer as the manager. For the remainder of the spring, the military struggled to bring order and safety to the plant, constructing concrete barriers and undertaking studies to obtain better yields. They also targeted the waste products of the plant, which posed a particularly serious hazard. An unlucky civilian was gassed when a small amount of chlorpicrin escaped with the waste into Stamford's public sewers, and the military supervisors sought safer alternatives, such as piping the material into the Long Island Sound. Any gas in the discharge would sink to the bottom, they believed, although they suspected the neighbors along the adjacent shoreline and oyster harvesters might wage legal protests at the remainder of the effluent, which the officers were not so sure would sink. For the longer term, they hoped to develop a process that produced less troublesome effluent.[59]

The military officers invested in the Stamford plant initially because they believed they had contracted with technically competent manufacturers. When that proved not to be the case, they continued to operate the plant themselves in case the chlorpicrin plant at Edgewood was incapacitated through accident or sabotage. In the end, the Stamford plant convinced the top officers they had not erred in commandeering the Stamford plant or in centering operations at Edgewood where the military could control production more tightly. They learned there was no advantage in technological expertise at the private firm. Records suggest the plant ran more smoothly under military control; they also suggest demand from the front was not expansive enough to drive the plant to full capacity. For all of the plant's problems, it still supplied 60 percent of the U.S. production of chlorpicrin, with the remainder coming from the plant at the Edgewood Arsenal.[60] In the end, however, the case was a failure with respect to the private manufacture of gases.

Dow Chemical made its primary contributions to gas warfare through their work with bromine lachrymators (tear gases), but they had also undertaken initial investigations of chlorpicrin. Their work on mustard gas rarely made it into official military accounts, but the records that describe the Dow Chemical Company's experimental mustard gas plant reveal some of the complexities in scaling up production of a hazardous material. Not all of these challenges were related strictly to manufacturing. The electrolytic chlorine plant at Dow Chemical made the Michigan company a likely candidate for the production of mustard gas. Albert W. Smith, the chemist from the Case School of Applied Science who frequently consulted for Dow

Chemical, was one of the chemists conducting poison gas research under the Bureau of Mines program. Beginning in October 1917, Smith spent half his time researching chloropicrin, arsenic chloride, and several bromine-based tear gases. By February 1918, the chemist turned his attention to mustard gas and received permission from Washington to expand his research in facilities at Dow Chemical. The Dow group scaled up its bench process to produce small 300-pound batches and then built a commercial plant with four units to become the first full-scale manufacturer of mustard gas in the United States. Under Smith's direction, the first unit of the plant, with a capacity of fifteen tons per day, manufactured mustard gas in the middle of June until an accident at the end of the month interrupted operations. In mid-July, both the first and second units began production but were hampered by an inadequate supply of ethylene, which Dow produced at its Midland site from alcohol provided by the government. Alcohol shortages and an undersized ethylene plant contributed to Dow's shortage of ethylene, the only organic component in mustard gas. However, at the end of July, Colonel William Walker of the Edgewood Arsenal issued an order for Dow to build a larger plant that would produce twenty tons of mustard gas per day.[61]

Until early June, Dow Chemical manufactured the mustard gas with its own employees and four technical assistants sent by the Bureau of Mines, all under the supervision of Albert Smith. But when the first unit of the commercial plant was ready to operate, the army, for security reasons, sent twenty-three enlisted men to work in the plant alongside the Dow civilian employees. The soldiers answered to their lieutenant, but the lieutenant followed orders from Smith. Initially, this chain of command caused friction, but tensions receded with time. Dow and the army kept the mustard gas program secret, also in the name of security.[62]

Like many industrialists in the Progressive Era, Herbert Dow believed, or at least espoused the belief, that most industrial accidents were the result of worker carelessness. Although he had not shown much remorse about earlier plant accidents, Dow seemed truly distressed by the rate of injuries among his employees in the mustard gas plant. He noted at least two minor injuries every day, and the accidents led Dow to establish hospital facilities for the victims of the gas burns. The hospital consisted of two refurbished rooms in the Dow Company's education building. A local doctor attended the hospital until an army doctor, Lester L. Roos, arrived at the end of May. Local nurses staffed the hospital rooms throughout the war.[63]

The hospital soon received its greatest challenge in late June when a serious accident occurred at the first unit of Dow's mustard gas plant, sending seven men with severe chemical burns to the hospital. The accident shut down the plant immediately at a time when the army called for maximum output. Subsequent events in the hospital affected the relationship between Dow Chemical and the army and consequently the way that poison gas manufacturing would be organized in the United States. At the hospital, the staff worked long hours to treat the patients, but two men died within a month after suffering through the burns and blisters caused by mustard gas. On July 30, Roos, the army doctor, came down with the same symptoms as his former patients—swelling and burning on his foot and leg. Observers quickly realized that someone had put mustard gas in the doctor's shoe.[64]

The crime quickly brought to the scene army investigators in search of German sympathizers. Almost immediately, Winnifred Murphy, the head nurse at the hospital, confessed to the deed. News of mustard gas production and the accident reached the public when the patients were transferred to the local Bay City hospital. Roos also went to Bay City and while there for a few days accused Dow Chemical of mistreating the patients, in addition to sabotaging his shoe. Roos and two medical doctors from the University of Michigan who had been invited to observe the cases implicated not only Murphy but also the entire staff of the hospital and even the leadership of Dow Chemical Company in disloyal activities. For the rest of the war, the army feared for its enlisted men's safety at Dow.[65]

Dow's own investigations painted a very different picture. Interviews with the nurses, patients, other army officers, and Dow employees revealed Dr. Roos to be "unprofessional, vulgar, profane and unbecoming an officer and physician."[66] The nurses especially found the doctor difficult. They accused him of roughly handling dying and suffering patients, including subjecting one man to such an unusually long period of time in a chemical bath that his skin tissue died. On other occasions, the nurses stated, Roos would allow the university doctors to move the patients unnecessarily to take photographs of their injuries. But beyond the questions about medical treatment, the doctor and nurses developed an extremely antagonistic relationship. Instead of renting lodging, the doctor resided in the hospital rooms. He placed a cot in one corner of the room and, according to the nurses, sent his laundry out with the hospital's, ate food purchased and prepared for the patients, walked around the hospital half-dressed, made at least one unwanted advance toward a nurse, and expected the nurses to prepare his meals at unusual hours and even get him dates. Head nurse Murphy lost her patience and was able to obtain a small amount of mustard gas from a friend working in the plant.[67]

Murphy's action had several consequences. First, Dow fired her, and the army transferred Roos to another location pending completion of its investigations. Murphy's step also had repercussions for the nation's mustard gas program. In mid-August, the commander of the Edgewood Arsenal ordered Dow Chemical to cease production of mustard gas and suspend construction of the larger plant. The telegraphic order explicitly cited "difficulties and complications arising from action of your [Dow's] employees relative to Captain Roos" as the justification for shutting down Dow's mustard gas production. Dow officials could not quite believe that an "entirely trivial" incident had stained the company's reputation. They also could not understand why the government halted production of a high-priority war gas. Dow's small plant remained idle when the armistice arrived, and the larger plant stood incomplete. Published records rarely mention Dow's mustard gas production.[68]

The organizational relationship between industry and government in the gas program continued to evolve until the armistice. The future trajectory of the program is illustrated in the way the military developed a new gas in 1918. Liking neither the strategic weaknesses that attended the concentration of facilities at the Edgewood Arsenal, nor the complications of coordinating with private civilian plants, the military officers chose a third way. They established a factory in Cleveland designed and run entirely by the military. The new gas was Lewisite, a chlorinated arsenic compound initially pursued by a chemist from Northwestern University, Winford Lee Lewis. After the work by Lewis and his research group at the Catholic University branch laboratory, the project moved through the developmental stages, in which the young Harvard chemist, James B. Conant was a key participant. The military planners believed that, to maintain secrecy and discipline, the workforce must be "strictly military," with no civilians, allowing only such liberties as necessary to keep up the men's morale. In planning the location and layout of the factory to produce the new gas, the planners were acutely conscious of the "very toxic" nature of their chemicals and the consequent risks to workers. "A fair estimate is that from 10 to 25 percent of the entire working force will be on the sick list all the time," wrote the planners, who wanted the workforce to be 50 percent larger than required in some jobs to accommodate the high casualty rate. The entire project remained top secret for the duration of the war and created a stir in the trade press after the armistice when it was made public (the new gas was "seventy-two times more powerful than mustard gas," suggested *Drug and Chemical Markets* rather fantastically). At the armistice, the first batch of Lewisite was on the way to Europe; the crew dumped it overboard.[69]

Through 1918, the American gas program grew fast and furiously; plans existed for even larger production in 1919. In fact, after the armistice,

Table 3.5. American War Gas Output (in Tons), 1918

	Chlorpicrin	Phosgene	Mustard Gas
Jan.	10	—	—
Feb.	27	—	—
Mar.	59	—	—
Apr.	33	15	—
May	130	18	—
June	263	23	6
July	499	100	21
Aug.	646	314	36
Sept.	564	327	144
Oct.	445	664	361
Nov.	100	155	143
TOTAL	2,776	1,616	711
Shipped to Europe	1,903	420	190
11/1/18 monthly capacity	1,500	1,050	900
1/1/19 monthly capacity	1,500	1,650	4,000

Note: Haber provides numbers for the major belligerents from 1915 to 1918. All amounts are listed in tons as chlorpicrin, phosgene, and mustard gas, respectively: Germany, 4.1, 18.1, and 7.6; Britain, 8.0, 1.4, .5; France, .5, 15.7, and 2.0; United States, 2.5, 1.4, and .9 (Haber's figures on the United States are higher than Crowell's). Germany also produced 11.6 tons of diphosgene. Haber, *The Poisonous Cloud*, 170.
Source: Benedict Crowell, *America's Munitions*, 409. The monthly numbers also appear in the archival records, although Crowell's numbers for amounts shipped overseas are generally lower than the archival statistics. Fries, "Gases Produced by the Chemical Warfare Service," August 13, 1919, National Archives and Records Administration, RG 175, entry 7, box 16.

William Walker of the Edgewood Arsenal described chilling plans to fly over enemy territory like Coblenz or Metz with one-ton containers of mustard gas, exploding them in the air to destroy at once an acre or more, in which "not one living thing" would survive.[70] However, for all the growth and plans for the future, the armistice arrived before American gases made any sizable contributions on the front. Development time, shortages of shells and other supplies, and the challenges of transporting gases all meant that the American program began to make an impact just as the war came to a close. The three main European belligerents, making gases since 1915, all outproduced the United States. After the war, the military officers celebrated their accomplishments, though they rarely noted publicly that output had been considerably below expectations and capacity.

A month after the armistice, Harvard graduate Edward O. Holmes Jr. felt confident that the skills he had learned at Du Pont during the war would serve him well. For eighteen months, he "spent [his time] entirely on

Table 3.6. Gas Output of Outside Plants and Edgewood Arsenal, 1917–1918 (in Tons)

Gas	Plant	Output
chlorpicrin	American Synthetic Color Company (Stamford, Conn.)	1,613
	Edgewood Arsenal, Md.	1,160
phosgene	Oldbury Electrochemical Co. (Niagara Falls, N.Y.)	476
	Frank Hemingway, Inc. (Bound Brook, N.J.)	205
	Edgewood Arsenal, Md.	935
mustard gas	Zinsser & Company (Hastings-on-Hudson, N.Y.)	0
	National Aniline & Chemical Co. (Buffalo, N.Y.)	0
	Commercial Research Corporation (Flushing, N.Y.)	? (semi-scale amounts)
	Dow Chemical (Midland, Mich.)	? (semi-scale amounts)
	Edgewood Arsenal, Md.	711
brombenzylcyanide (a tear gas)	Federal/Union Dye and Chemical Co. (Kingsport, Tenn.)	5

Note: The Zinsser and Co. and National Aniline and Chemical Co. plants were incomplete at the armistice. Dow's brine wells and Federal's plant were just ready at the armistice, whereas others had been shut down by then. The CWS canceled the contract with Commercial Research to make mustard gas by the ethylene chlorhydrin process after they worked out the sulphur monochloride process. Toward the end of the war, the CWS had also built other plants outside of the large Edgewood Arsenal plant in Maryland not having any role for private firms.
Source: Benedict Crowell, *America's Munitions*, 401–4, except that Dow Chemical's mustard gas program is excluded.

organic research on new explosives, the formulars [sic] of which can't even be found in the literature and so I have become acquainted with an abundance of organic chemistry." With additional experience in the semi-works plant, Holmes approached his postwar future with optimism.[71] Certainly not every chemist in the explosives industry gained the same experience as Holmes, but enough of them did to contribute to the production of dyes, pharmaceuticals, and other peacetime products of the growing U.S. synthetic organic chemicals industry. Although by no means sufficient to sustain a highly developed commercial industry, the knowledge gained by explosives manufacturing and gas research contributed to an accumulating base of knowledge. In the war crisis, the usual guards against sharing know-how with competitors were relaxed considerably, and the war agencies provided new avenues for diffusing the knowledge. As the firms and the military demobilized, many chemists voluntarily and involuntarily took their technical skills elsewhere—to (or back to) industry and academia.

Beyond expertise, the massive production of chemicals for war also increased the supply of raw materials for domestic synthetic organic chemicals production. Dyes and pharmaceuticals and all of the peacetime product would use a mere fraction of the toluene, benzene, and phenol consumed in the wartime production of explosives, but the war gave steel firms the incentive to expand their use of by-product coke ovens, ensuring that raw materials would not be a critical barrier to the survival of the domestic industry. Third, both the production of explosives and war gases pushed and expanded Americans' abilities to make chemicals in volume, building on and expanding expertise in chemical engineering, which contributed to the American manufacturers' abilities to take the synthetic organic chemicals industry in new directions in the future. Finally, the war also generated enormous profits for manufacturers, large and small, who would invest in the production. Du Pont is the obvious case, but the trade press already in 1916 reported that Aetna planned to convert their explosives plants to dyes manufacturing. Although the U.S. government cut firms' profits through higher tax rates once the United States entered the war, the scale of operations continued to expand, and competent firms still enjoyed a high rate of return on their investments.[72] As the firms attempted to diversify, the war revenues helped to cover the steep costs of learning to make synthetic organic chemicals for the civilian market.

4 / Ideology and Institutions

American Chemists Respond, 1914–1918

In the U.S. effort to build a domestic synthetic organic chemicals industry, the skills and commitment of American chemists would be crucial. American manufacturers needed their in-house knowledge, but they also believed—based on German experience—that their ability to innovate depended on universities' ability to expand their research in synthetic organic chemistry and to train employees for the industry. Historians of business and technology have increasingly highlighted the networks and nodes that construct an infrastructure of knowledge to support science and technology-based industries. When the war began, American chemists and the larger chemical community showed great ambivalence about the European war, and their public discussion tracked the broader debate in the nation as a whole. Many American chemists had trained in Germany, had close friends in Germany, and had profound respect for the achievements of German chemists and the German chemical industry. In the scientific and trade literature, several members of the chemical community suggested that Americans should simply leave the synthetic organic chemicals field to the Germans—the U.S. market could survive the war shortages with temporary solutions—and invest their time, energy, and capital in other branches of chemistry and industry. As the previous chapters have indicated, however, chemists became deeply involved in fighting the war, and the patriotic zeal that gripped the nation manifested itself among chemists in part by an ideological commitment to "freeing" synthetic organic chemistry from German dominance. In their work and lobbying, American chemists formed a vision for their community that mirrored the autarkic isolationism that shaped the political economy of the war and interwar. By the end of the war, the American chemical community was strongly behind the effort to build a domestic synthetic organic chemicals industry.

Like World War I, science itself was an enterprise that was at once national and international. Scientific ideas flowed across the Atlantic with increasing ease before the war, and American chemists felt they belonged to the larger scientific community of the West; international trade in chemicals additionally bound the American and German chemical communities. As the war crisis worsened, American chemists cut their German ties—scientific, commercial, and personal—and embraced nationalism in ways particular to their goals and roles in the war. They asserted that American chemists could be intellectually independent of Germany and that they were able to support an American industry. They grew closer to the British and French chemists through military and commercial collaborations, but the war began an isolation of the German scientific community that would continue through the interwar years.

Besides weakening ties to the international science community, the war accelerated chemists' growing identification with the nation and a national community of science. The growing anti-German sentiment in the country permeated the chemical community and shaped the response to war. The voices of the chemists emerged most clearly and loudly through the American Chemical Society (ACS), their primary professional society, and the ACS journals. The shortages put chemists in a national spotlight, sometimes critically, and they sought to use the attention to build appreciation for chemistry and support for the industry. The American chemical community openly and regularly speculated about the war's long-term impact on the international chemical market; they increasingly wanted the changes to be permanent and supported policies designed to prevent the Germans from reclaiming the American market after the war. Fighting and mobilizing for war brought the chemists into contact with several agencies within the federal government, increasing their collective familiarity with the political mechanisms in Washington. Over the course of the war, the ACS leadership became more deeply involved in promoting American "independence" from Germany in synthetic organic chemistry.

As academic chemists moved from universities to war work, they reoriented their research and teaching from the campus to the nation. Universities sat in the difficult position of trying to train new chemists at the same time they were devastated by military conscription and absent faculty, in addition to the demands made by industry. Although they tried, the military and universities never reached a stable balance between their respective needs, and higher education suffered as a result. For universities, the war meant coping with shortages of students, faculty, and even laboratory supplies for the chemistry departments. Despite the toll on campuses, the universities' faculty and students could benefit from the unusual opportunities offered by war. Often completely absent from campus, the top chemists researched and shared knowledge in settings that rearranged their professional and personal networks, creating new paths for the flow of knowledge, including into the nascent synthetic organic chemicals industry. The war gave chemists an ideological mission and new public recognition, but it also served as an enormous practicum that expanded education, experience, and networks for an entire generation of American chemists.

Nationalism and Internationalism

The fight among Europeans for the empathy and support of neutrals—including scientists of neutral countries—escalated as quickly as the military attacks. The German army moved rapidly into neutral Belgium, blurring distinctions between military and civilian, and shocking western Europeans in the way Belgian civilians suffered in the initial stages of a modern total war. About 6,500 Belgian civilians died, the majority in August 1914, and the Germans deported others back to Germany or used them as human shields. There were 129 cases of Belgians being shot in groups of ten or more. The Germans justified the violence as necessary in the face of Belgian resistance, but the Germans exaggerated the threat, sometimes intentionally and sometimes simply as the result of nervous soldiers sensing danger where none existed. The invaders burned about 20,000 buildings, most famously the university library at Louvain with its 280,000 volumes and medieval manuscripts. For many in the West, the destruction of the library stood symbolically as an attack on civilization and a sign of German barbarity. Very quickly, the documentable acts of violence flamed rumors of greater atrocities such as mass rape, severed hands, and smashed babies; each side dehumanized the other and vilified their culture.[1] The nation of *kultur*—of philosophy, music, and science—had abandoned the values of modern Western society, in the opinion of other western Europeans. Rumors and propaganda became a deeply ingrained part of total war, and they would have an impact on the course of the Western scientific community, including American chemists.

Like other Americans, chemists as a group maintained a neutrality of sorts upon the outbreak of the war in Europe; concern and uncertainty dominated the reactions. In September 1914, Eugene F. Roeber editorialized in *Metallurgical and Chemical Engineering* that "all Europeans have suddenly gone crazy," and he likened the war to an enormous chemical reaction in which individual constituents had no more control "than the 'free' ions of the dissociation theory." A British journal, *Mining Magazine*, responded with an irate editorial, appalled that Americans seemed incomprehensibly oblivious to the atrocities the Germans had wrought on Belgium. From the British point of view, only the Germans in their "wanton barbarity" had "gone crazy." Indeed, they argued, Roeber's editorial came close to "that pestilent Prussian notion that war is a 'biological necessity.'" Over time, however, Americans were drawn into the war emotionally as well as economically and militarily, and after the United States entered the war in 1917, American hyperbole and propaganda—and censorship—caught up with and then surpassed the European belligerents.[2]

More than most American chemists, Leo Baekeland agonized publicly and privately over the horrors of war and chemists' role in it. Born in Belgium, he grieved for his native land, but he mourned even more for the science of chemistry as the Western world turned the chemists' "proudest discoveries . . . into means of relentless destruction and human slaughter." Nor did he excuse chemists from partial blame: "Our very successes will threaten to devour us as long as all of us have not yet become imbued with the truth that greater knowledge, like greater possession of wealth or power, demands a greater feeling of responsibility, greater virtues, higher aims, better men." To his diary, he confided a more extreme version of his faith in science, "which if turned to the purposes for which it is intended would make a paradise of this Earth and make the human race a race of angels." In war, however, "militarism [and] despotism, . . . [have claimed] for their own selfish barbarian purposes all the results of Science, education, intelligence. . . . I have never felt as discouraged in my life!" Baekeland fell back on the hope that the "nakedness of [the war's] nasty horrors" would subsequently lead societies to abandon war altogether.[3] Dying in 1944 at eighty, Baekeland lived long enough to learn that World War I would not be the war to end all wars.

Baekeland's distress and the reaction to Roeber's editorial reflected the profound challenges World War I posed to the ideology of scientists, particularly to their identity as part of an international community. On the one hand, scientists believed their work collectively contributed to an ever-accumulating body of "universal knowledge" in which national borders were irrelevant. On the other hand, almost all scientists practiced their trade in national or nation-bound institutions, and competition among nations helped to drive scientific exploration. During times of peace, their national and international identities co-existed—international recognition brought "fame and honor" to the scientist's nation, and some colleagues collaborated across borders.[4] But the war exaggerated nationalism in science, including chemistry, and scientists accused their colleagues in enemy nations of betraying ideals by working toward illegitimate national goals.

The ideological skirmishing among scientists began early in the war as European scholars lobbed nationalistic declarations back and forth across the English Channel. Most incendiary was an open letter, "To the Cultural World," signed by ninety-three prominent German scholars, defending their nation's invasion of Belgium and denying the destruction there. Among the signatories were thirteen chemists, including such luminaries as Emil Fischer, Adolf von Baeyer, Wilhelm Ostwald, Fritz Haber, and Richard Willstätter. Although a few signed without knowing the declaration's contents and few would have been in a position to know the reality in Belgium, they accepted their government's account of events. While both sides issued multiple declarations, this October 4, 1914, German proclamation made particularly clear that total war would include the scholars, too.[5]

After the initial salvoes of declarations, scientists from all belligerents engaged in their nations' ongoing retelling of history to disparage their enemies' culture. All sides challenged the national origin of scientific developments, a task that scientists with a flair for propaganda tackled with dedication. German physicist Wilhelm Wien urged his colleagues to cite the work of German research more frequently; British work was overly represented. A fellow physicist, Philipp Lenard, "alleged that the English habitually represented foreign discoveries as their own." A Bayer executive ridiculed British attempts to make dyes: "The nation is incapable of the moral effort . . . which implies study, concentration, [and] patience." The British retaliated in kind, with chemist William Ramsay suggesting that the war's "restriction of the Teutons will relieve the world from a deluge of mediocrity."[6]

By 1917, American scientists had joined the war of words, often echoing the British or French themes. Leigh R. Townes, a professor of chemistry at Georgetown College, penned a frequently cited article, "Germany's Stolen Chemistry" (1918), which dismissed forty years of German chemical development and argued that German chemists merely "borrowed" and imitated the great discoveries of other nations and erroneously claimed superiority in chemistry because of the "exaggerated ego" of a "Teuton." Synthetic dyes were originally British; even Justus Liebig and Friedrich Wöhler, two of the nineteenth-century's most notable chemists—particularly in organic syntheses—received their training in France, he alleged, mostly incorrectly. True, Germans had identified some fundamental elements on the periodic table, but only a "very small number of useful ones."[7]

These kinds of attacks reflected the cultural side of the war, waged to motivate all parts of the population to carry on the burdensome fighting. The shared sense of purpose, the underlying horror of war, and the fear and mistrust generated by the war all permeated the American chemical community. Prominent spokesmen increasingly imagined the German chemical industry had conspired with their government for years in preparation for war. "It is . . . not too much to infer," suggested William Nichols, that Germany waited until it had the Haber-Bosch process for synthetic nitrogen before embarking on war. "This is one instance of many which might be cited to show the extreme preparedness of the German nation." At home, Bernhard C. Hesse, a prominent Michigan-born chemist, felt it necessary to defend his patriotism against the anti-German hostility directed at so many Americans of German ethnicity. In early 1915, Hesse gave a speech in which he pointed out that he and his parents were citizens (naturalized, in the case of his father) and that his scholarly degrees were from American universities (B.S. from the University of Michigan, Ph.D. from the University of Chicago). "I am sure that this will relieve me of any suspicion of being a German spy or of being in the pay of German interests. I am as American as they make them."[8]

Synthetic organic chemistry most clearly illustrated the kind of scientific autarky that Americans embraced during and after the war; the association of synthetic organic chemistry with Germany made a profound difference in the chemical community's attitudes. Whereas few before the war worried about the relatively small if technically advanced segment of chemistry, the war prompted a debate—even as entrepreneurs responded to opportunity—over whether American chemistry and industry needed to have expertise in all fields. Academic chemists had begun to show significant achievements in other branches of chemistry, most notably Harvard's Theodore Richards, who won a Nobel Prize for his work on atomic weights. Like a growing number of German chemists, American chemists before the war believed that newer fields of chemistry could offer more opportunity and intellectual excitement than synthetic dyes and pharmaceuticals. As the war drew Americans more deeply into its snares and anti-German hostility increased, however, the professional discussion shifted, and more chemists argued that synthetic organic chemistry formed a vital and necessary part of a well-rounded chemical industry and academic discipline. They embraced the cause and increasingly linked synthetic organic chemicals with expanding university chemistry education, public health, economic development, and national security.

Even as they pursued a course of increasing national independence in chemistry, both in industry and academics, Americans continued to analyze and engage in international developments. Indeed, to aid their independence, they looked for strategies elsewhere, including Germany, however distorted the understanding of Germany that some Americans held. In particular, several chemists jealously eyed the Kaiser Wilhelm Institute and lobbied for a similar research center in the United States. Professor Marston T. Bogert of Columbia University maintained that such a center was necessary to train all the graduate chemists that the United States would need for a domestic organic chemicals industry. C. Alfred Jacobson of the University of Nevada envisaged an immense complex with 50 congregated buildings containing departments conducting research in 150 chemical subjects, employing more than 2,000 people, and including 1,350 chemists. Jacobson placed the initial expense at $27 million and annual costs of $5.6 million, "one-half the cost of a modern battle cruiser," perhaps funded in part by the chemical industry. Bogert and Jacobson hoped for a centralized government research institution that would outlast the war, executing fundamental chemical research that would lead to a variety of chemical products beneficial to society.[9] (In size, the war gas program came close to matching such visions.) Americans continued to use such explicit comparisons with Germany as they pursued their independence in synthetic organic chemicals.

Nationalism and the American Chemical Society

While weakening international scientific ties, the war sharpened nationalism among scientists, and the American Chemical Society (ACS) notables stepped forward as the most prominent voices for American chemists. When World War I began, the professional society stood in a better position to assume a national leadership role for chemists than it had occupied a decade earlier. The war helped the ACS by raising the status and visibility of chemists, mitigating its prewar internal disputes with a common and urgent mission, and dramatically expanding membership. From 7,170 in 1914, membership grew to 12,203 in 1918, with the greatest wartime increase occurring in 1917 and 1918.[10] But as the leading organization for chemists, the ACS stood in the spotlight, forced to account for and explain the crisis and the apparently underdeveloped condition of chemistry and the chemical industry of the United States. The ACS leadership responded vigorously with publicity and political engagement; they defended American chemists' abilities and actively lobbied for federal legislation and policies favorable to chemists and the industry.

The ACS leadership used its journals, particularly the *Journal of Industrial and Engineering Chemistry*, to keep members apprised of the contributions of chemistry to the war. The *Journal of Industrial and Engineering Chemistry*, along with the more commercial *Drug and Chemical Markets*, was the leading print forum for discussion of the domestic synthetic organic chemicals industry during World War I, and it continuously published articles during the war exploring the dimensions of the crisis and the possible solutions. The editor of the *Journal of Industrial and Engineering Chemistry* from 1911 to 1916, Milton C. Whitaker, had at age forty-one already spent years in both academia and industry; he served as editor almost concurrently with his time as the founding department chair of Columbia University's chemical engineering program. Although the promotional rhetoric of Whitaker's editorials pales in comparison to subsequent editors, he firmly encouraged measures to develop the industry. "Never," wrote Whitaker about the war's impact, have chemists and chemical engineers "been confronted with greater responsibilities nor greater opportunities."[11]

Whitaker was among the chemists who believed the chemical community needed to enlist the support of the broader American public "to sustain the efforts and share the risks of the pioneer . . . [because] now is the time to build great American chemical industries," he wrote from his editorial post.[12] Convinced that they could meet the challenge of supplying technical skill to a domestic industry, the chemists' task lay in persuading Americans outside of the chemical community that a domestic industry would be worth fostering. While the ACS publications created better communications among the chemists, the ACS leadership looked for ways to reach a broader audience, a task helped by the war because of chemistry's increased prominence, but the chemists also wanted to respond to criticism that came with increased scrutiny.

One way they preached the gospel of chemistry to the unconverted general public was through the new annual National Exposition of Chemical Industries, which drew an average daily attendance of 10,000 in 1915 and 13,000 in 1916. Over sixty exhibitors, drawn from nearly all branches of the chemical industry, professional chemical societies, and several government agencies, displayed their products and projects in Grand Central Palace in New York City. Domestic manufacturers of synthetic organic intermediates and fine chemicals especially seized the opportunity to display chemicals previously made only in Germany. The *New York Times* noted that Thomas Edison, who had become a chemical manufacturer during the war, attracted some of the largest crowds for his production of phenol for phonographic records.[13]

Creating familiarity with, and support for, chemistry among a wider public dovetailed with the more specific goal of convincing policymakers to pass favorable legislation. At the outbreak of war, the New York section of the ACS (ACS-NY) took the lead in assessing the depth of the dyes crisis.

Although the ACS national office had moved to Washington, New York City was the crossroads in the dyes trade. In response to the "hue and cry of the superficial hack-writer," as Whitaker described publicity over the shortages, the chair of the section, Allen Rogers (of the Pratt Institute), formed a committee of seven with representation from manufacturers of heavy chemicals, coal tar chemicals, dyes, and textiles, plus Herman Metz as the importers' representative, Rogers himself, and Bernhard C. Hesse as committee chair. The committee addressed several charges leveled at the chemists and chemical industry, many of which castigated the chemical community for having neglected such a vital specialty and for responding slowly to address the shortages.[14]

Particularly because the report came in 1914, before many people expected a long war, the committee members were deeply concerned about competing against the German industry and advocated stronger tariff protections. With so many chemists drawn from industry, the committee had a respect borne of familiarity with the complicated workings of the German industry, or, alternatively, familiarity with losing money trying to compete with the German industry. Playing on American preoccupations of recent years, the committee noted that German firms were free to form cartels, an advantage denied to Americans by antitrust laws. Wrote the committee: "The American chemical industry is expected to cope with the foreign industry while both its arms are tied behind its back." In no substantial way could an American dyes industry survive without greater trade protection, they argued. First, convinced that Germans had sold dyes and intermediates below cost when Americans launched rival products before the war, the committee recommended effective anti-dumping laws and their enforcement. Second and most important, Congress must raise tariff rates, they argued. After reviewing the years before the tariff of 1883, the supposed golden age of American synthetic dyes tariffs, the committee recommended a 7½¢ per pound specific duty as well as a 30 percent ad valorem. The committee also recommended that intermediate products receive protection at half the rate of the dyes.[15]

The committee members understood perfectly that protectionist legislation helped dyes manufacturers at the expense of consumers. This, however, was justified because "private enterprise and private capital have gone their limit. . . . The nation as a whole must bear the burden . . . if . . . our industries requiring coal tar chemicals such as dyestuffs shall forever be protected and made independent of foreign nations for the supply of those materials." The committee believed that the domestic chemical industry had developed appropriately to American resources and markets, "utilizing and exploiting every reasonable opportunity to its full extent." The main obstacle lay with the large and politically connected textile industry, the primary consumers of dyes. Textile manufacturers had lobbied for decades before the war to keep tariff rates at 30 percent, which, the committee argued, had obviously been too low to support that sector of the chemical industry. Committee members also asserted that the chemical industry had reacted quickly to the shortages that fall, reactivating mothballed plants and building new capacity as quickly as possible. In a final point, the committee strongly and unanimously recommended against changing U.S. patent laws, such as to include a compulsory working clause. The report, submitted in November 1914, initially received wide attention within the chemical community.[16] Some dyes proponents among the ACS leadership went on to use the report as a guide to their lobbying efforts in subsequent months.

The analysis and proposals in the ACS-NY report strongly reflected its primary author, Bernard C. Hesse. Hesse's knowledge of the field explains why ACS-NY made him the chair of their committee; his became a frequent and familiar by-line closely associated with analysis of the synthetic dyes crisis. One of the most intriguing chemists to rise to prominence during the war, Hesse understood and could explain the synthetic dyes industry better perhaps than any other chemist in the United States. After his doctoral education at the University of Chicago, Hesse moved to Germany to work in the research laboratory of BASF in Ludwigshafen, where he learned the dyes business. He returned to the United States to supervise BASF patents and patent applications in the United States for several years before becoming an independent chemical consultant in 1906.[17] In the ensuing years, Hesse carefully observed the dyes market in the United States, and, more than most, he stood in a position to analyze the prospects for an American synthetic organic chemicals industry once the war broke out.

Between 1914 and 1917, Hesse wrote several informed and articulate essays for professional and trade journals, articles that still ring with clarity and authority. His analysis of the dyes crisis and the German industry influenced Americans' understanding of them and helped to identify and shape major points of debate over the coming years. Particularly in 1914 and 1915, he provided a wealth of data on the German industry, citing especially the elaborate intra-firm and intra-industry exchange of chemicals—by-products of one process created inputs for another. His articles were invaluable in explaining the scope and nature of the dyes industry to specialized and general audiences, particularly those who wanted to promote a domestic industry. Interestingly, however, Hesse himself doubted the value of an American dyes industry. In his earliest war articles, Hesse suggested that the United

States might be better off without undertaking the costly support for a domestic industry. "The truth seems to be that the whole of this industry cannot be successfully transplanted ... and if it could be transplanted as a whole the net result would not be commensurate with the expense, effort and risk connected with it."[18] Unlike most of the other characters in this story, Hesse believed in free market economics, staying skeptical throughout the war about the wisdom of investing public money to establish a dyes industry.

In the face of broader support for a domestic industry, Hesse only lightly questioned the wisdom of such a goal after September 1914, at least publicly, and concentrated on articulating policy measures he thought would help create a viable industry. If Americans wanted the industry, he saw no option but to turn to help from outside the chemical community, including from textile manufacturers and the federal government. Unless the government altered market conditions through policymaking, the country should not expect a significant domestic industry, an attitude reflected in the ACS-NY report. Americans and their universities certainly could supply sufficiently trained chemists, Hesse argued, but if the United States wanted more chemists trained in synthetic organic chemistry, then capital would have to flow to the industry to lure more chemists to the field.[19]

Nowhere was Hesse's imprint on the ACS-NY report more apparent than in the committee's rejection of patent reform. The war sharpened the periodic debate about the American patent system, and some Americans, not all in the chemical community, believed that the proportion of U.S. dyes patents owned by Germans indicated the need to overhaul American patent law. In particular, they revived a prewar public discussion about a "working clause" in patent law that would require patent owners to manufacture or license the patented products and processes in the United States. Since the 1870s, Germans filed most of the patents for dyes in the industrial countries, leaving industry newcomers in a poor position to produce legally protected dyes. Ironically, given subsequent events, few people acquainted with the industry viewed the percentage of patented dyes significant enough to thwart or restrict development of a domestic dyes industry. Once again, Hesse's numbers-based analysis, so characteristic of him, influenced the debate. First, he pointed out, patents ran just seventeen years. Of the 921 dyes that had been patented through 1912, 714 no longer remained under patent. By the end of 1915, another 44 dyes patents had expired, leaving in effect 163. Hesse concluded: "At no time was the American industry throttled or even handicapped by U.S. patents held by foreigners to such an extent that it could not offer successful substitutes for the great majority of patented articles." Instead, he argued that Americans should pay close attention to the U.S. patents issued to foreigners and use that information to stay abreast of new innovations and developments.[20]

At the heart of Hesse's analysis was an explicit comparison of American potential against the German industry; implicit was the assumption that an American industry would need to follow the German example. Nearly all of Hesse's articles focused on dyes, the oldest and most prominent among the synthetic organic chemicals and the products most recognizably German. Like many chemists, he had throughout his career viewed the German dyes industry as the pinnacle of industrial chemical expertise. In 1914, he wrote respectfully of the powerful lead in synthetic organic chemistry the Germans had developed over the decades, suggesting Americans could more easily focus elsewhere. A few American chemists rejected the unfavorable comparison to Germany's chemists. For the industry to succeed, wrote one, it "must be an American one, developed under American conditions and by American chemists." MIT chemical engineer William Walker suggested that "the magnificence of German chemical industry has been compared to the chemical industry of America almost ad nauseum. Generally the inequalities are greatly exaggerated and many may be explained without any discredit to the American profession."[21] As the war progressed, Hesse's descriptions modified, and the cosmopolitan chemist distanced himself from the German industry, which he knew so well. By 1917, the impact of three years of fighting and the American entry into the war had led Hesse, like so many others, to take a stronger nationalist position.

Hesse's commitment to a rationally planned economic project led him to become a strong advocate for collecting and generating market information. In his own articles, he provided compilations of statistics about costs, explained the complexity of production, and outlined the broad-based scientific and economic infrastructure necessary to develop a domestic industry. If he were to support a tariff, such a tariff, he believed, must be based on good data, and he threw himself into initiatives to collect such information, convinced it would aid American entrepreneurial decision making. The textile manufacturers could most help the dyes crisis by providing and tabulating their past consumption of dyes, particularly German dyes, he argued. In this way, chemical entrepreneurs could base their decisions about manufacturing on more solid market information. While Hesse's entire approach to the crisis fit the rational, scientific ethos of the early twentieth century, it underestimated the degree to which market competition and rivalries inhibited the collaboration and sharing he advocated.

By 1917, Hesse's data-driven, precise, and concrete analyses were overshadowed by more passionate arguments. The war stirred powerful emotions, and one consequence in the chemical community was to redefine and repackage the industry in ways suited to American resources and designed to garner more public and government support. The public conversation increasingly placed dyes in a more broadly defined synthetic organic chemicals industry that included military explosives and gases and pharmaceuticals, among others. Hesse gave way to other industry leaders, particularly Charles H. Herty, who brought emotion, a more intense nationalism, and considerable personal and public powers of persuasion. The shift in the relative prominence of the two men reflected a larger dynamic—the community of chemists moved more strongly behind the promotion of synthetic organic chemistry and its industry over the course of the war. In 1917, Hesse took a position with General Chemical and moved to National Aniline & Chemical Company with the merger, and he appeared in the public record much less often.

Charles Holmes Herty became a distinctive and influential voice for chemists and the larger chemical community in World War I, particularly with respect to the domestic synthetic organic chemicals industry. Over his life and career, Herty crossed boundaries, both geographic and metaphorical: from the American south to north, from a childhood in rural Georgia (during which both parents died) to the meccas of advanced chemistry in Europe, and from "pure" chemistry to commercial applications. Herty's professional education followed a common trajectory for American chemists of his generation. He took his undergraduate degree at the University of Georgia and then received his Ph.D. in 1890 from Johns Hopkins University under Ira Remsen, one of the nation's most prominent chemists and an academic known for training students rigorously in research. After a few years teaching back at the University of Georgia, Herty packed up his family to spend 1899–1900 in Germany and Switzerland, learning from prominent chemists such as Walther Nernst, Georg Lunge, Otto Witt, and Alfred Werner. On this and subsequent trips, Herty was particularly interested in the industrial applications developed by Germans and other Europeans, even as he initially made his scientific reputation studying the halides of lead and alkali metals. Engaged with the increased attention to conservation of natural resources in the early twentieth century, Herty worked with the southern pine industries to develop an effective "cup and gutter" device to collect the sap used in turpentine without permanently damaging the trees. In 1905, the University of North Carolina lured Herty away from Georgia, and he gained a reputation as a diplomatic mediator in academic politics; during the war, he applied his organizational and people skills on a national scale.[22]

The ACS recognized Herty's achievements by electing him to consecutive terms as president in 1915 and 1916. His presidential duties frequently required him to make speeches before different chapters of the society, and in 1915 his favorite topic became the domestic synthetic dyes industry; he urged his audiences to build the new industry while the "time is ripe." As Herty completed his time as president, the society elected him as the full-time editor of the ACS's *Journal of Industrial and Engineering Chemistry*; he left academia for good by age fifty. Once his editorship began in January 1917, Herty kept the cause of the domestic synthetic organic chemicals industry prominently in the journal's pages. From these two ACS posts, Herty became the most recognized and vocal spokesperson for the chemical community on the topic of the tariff and the U.S. dyes industry.[23]

Whereas Hesse wrote with great familiarity and respect for the German industry, holding it up as an instructive guide for Americans, Herty saw the German industry as wielding "enslaving power" and as a dangerous "monopoly" against which Americans needed to be protected. Hesse cited the Germans' technical savvy and experience; Herty dwelled on the "humiliation" that came with dependence on a foreign power. In his occasionally folksy, occasionally confrontational style, Herty ardently advocated policy measures to promote the industry. He argued that "Industrial Freedom" was integral to preparedness, emphasizing the chemical relationship between dyes and explosives. The Germans, he argued, competed as ruthlessly in commerce as in war, and a nascent American dyes industry required government protection from the "nation which first introduced into warfare the use of asphyxiating gases." In addition, the British and Japanese had taken steps to ensure their domestic needs, leaving only the United States at the mercy of Germany.[24]

Herty mixed his pointed rhetoric with persuasive appeals to many audiences, and he became a "coordinating influence," a goal he announced for himself in his first editorial. Using his ACS positions, Herty made a wide range of contacts in the federal government, including members of Congress, the Chemical Warfare Service, and the Department of Commerce. As a southern Democrat, Herty also had purchase with the southern textile manufacturers, a group more reluctant than northern manufacturers to back higher tariffs for synthetic dyes.[25] The world's largest professional society of chemists effectively advocated on behalf of their expanding constituency, representing the chemical community in a wider public forum, a task at which Charles Herty especially excelled.

Nationalism and the Universities

To win the war, academic chemists redirected their research and teaching to the national emergency, generally at the expense of universities. The chemists' everyday routines that bound them to their universities turned upside down, and the emotions of the war challenged some of the values that shaped academics' identities, whether the pursuit of pure science or the defense of academic freedom or tolerance of dissent. The stresses on universities took many forms, including financially and their own shortages in chemicals and equipment. Faculty and students left campuses in droves, either temporarily for war service or more permanently for industry; those left behind watched military training increasingly displace academic studies. Despite the impact on universities, the strategic and symbolic significance of synthetic organic chemicals in World War I created opportunities for chemistry faculty and students, who not only developed different relationships and expertise working with the military and industry but also could imagine a future in an American synthetic organic chemicals industry. Personal and institutional networks among chemists shaped the flow of knowledge and expertise, and the war's impact on university chemists and chemistry departments helps to illustrate how such a trauma can turn institutions and individuals down different paths, either voluntarily or involuntarily. Chemists identified more with national networks, whether the ACS, defense, or industry. While much of the nationalism calmed down after the war, the strengthened national institutions and cross-country relationships persisted through the interwar years.

Patriotic fervor captured people and institutions, including university communities, when the United States entered the war. University presidents declared their campuses and personnel to be at the service of the nation. Faculty from every discipline felt the war's attraction: psychologists tested theories about mental and emotional capacities on thousands of military recruits, foreign language scholars offered their services in translation, and historians sought relevance through active propaganda work. Scientists understood they held the public's attention, the war having given them an exceptional opportunity to demonstrate the practical contributions of their research. Scholars also succumbed to the xenophobia and super-patriotism that affected the rest of society. As a professional value, academic freedom was new and fragile. World War I overwhelmed it, and only after the war did academics and their critics acknowledge how much war passions had swept up institutions of higher learning. Indeed, writes historian Carol Gruber, "as a group, American professors were among the most enthusiastic supporters of the cause," and they "assumed that the universities . . . had a special contribution to make." Before the war, universities embraced a "service ideal," a commitment to contribute to the social good; during the war, service meant contributing to winning the war.[26] In this new total war, the universities provided desired knowledge and learning, but redirecting faculty and students to military purposes undermined the universities' ability to teach and sustain expertise.

The war's demand for young educated men threatened campuses and plunged the universities into the politics of mobilization, particularly military conscription and enlistment. Chemists knew that some of Europe's brightest scientists had died on the battlefield early in the war, denying countries their expertise. To the scientifically minded Americans of the Progressive Era, this kind of voluntarism was blatantly wasteful, a situation that called for a coordinated organization of human expertise. The abstract efficiency of a bureaucratic plan, however, could never tame the intensely human and political passions tied to war and sacrifice. The politics of conscription—shaped by scarcity of expertise and longstanding U.S. political currents—played out in congressional debates, among firms and industries, and on college campuses. President Wilson, struggling to stay out of the war, resisted attempts prior to 1917 to institute a military draft and only on the brink of war threw his support behind a national, centralized plan.[27]

As Wilson and Congress shaped the draft, their primary selling point centered on "selective service," the component that protected experts and labor in fields or industries with particular relevance to prosecuting the war. Americans wanted to conscript an army without disabling industry, which was also necessary to winning the war. All soldiers served until the end of the war, but the 4,500 local draft boards could exempt men from conscription if individuals or firms appealed to them with qualifying reasons. The "rationality" of the plan appealed to manufacturers, although in practice they were often frustrated with labor shortages. George Merck, for example, appealed regularly to the local draft board to excuse essential employees from the draft. In his letters of appeal, Merck emphasized the significance of his firm to the war effort. Thirty percent of his business was in government contracts, he wrote in mid-1917, and he expected within months that 80 percent of his firm's output would go to "the Government, Allied Governments, or the Red Cross." Merck & Company had thirty-four chemists on staff and twenty-six of them were registered for the draft. If the board drafted the "indispensable" chemists, they "endanger[ed] the lives of the sick and wounded." George Merck was not alone; around the country, employers struggled to hang on to their technical talent. And the Department of War leadership fretted by 1918 that "the process of classifying" men took too long; some of the applications for exemptions for industrial positions particularly might require time-consuming review.[28]

As the demand for soldiers increased and the number of students decreased, the military and universities arranged to place students on campuses in the Student Army Training Corps, partly to avoid the "virtual breakdown of the entire educational system" that occurred in Great Britain and Canada, but neither side believed the SATC worked well. When the army inducted 140,000 soldiers into the Student Army Training Corps on October 1, 1918, the military officials explicitly aimed to get soldiers through military training within a year, and the universities looked more and more like boot camps. Secretary of War Newton D. Baker told a House committee that while it would be "atrocious" to sustain universities "at the expense of our military efforts," higher education was a "national asset" to be protected if possible. Baker knew, however, that balance eluded them; "the maintenance of academic education in America," he admitted, had become one of the serious "unsolved problem[s]" of the draft. After the war, one legacy of the SATC was the introduction it gave many men to higher education, which contributed to increased enrollments, but the immediate verdict on the SATC was mostly negative. "Then began a veritable academic nightmare," wrote the 1920s chronicler of Illinois's chemistry department about the SATC. The university survived the drain on personnel in 1917—Illinois students served in "Ordnance, Quartermasters, Chemical Warfare, Medical and Sanitary Corps" in the U.S. Army, plus positions in the U.S. Navy and civilian government agencies—but the SATC severely undermined the regular curriculum. "The difficulties were insurmountable and the quality of academic work probably never sank lower." The students lived in barracks on campus, subject to military discipline, and were placed under "officers . . . of no military experience, and inferior in education and younger" than the students. The university faculty bid farewell to the SATC with relief at the end of the war, and the department set up ad hoc means to grant partial credit to returning students who learned particular skills in the service.[29]

For much of the war, the universities continued to pay the salaries of their absent chemists. At Columbia University, President Nicholas Murray Butler wrote to chemistry professor Marston T. Bogert expressing his intention to pursue "full cooperation between the University and the Government in every way that is possible," and said that Columbia saw Bogert's government duties as work of "great importance" and a "matter of pride" for the university. However, "the resources of the University . . . are being swept away in a fashion that is staggering." Just a "stroke of the pen" could and should provide government salaries for those professors engaged in government work. "If the Government is not levying its taxes and issuing its bonds for just these purposes, then what is it issuing them for?" Marston

Bogert was one of the prominent academic organic chemists who left campus to take on significant responsibilities in Washington. Already in the spring of 1916, Bogert asked permission from Columbia to spend half his time to serve on the newly formed National Research Council (NRC); a year later, he asked for full-time paid leave because his mobilization work consumed all his time. With no other source of income beyond his Columbia salary, Bogert seriously worried that he might lose his job as Columbia's finances deteriorated, and he told President Butler he would return to campus if necessary.[30]

But with government funding in spring 1918, Bogert stayed in Washington, and his experience illustrates the way in which the war channeled the expertise of academic chemists away from campus and toward national mobilization. As Columbia's senior organic chemist, Bogert had led efforts to expand training early in the war to meet shortages. "We owe it to the community," he argued, adding that Columbia was the nation's largest university and one of the few in a position to train graduate students in synthetic organic chemistry for industry. He pushed efforts in the chemistry department while Milton Whitaker oversaw significant expansion of the chemical engineering program. Bogert quickly developed ties to local firms, including dyemaker Zinsser & Company, to obtain donated chemicals, and he asked the university for permission to establish an advisory committee of manufacturers. Within a month of that April 1916 request, however, Bogert became heavily engaged in war work and increasingly reduced his efforts in the chemistry department. In Washington, Bogert chaired the NRC's chemistry committee, which helped to identify relevant research problems and find academic and industrial chemists to address them. But he also advised the War Industries Board, the War Trade Board, and ranked colonel in the Chemical Warfare Service by the end of the war.[31]

Like Bogert, Elmer P. Kohler of Harvard University applied his chemistry and administrative skills to war work, primarily in war gases, and the records of the Harvard University chemistry department suggest the impact of the war on faculty and recent graduates alike. Although not the largest chemistry department in the country, it employed or produced some of the most significant synthetic organic chemists of the war generation. Kohler was one of the most prominent organic chemists in the country and the Ph.D. advisor to, among others, Roger Adams and James B. Conant, both of whom rose to the top of their profession in the 1920s. The Bureau of Mines war gas officials initially negotiated with the university president, A. Lawrence Lowell, to permit Kohler to serve two masters at once. In the fall of 1917, Lowell offered to provide laboratory space for Kohler to conduct military research while staying present in Cambridge to fulfill the most necessary duties at Harvard. At the time, the arrangement suited the Bureau of Mines well because of insufficient working space in Washington, particularly for organic chemistry. As part of the arrangement, Harvard set aside laboratory space "somewhat apart" from the main laboratories and stationed guards twenty-four hours per day. The war gas program's directors also assigned "several drafted men" to assist Kohler in the laboratory work. It would not be long, however, before Kohler moved to Washington. At least one of the chemists in the war gas effort, Lauson Stone, believed Kohler to be "without doubt the best all around research organic chemist in the United States." He had become the "right-hand-man" to George Burrell, the head of the Bureau of Mines war gas program, where he was "unquestionably one of the most valuable men in the organization." In arguing to obtain a raise for Kohler, Stone wrote that Kohler possessed "a combination of chemical knowledge, executive ability and good personality . . . exceedingly difficult to find among chemists." With Kohler and many colleagues absent, one of the department's own recent Ph.D. recipients shouldered the burden of teaching many of the elementary organic classes.[32]

As much as chemists focused on war work, most never lost sight of postwar goals for themselves and for their discipline. When they could, they attempted to address war-induced problems with solutions that would continue to build synthetic organic chemistry in the United States after the war. In addition to the shortages of faculty and students, university laboratories—and industrial and government laboratories—felt the impact of the war in shortages of scientific literature, chemical apparatus, and reagent chemicals.

By late summer 1918, as mobilization of laboratory and industry shifted into a higher gear, chemists felt the scarcity of common reference works and journals more acutely. In one case, Major William Lloyd Evans of the Edgewood Arsenal laboratory appealed successfully for help from industry and academics: both Ira Remsen of Johns Hopkins University and V. H. Gottschalk of the University of Missouri loaned journal collections and reference works to the laboratory for the duration of the war; Du Pont had helped to advertise the lab's request. The ACS leadership—the usual suspects—rallied to the cause. Charles Herty and Bernhard Hesse agreed that American chemists needed to produce their own reference works to demonstrate independence from Germany. As Herty put it, "American chemists must seriously face the necessity of compiling in an authoritative manner great reference works for all fields of chemistry, books written in our own language and adapted to our own methods of work." This, however, was a task for after the war. In the meantime, American chemists could reproduce existing copies of German reference works, and Herty particularly had his eye on the *Handbuch der Organischen Chemie*. Compiled initially in 1880 by Russian chemist Friedrich Konrad Beilstein, the *Handbuch* gave the known properties of thousands of chemicals. Herty reported that it would cost $30,000 to reproduce Beilstein, as it was known, and suggested a wealthy patron might take on the expense. The federal government could justifiably override the copyright, Herty argued. "Do we feel any qualms of patriotic conscience about such a reproduction? Well, we should worry! Germans are daily profiting in the conduct of the war through the utilization of American inventions, the submarine, the telegraph, the telephone, the machine gun and what not."[33] The war ended before Herty's Beilstein plan could be implemented, but the wartime scramble for chemical literature gave way to expanding numbers of chemical publications after the war.

Shortages of lab equipment—also made primarily in Germany—required domestic substitutes. By most accounts, American laboratories and dealers carried a sizable stock of equipment and apparatus at the war's outbreak. Test tubes, beakers, pipettes (similar to medicine droppers), burettes (vertical measuring cylinders), porcelain crucibles, and specialized thermometers were among the apparatuses that Americans began to make during the war. Several of the big glass manufacturers in the United States, who had not bothered with the high-cost specialized glass before the war, found the wartime prices a lure. The Coors brewery, facing Colorado's 1914 decision to go dry, manufactured chemical porcelain as a partial substitute for lost beer revenue; they had eighty employees by the end of the war. Corning introduced Pyrex in 1915, which scientists liked for its exceptional durability. And while some laboratory apparatus used mechanically made glass, others depended on hand-blown glass, and the industry scrambled to find or train enough skilled laborers. The University of Illinois, among others, kept a glassblower on staff to support their chemists. With help from the glass and ceramic industry and improvisation, the laboratories found the equipment to continue research.[34]

Reagent chemicals, used in laboratories for experiments, had also come from Germany, and developing American substitutes involved a university-industry collaboration that carried on after the war. Even before the war, the reagents could be scarce and expensive, prompting chemists at the University of Illinois to find an alternate solution. Chemist Clarence Derick and his successor at the university, Roger Adams, initiated and ran a summer program for chemistry students to synthesize organic chemicals for research. The university acted as a clearinghouse for chemicals. Companies and laboratories that produced a surplus of certain types of chemicals sent them to Illinois for processing and in return could buy small amounts of a variety of laboratory chemicals. By 1918, however, the project outgrew the facilities at the University of Illinois, and Roger Adams encouraged the American Chemical Society to enlist industrial support, which came from Eastman Kodak of Rochester, New York, in 1918.[35]

In the reagent project, Kodak demonstrated the possibility of broadening the pool of potential American chemists to include women. Kodak's photographic processes required small amounts of synthetic organics, such as metol and hydroquinone, and the company's research director, C. E. K. Mees, viewed the project as a contribution to the war effort. However, Mees and George Eastman felt they had no male chemists to spare and organized a lab staffed by women. The young women, who filled the roles of chemists, assistants, and clerical staff—under the direction of a male chemist—worked during 1918 to produce the research chemicals. Kodak management cited two disfiguring accidents to support their decision to replace the women after the war when male chemists returned, although as late as 1920 three of the women—"girls of exceptional ability whom we should be sorry indeed to lose"—remained.[36]

Other firms and even military officers considered using women to help fill the demand for chemists. The Monsanto Chemical Works hired the most women, about 125 "in the various mechanical departments," but few if any were chemists. More typically, employers worried about violating cultural norms. Herbert Dow considered employing women chemists but dismissed the idea because he believed it inappropriate for women to work the night shift. In addition, he argued that the company could ill afford to construct the separate laboratory that he believed necessary to segregate the women and men. Du Pont hired women as analytical chemists in the government plant in Nashville that the company built and operated. The Du Pont recruiter noted that "we have provided safe accommodations for women at that plant, and have employed a young lady with experience to take charge of their work." As in most cases, their work was in analytical chemistry, rather than organic or physical chemistry. This relieved people like industrial health researcher Alice Hamilton, who worried about the impact of the synthetic organic chemicals on the women's reproductive systems.[37] Despite the demand for chemists, women remained a relatively untapped source; the synthetic organic chemicals industry mostly left women out of the American solution to the nation's chemical crisis and postwar future.

The universities and industry built multifaceted relationships that were usually mutually beneficial. Firms hired chemistry graduates, but the relationship also included academic consultants to industry, firms' support of university fellowships, and the development of close ties between particular firms and universities.[38] In each case, they helped to construct a national knowledge infrastructure—the formal and informal human channels through which knowledge and know-how passed. Although tension arose when firms competed with university faculties for talent, the synthetic organic chemical sector's demand for highly trained graduates helped to spur growth in universities' chemistry and chemical engineering departments. While the alliance between industry and universities in the interwar period was weak compared to the Cold War years, the two found common ground in the pursuit of a U.S. synthetic organic chemicals industry.

Merck & Company's archives contain an intriguing trove of contracts or "employee agreements" for new chemist hires, primarily from 1914–1924, which provide a significant look at the way one firm attempted to meet its needs for new technical help. The records do not clearly indicate the number of total hires or whether all new hires are represented in the extant contracts, and consequently the records are more impressionistic than definitive. From 1914 to 1918, the contracts show twenty-nine new hires by Merck & Company. Besides the fact that all were men, the single most striking characteristic of this group is the relative youth of its members. Of the twenty-nine, twenty-two were in their twenties and three in their thirties; the ages of the remaining four are unknown. A few had work experience prior to coming to Merck & Company, such as in pharmacies or a rubber company, but the majority appear to have come directly from an educational institution. Some records indicated the new hires' courses of study, with at least thirteen trained in chemistry, seven in pharmacy, and five in chemical engineering (some men had education in more than one field). Among the twenty-nine, the most frequently attended institutions were the University of Michigan (six), Harvard University (four), and MIT (four). Collectively, the men attended nineteen different educational institutions, ranging from the Ivy League and state universities to colleges of pharmacy, small colleges, and technical institutes. Only one man arrived with a Ph.D. (from Harvard), and one with a terminal M.S. (from Allegheny College), although many indicated they had taken at least some graduate training without completing a degree. With the exception of one hire, a high-level manager brought in with Merck & Company's purchase of the Rahway Coal Tar Products, the starting salaries ranged from $85 to $175 per month.

The Merck records also show that many of the men left the firm, either voluntarily for new opportunities, the result of some "incompatibility of temperament," illness, or the sharp economic recession in 1921. Merck fired one man because he had conspired to leave to become a partner in a start-up firm and unsuccessfully encouraged another Merck employee to join him. In Merck's eyes, his crime lay in his plans to use the skills and know-how he gained at Merck & Company to become a rival. Normally, when Merck & Company and the men parted ways, the firm paid the men an ongoing retainer, such as $10 per month, to keep them from revealing any Merck & Company secrets for a term specified by the firm. Two men left for the Chemical Warfare Service, two for the Ordnance Department on high explosives, and two for the Aetna Explosives Company (the chemist with the Ph.D. worked at both Aetna and the Ordnance Department and then ended up at Zinsser & Company, a small dyes firm in New York).[39] As much as Merck and other firms discouraged them, chemists gained experience in one venue that they could carry to another, which could strengthen the industry collectively even when a departure might hurt particular firms.

Firms hired faculty as well as students, which posed more of a challenge to universities during and after the war. The difference in salaries between academic and industry jobs could be stark. For junior faculty, especially, the prospect of an industrial salary held a temptation too great for some to resist. Clarence Derick, Ernest Volwiler, and Lee Cone were among many top young organic chemists who permanently abandoned academic careers for industry. Derick, a promising organic chemist at the University of Illinois, earned $5,000 per year when he accepted the position to head the research laboratory of the Schoellkopf Dyes factory in 1916. Even ten years later, the University of Illinois offered a salary of $1,800 to a new junior professor, and the chemistry department chair, W. A. Noyes, begged his superiors "to recognize the competition with the industries which must be met in some degree if we are not willing to content ourselves with mediocre men."[40]

Continuing a prewar tradition, many university chemists consulted in industry from their academic posts rather than completely embrace industrial employment. Manufacturers typically called on exceptional academic chemists to serve as consultants in industry, expecting the chemists to help shape industrial research programs by providing ideas and students. In the 1920s, the use of consultants would increase, but notable examples existed during the war as manufacturers struggled to learn the science and production of synthetic organic chemicals. As historian John P. Swann has written, Roger Adams of the University of Illinois was an extraordinary chemist who began a lifelong cooperation with Abbott Laboratories in 1917. Likewise, throughout his career Albert W. Smith of the Case School of Applied Science aided the Dow Chemical Company with both research and Case students, and Du Pont worked with academics at Harvard and Illinois, among others, and also sponsored fellowships at several.[41]

Yale is another case in which a university developed particularly strong ties to one company. Yale's relationship to the Calco Chemical Company initially derived from the same shortage of reagent chemicals that prompted the University of Illinois and Kodak to respond. In its annual report of 1914–1915, Yale University proposed building a "small unit plant," in which "advanced students" could simultaneously manufacture the scarce chemicals while gaining "invaluable" experience. Without stating matters too explicitly, the report's authors hinted that the department of chemistry lacked only a donor to provide the equipment. If the proposal could be realized, the production of reagents would "lead to a closer cooperation with chemical industries," which at the time was deemed "something much to be desired." Yale's challenge generated a response from Robert C. Jeffcott, an 1897 graduate of Yale's Sheffield Scientific School and head of the Cott-a-Lap Company. The company, which he inherited from his father in 1906, produced wall fabrics. In 1915, however, Jeffcott became one of the capitalists venturing into synthetic organic chemicals, for which he formed the Calco Chemical Company of Bound Brook, New Jersey. Calco's comparative success over the next few years derived in part from the close relationship Jeffcott developed with the Yale chemistry department. During the war, Jeffcott purchased the equipment desired by the department, but he also created two fellowships to support graduate students in organic chemistry. In exchange, Jeffcott could expect Yale to produce a greater number of more able chemists, many of whom would be familiar with his company as they sought employment after completing their degrees. One former Calco employee recalled having the impression that the company employed Yale graduates in "nearly" every position in the "administration and operating departments" during the war.[42]

The growing synthetic organic chemicals industry was one reason university chemistry programs expanded after World War I. During the war and the following decade, several universities gained recognition for the strength of their chemistry departments. Contemporaries placed Illinois, Harvard, Yale, Columbia, and Cornell at the top of the list, with the Universities of Michigan and Chicago also sometimes included. One survey in 1925 suggested that about 25 percent of American graduate students in chemistry specialized in organic chemistry. (By comparison, at least half of German chemists appear to have specialized in organic chemistry.)[43]

By the end of World War I, American chemists had developed a strong sense of national community. Challenged by the shortages and military mobilization, the chemists became a crucial part of the nationalist impulse

Table 4.1. Chemistry Ph.D.s in Germany and the United States, 1910-1930[†]

Year	Germany	U.S.	Univ. of Chicago	Columbia Univ.	Cornell Univ.	Harvard Univ.	Univ. of Illinois	Yale Univ.
1910	?	78	7	2	5	2 (1)	6	3
1911	381	87	8	10	5	8 (?)	2	5
1912	?	94	6	7	10	4 (3)	6	7 (4)
1913	?	87	1	6	9	2 (1)	4	8 (5)
1914	329	97	4	6	7	6 (4)*	6	3
1915	232	107	8	9	7 (0)	6 (0)	6	7
1916	85	115	9	9	8 (5)	4 (2)	14	9 (6)
1917	?	121	6	9	11 (5)	10 (3)	8	5
1918	57	96	13	5	12 (1)	3 (1)	12	—
1919	103	65	7	5	6 (0)	3 (1)	7	1
1920	?	104	12	7	4 (0)	3 (3)	8	7
1921	?	137	15	15	4 (2)	4 (2)	7	10
1922	864	147	13	8	8 (2)	3 (2)	18	11
1923	949	167	11	25	7 (2)	3 (2)	19	14
1924	?	218	12	31	15 (4)	16 (8)	17	7
1925	785	220	15	23	12 (5)	11 (5)	18	9
1926	?	248	19	29	10 (3)	9 (6)	18	12
1927	669	211	17	19	17 (7)	9 (4)	13	9
1928	?	232	9	15	21 (5)	6 (2)*	15	14
1929	557	229	13	16	16 (6)	6 (2)	14	10
1930	496	332	25	20	14 (4)	6 (3)*	23	3

[†]When known, the number of Ph.D.s in organic chemistry is given in parentheses.
Sources: Germany: Jeffrey A. Johnson, "German Women in Chemistry, 1895-1945" (in two parts), 8, 11, 68; I have combined two of Johnson's charts—one covering Ph.D.s per academic year and one covering Ph.D.s per calendar year, making the table easier to compare in a general way but less accurate (the numbers include men). U.S.: Thackray, Sturchio, Carroll, and Bud, *Chemistry in America, 1876-1976*, 265-66; Univ. of Chicago: "Doctors of Philosophy, June 1893-April 1931," *Announcements* 31, no. 19 (May 15, 1931): 77-94. Columbia Univ.: *Annual Reports of the President and Treasurer to the Trustees with Accompanying Documents*, 1910-1930 (New York: Printed for Columbia University, 1910-1930). Cornell: Compiled from commencement programs (Collection 37/8/346, box 2) and "Official Publications of Cornell University," vol. 1 (1910-11) to vol. 23 (1931-32), Cornell University Library, Rare Books and Manuscripts Collections; because the numbers were difficult to compile, they are approximate. Harvard Univ.: Compiled from "Commencement Class Day Exhibitions," 1910-1930, HUC 6910.112 to 6930.112, boxes 182-88, and from the "Official Register of Harvard University" from 1910-11 to 1929-30, Harvard University Archives. The asterisk indicates disagreement between the two sources. Univ. of Illinois: Department of Chemistry, *Developments During the Period, 1927-1941, Doctor of Philosophy Degrees in Chemistry* (Urbana: University of Illinois, 1941), Liberal Arts and Sciences, Chemistry and Chemical Engineering, Departmental History, 15/5/801, box 1. Yale Univ.: *Doctors of Philosophy, 1861-1960* (New Haven, Conn.: Yale University Press, 1961).

to build a U.S. synthetic organic chemicals industry. Rejecting long ties to German chemistry, American chemists turned inward and emerged as the central cogs in the development of domestic scientific and technological expertise. As important stakeholders in the growing U.S. synthetic organic chemicals industry, the chemists also became vocal advocates for federal promotional policies.

5 / Xenophobia, Tariffs, and Confiscation, 1914–1918

The wartime crisis in dyes and pharmaceuticals shaped public policy in ways that had significance beyond the German chemicals. World War I left Americans feeling vulnerable from interruptions in international trade, and they sought to rectify the perceived weaknesses. Synthetic dyes and pharmaceuticals became a harbinger of the shift to autarky and isolation that gripped the United States during and after the war. In the relatively small sector of synthetic dyes, Americans rejected the globalized economy of the long nineteenth century, and the rejection came most explicitly and tangibly in promotional policies adopted by Congress and enforced by executive agencies. Americans backtracked from prewar lowering of protectionist barriers and pursued other policies that more aggressively attacked the bonds of international trade. The war's disruption of trade created economic incentives for Americans to manufacture synthetic dyes and pharmaceuticals, but fear instilled by the war led to widespread political support to use the tools of the federal government on behalf of the domestic industry.

The exaggerated anti-German sentiment affected the goals and severity of policy measures. Distinctions among Germans in the imperial army, Germans in chemistry, and German American citizens grew hazy in the eyes of many Americans. The hostility to Germany helped advocates to convince key Americans and policymakers that the ability to produce synthetic dyes and pharmaceuticals, this seemingly small cluster of chemicals, had become not only an urgent priority for the economy but also a patriotic mission. The political pressure to address the crisis became acute.

Early calls came for higher tariff rates to protect the industry, one of the most traditional of policy tools and a more predictable protection than the war's blockades and economic disruption. The partisan positions over protective tariffs had a long history, but the fight was particularly intriguing as the Democratic White House and Congress had managed to lower average tariffs in 1913 after decades of rising rates. They sought to preserve the 1913 rates and implement alternative measures, including enlisting the Department of Commerce to provide market information to aid the manufacturers. Concerned about upcoming elections, however, Democrats brought out a tariff in 1916 that reflected political compromise—making the first crack in the 1913 reforms.

More dramatic than tariffs, legislation to address German-owned property in the United States aided those policymakers bent on permanently driving the German chemical industry out of the country. The Trading with the Enemy Act (TWEA), passed in October 1917, was among the most dramatic of war laws. Although initially conceived as a liberal law intended to protect private enemy property from the passions of war, it became instead a harshly punitive measure aimed at, above all, the German chemical firms. The Alien Property Custodian (APC), a new executive position, carried out the TWEA provisions related to supervising German-owned property in the United States. Under the leadership of A. Mitchell Palmer, the APC aggressively targeted the German chemical firms and their American partners, attempting to destroy the international ties and "Americanize" the industry. Using techniques hard to imagine outside the context of war, the Office of Alien Property contributed the harshest component of the federal government's multifaceted promotional industrial policy.

The Tariff Fight

The manufacturers of synthetic organic chemicals, as well as other advocates in the chemical community, very early reached the consensus that they needed a protective tariff to develop the industry. The war and transatlantic blockade provided the nascent industry with an extraordinary amount of protection, but the manufacturers feared for the postwar return of German competition and turned to Congress to reduce the risks of investing in the industry.[1] In their quest for a tariff, however, the chemical community needed to convince powerful interests of the value of their industry and the desirability of a tariff. Two powerful groups potentially posed the largest obstacles: Democratic politicians, who held the White House and both houses of Congress, and textile manufacturers, who exhibited ambivalence about committing to higher dyes prices after the war. Although the Department of State and Department of Commerce had almost inescapably become involved in the synthetic organic chemicals crisis, a move by Congress to establish a protective tariff on the chemicals would mark a firmer commitment to the industry. The chemical community lobbied vigorously to obtain such recognition and aid from

Congress, although the questions about the long-term viability of the industry remained.

Chemists and manufacturers turned so rapidly to a tariff remedy because for forty years manufacturers of all stripes received increasingly high tariffs and had grown to consider protectionism a normal condition. They and their congressional allies, primarily but not exclusively Republicans, believed the federal government should offset the lower costs of production elsewhere, which caused higher prices for consumers but aided industrial development. That American industry had expanded dramatically during those decades lent credence to the protectionist approach, especially in an era when free-trade economists remained largely confined to their ivory towers. In 1913, however, reformist Democrats came to Washington in force, taking the White House and Congress. Almost immediately, President Woodrow Wilson and congressional Democrats overhauled the tariff schedule. Although not advocates of entirely free trade, the Democratic reformers aspired to what they deemed a more "scientific" tariff that dropped rates to the minimum that would sustain domestic industry. They succeeded in lowering rates on almost all products, and they installed the direct income tax to generate lost revenue.[2] After finally achieving a "Democratic" tariff, Wilson and his allies were loath to contemplate higher rates again. The chemical lobby needed to convince these Democratic politicians to raise rates.

In the executive branch, the Department of Commerce first addressed the chemical shortages, and Secretary of Commerce William C. Redfield became President Wilson's point man in the administration's response to the shortages in synthetic organic chemicals. A New Yorker, Redfield spent thirty years in machinery manufacturing and moved into banking and finance in the 1920s. Prior to his six years as secretary of commerce (1913-1919), he served as Brooklyn's Commissioner of Public Works and spent a term in Congress (1911-1913). Perhaps not surprisingly, Redfield rejected the characterization of business as the province of robber barons, arguing that "commerce is the ally of progress and develops, not destroys." But Redfield also embraced the reform impulses of the Progressive Era, and in 1912 he published *The New Industrial Day*, in which he implored manufacturers to invest heavily in their workforce. "The man can grow, the machine can not, and we must be sufficiently scientific in our management to avail ourselves of the *growth of the man*. We must deal with inefficient labor by teaching it and by paying it enough to stimulate it into efficiency." In addition, Redfield believed that the United States needed to embrace global markets—not tariff protection—and that increased productivity from a happy, trained, and motivated workforce was the way to compete abroad successfully. In the

1920s, when isolationism appealed to so many, Redfield continued to advocate international engagement.[3]

Redfield endorsed the goal of establishing a domestic synthetic organic chemicals industry, but his antagonism toward tariff protection and his embrace of the "scientific" study of markets heavily shaped his response to the dyes crisis. As Democrats came under increasing pressure to raise tariffs, Redfield resisted the attempts to roll back the reductions in the 1913 tariff. Among other things, Redfield privately lobbied capitalists like John D. Rockefeller and Eugene Meyer to invest in the nascent domestic industry. Already by 1915, Redfield had decided that the "German stoppage" made an American industry "morally certain," a positive development that would "relieve us from the rather humiliating dependence upon Germany." He wrote to Secretary of State William Jennings Bryan that he simply could not "feel any deep enthusiasm over the difficulties" of the importers. Redfield committed himself and the resources of his department to promoting the domestic synthetic organic chemicals industry, but he opposed a tariff increase. A higher tariff "might seem to be the simplest and easiest way," he argued, but it "merely results in American people paying more money for the product." More than Wilson, Redfield received the brunt of the criticism for upholding the Democrats' tariff law of 1913, and the chemical community never accepted him as one of their allies.[4]

The one potential "danger" Redfield admitted concerned the possibility that the German firms might engage in unfair trade practices. First, American manufacturers regularly accused the German firms of dumping chemicals into the prewar U.S. market, perhaps aided by subsidies from the German government. Dumping, or selling below cost, was an illegal trade tactic that Redfield promised to eliminate with effective consular and customs administration and, if necessary, by supporting new anti-dumping legislation. Second, if a German firm threatened to withhold dyes from any customer who refused to buy all of his dyes from the same German firm, a practice that earned the title "full-line forcing," then the Department of Commerce would attack that unfair trade practice as well. In fact, Redfield argued that a tariff would be ineffective in the face of either dumping or full-line forcing, because the Germans could "absorb" the increased costs. He also privately suggested that the "slaughter" of World War I would exceed the bloody U.S. Civil War, and at one point he wondered whether the destruction of the "managing brains . . . as well as the working hands" might hinder the German chemical firms' postwar prowess. In any case, for too long Americans had given in to their "tariff cowardice," and he yielded no ground to the supporters of tariffs.[5]

Redfield's department made its greatest impact on the domestic industry by systematically collecting and publishing market information. At the instigation of a Senate order, the Department of Commerce began an investigation of the dyes crisis and issued a series of reports, the first statements from the federal government to sanction the quest for a domestic synthetic organic chemicals industry. Through his Bureau of Foreign and Domestic Commerce, Redfield hired Dr. Thomas H. Norton, a chemist near the end of his career who had long experience in academic, industrial, and government employment, to analyze the dyes shortage. Norton's reports became some of the most important documents for the dyes industry during the war.[6]

Over 1915 and 1916, Norton produced three significant reports that systematically laid out the scope of the crisis. His reports argued clearly and concisely that the United States ought to develop a domestic synthetic organic dyes industry. Norton pointed to the abundant raw materials in the United States and the demand from its textile industries, and he also noted that the vast majority of the dyes consumed in the United States never were or were no longer protected by patents—all conditions conducive to promoting a domestic industry. While the chemical community applauded the support of the Department of Commerce and Norton's investigations, they disputed some of Norton's analyses, particularly his suggestion that Americans should cooperate closely with Swiss dyes firms to acquire technical expertise, which American chemists took as a slight. In addition, most felt Norton's numbers in the two reports of 1915 were overly optimistic, they criticized him for rejecting a tariff as part of the solution, and they even ridiculed him in private. Still, the reports carried an authority and imprimatur of the federal government that legitimized the domestic manufacturers' pursuit of promotional policies and helped to lure capital to the industry, as even Norton's critics conceded.[7]

Norton's most important report came in 1916 and provided market information that no private entity could have supplied. He supervised the production of a "census" of dyestuffs, a compilation of import statistics gathered from the U.S. Treasury's customs invoices from the year prior to the war. Based on a suggestion Hesse had frequently made in the trade literature, and probably on the failed British attempt to compile a similar statistical profile, the census revealed the quantity and variety of dyes consumed in the United States and provided a fairly comprehensive overview of the market. "In no other way," he wrote, "can the creators of such an industry try to avoid duplication, overlapping, waste, and blundering, tentative struggles to adjust [production] to a vague, indefinite demand." The importers protested Norton's report before it was released to the public because it violated laws governing use of the confidential information provided to customs, and the department revised the census to include only aggregate numbers, rather than for individual brand name dyes, but not before the *Journal of Industrial and Engineering Chemistry* published a section of the unrevised census. Again the chemical community disputed Norton's numbers, particularly regarding dyes that had been on the "free list" and therefore not carefully noted by customs agents. Over the next several years, Norton's census became, as the Department of Commerce intended, a guide to many in the industry and a standard reference work cited frequently in subsequent debates over the status and future of the domestic dyes industry.[8]

Redfield had hoped that solid market information would aid the development of the industry without new tariff protection, but the pressure mounted on the Wilson administration and on Congress to encourage the domestic industry with new tariff legislation. The administration heard, often in very blunt terms, from the interested parties. One old dyer complained about the Democratic representatives knowing "more about chewing tobacco than Tariff Measures" and that unless the Democratic administration backed the tariff, it would be replaced with one that had more "Horse Sence" [*sic*]. That tariffs should be only for revenue rather than protection struck A. R. L. Dohme of Sharp & Dohme, pharmaceutical manufacturers in Baltimore, as contrary to American experience, not to mention a "heresy" that "wrecked our national development." In Dohme's view, the Wilson administration's opposition to higher tariffs for a domestic synthetic organic chemicals industry was symptomatic of a failed Democratic party ideology. Quite "simply," he believed, a "low tariff spells financial and commercial disaster," a policy that satisfied only importers and Socialists.[9] Dohme expressed his vehemence in more colorful language and partisan vitriol than most, but he represented the majority opinion among manufacturers.

At the end of 1915, Representative Ebenezer J. Hill, a Republican from Connecticut, proposed a flat ad valorem rate of 30 percent, plus another 7½¢ per pound, or specific, rate to protect American dyes manufacturers from renewed postwar competition from the German firms. Hill wrote to Redfield arguing that while "it is beautiful to maintain a theory," Democratic opposition to tariff legislation would prevent the investment necessary to make the domestic industry viable in the long run. Hearings on Hill's bill opened on January 14, 1916, before the House Ways and Means Committee, chaired by Representative Claude Kitchin, Democrat of North Carolina. J. F. Schoellkopf Jr., William Beckers, and Herbert Dow all testified that their firms would be unlikely to prosper in the face of renewed German competition after the war. Schoellkopf, with perhaps the most to gain from the tariff legislation, unabashedly remarked that the committee might consider even higher rates. The Hill rates might gradually attract capital, but higher rates promised more significant immediate returns. In his view, the business remained risky and "nobody is going into the business for the fun of the thing. Everybody goes into business with the idea of getting a return on his money." Dow argued that he had invested in his indigo operations "on the assumption" Congress would establish an effective tariff. Both Schoellkopf and Beckers expected their firms to grow rapidly if the tariff bill passed, and each claimed they could make as much as 50 percent of U.S. needs. Other observers thought they exaggerated, but the statements allowed Kitchin to question whether a tariff might encourage an American dyes monopoly.[10]

For the chemical community to win over the Democrats, the chemists and manufacturers needed to convince the public and policymakers that the significance of the domestic dyes industry went far beyond the seemingly tiny sector of the economy. As the war dragged on and the chemical shortages persisted, the advocates of the industry successfully played on emotions, highlighting the indignity of a great nation like the United States groveling before Britain and Germany to "beg" for German dyes. They discussed the nation's inability to supply red and blue for the red, white, and blue, and green for the greenbacks.[11] Most importantly, however, the witnesses in support of the tariff legislation argued two primary points: the impact of the dyes industry on the giant U.S. textile industry, and the contribution a dyes industry would make to national defense.

To make the case on behalf of the textile industry, the chemical community lined up witnesses to explain the significance of dyes to the textile industry. I. F. Stone of the Schoellkopf firm attempted to explain the larger relevance of the industry. By the time he finished calculating the ripple effect of the dyes industry through the textile industry, he concluded that "approximately one-fiftieth of the entire population of the country are directly concerned in their daily lives in the dye situation." But the chemical community knew that its most convincing witnesses would be the representatives from the textile industry themselves, particularly the manufacturers from the southern states who could persuade the southern Democratic members of Congress that their own states suffered from the dyes crisis.[12] Despite all the intensive lobbying from people representing the chemists and chemical manufacturers, many of them believed the primary influence on decision making lay largely with the textile manufacturers, as it had for decades.

Approximately 70 companies affiliated with the textile industry conveyed opinions in person or by correspondence to the House committee; 58 favored the bill. Over a dozen textile manufacturing trade associations went on record to support the tariff increase, and one textile chemist estimated that 95 percent of textile manufacturers supported the Hill bill. The war, they said, had led most of them to decide a domestic industry could provide a more stable supply of dyes and prevent similar crises in the future. Although quantities varied over specific textile products, dyes usually constituted only a small portion of a textile manufacturer's costs, and most of them decided they could afford to pay more after the war in higher prices for the "insurance" that came with a more reliable supply. One southern manufacturer, Fuller Calloway of Georgia, offered his support for the legislation and also a few jokes to the committee. He favored the dyes tariff, in part because the war had made him "more money" than he had "made in 30 years put together." Tellingly, only two of the textile manufacturers appearing on behalf of the dyes legislation came from the southern states, and the members of the chemical community felt the absence of the southerners.[13]

Witnesses also testified to the national security that a domestic synthetic organic chemicals industry would bring to the country. The shared chemical relationship between dyes and high explosives meant that a domestic dyes industry in peace could quickly convert to explosives factories in a future war, they argued. Schoellkopf and Beckers stated they could make the transition in a week because of common intermediates and also because both dyes and explosives consumed large quantities of similar inorganic chemicals, primarily nitric and sulphuric acids. The implication for national defense appeared obvious from the viewpoint of the chemical manufacturers and a few sympathetic textile industrialists. A domestic dyes industry would contribute to a strong national industrial base by providing peacetime markets for intermediates, which could be funneled into high explosives during war. Schoellkopf finished his testimony with the dramatic exclamation that "some day this Nation will cease to exist unless we have that industry."[14]

At least a few in the chemical community questioned the direct link between dyes and explosives, although not before the House committee. The critics compared the dyes and explosives production and noted the disparity in magnitude between the consumption of organic raw materials such as benzene and toluene and also the inorganic nitric and sulphuric acids. The entire amount of raw materials consumed by even a full-grown dyes industry amounted to a tiny fraction of the consumption of an explosives industry at war. The quantity of common intermediates would be even smaller.

Herman Metz argued these points, although the proponents of a tariff dismissed his criticism as self-interested because of his long association with a German firm. But Bernhard Hesse, the respected chemist who lived for statistics, also remained dubious of the usefulness of the links between the two industries, although he kept his opinion private. His performance at the hearings earned him public criticism in *Weekly Drug Markets*, which reported that "Dr. Hesse's testimony was regarded as unfavorable," largely because "he failed to take the position which many of those present at the hearing expected him to take." Indeed, the article accused him of protecting German interests. Had anyone taken a serious look at the state of the American explosives industry in January 1916, with its huge sales to Britain and France, the critique might have seemed obvious, but it carried little weight. The widespread belief that the German dyes factories produced high explosives carried more rhetorical power, particularly when the connection was exaggerated.[15]

When the hearings concluded, Herty took on the task of rallying support among textile manufacturers in his native South, and he brought more than the usual arguments on his lobbying tour. Within the chemical community, Herty was among the first and most vociferous to espouse anti-German rhetoric. As the political battle for the tariff intensified, the frequency and impact of fiery speeches linking a domestic industry to national defense and the horrors of the European war increased. In his speech before the American Cotton Manufacturers' Association in Atlanta, Herty castigated Secretary of Commerce Redfield and the Democrats on the Ways and Means Committee for playing party politics, for "trifling with a serious national disaster," when the crisis demanded "statesmanship and broad-minded Americanism." The bottom line for the cotton manufacturers, Herty told them, was their future ability to compete in the export market against countries with textile industries that could depend on a more certain supply of dyes. The association obliged him after his speech by passing a resolution backing the tariff legislation.[16]

In March and April, the Democrats in Washington started to crack and show signs of compromise, concerned about the fall elections; the compromise centered on the duration of the tariff provisions. The "whole trouble" with infant industries, wrote Henry T. Rainey, Democrat of Illinois, to Representative Hill, is that they "never grow up." However, if Hill and his allies in the chemical community would suggest a reasonable length of time, say, until domestic manufacturers could make about two-thirds of American demand, then Rainey would support tariff legislation that contained an expiration date. During the hearings, the committee members asked several witnesses how long they would require tariff protection to establish their infant industry. Charles H. Herty, representing the ACS, asked for "a reasonable tariff for a reasonable time" and identified "five or six years" as a period after which American manufacturers could "hold our own with the Germans," and then Congress could gradually reduce the rates. J. Merritt Matthews anticipated the same timeframe: "practical benefit" would come "inside of 12 months" and "full benefit" in "three to five years." Herbert Dow, as well, expected his firm would need "five or six years" of protection, long enough to "learn the tricks of the trade." After the learning period, the large scale of indigo production and consumption would allow him to compete without a tariff. In a trade journal, I. F. Stone suggested that with "proper protection," the United States could "supply all its own goods within a year."[17] Modified by the time limits, the Hill bill—now the Kitchin bill—passed the House of Representatives; the legislation then moved to the Senate.

The Senate's discussions of the bill were limited to amendments and adjustments to the House bill, and a key issue was synthetic pharmaceuticals. John F. Queeny, founder and president of Monsanto, had lobbied long and hard for Congress to consider not only synthetic dyes but also synthetic pharmaceuticals as they debated a new tariff. Neither the Hill nor Kitchin version of the bill included synthetic pharmaceuticals. He estimated that his firm had invested $150,000 to $200,000 in equipment to make from scratch synthetic pharmaceuticals and flavorings such as acetphenetidin, phenolphthalein, and saccharin. Queeny argued throughout the spring of 1916 that policymakers should recognize the close relationship between the synthetic dyes and pharmaceuticals and offer comparable protection. Like the dyes manufacturers, he invested under "the belief" Congress would "make the necessary changes" in the tariff. In fact, judging by price increases, he argued, the pharmaceuticals interests suffered more than dyes. If the final tariff included the proposed 15 percent rate on intermediates but gave no protection to synthetic pharmaceuticals, he would be worse off with the legislation. Similarly, Leo Baekeland argued that the 15 percent rate on phenol without protection for Bakelite put him at a disadvantage.[18]

The bill that became law on September 8, 1916, retained the imprint of Representative Kitchin. Coal tar crudes could enter duty free. Intermediates received a 15 percent ad valorem rate in addition to a 2½¢ per pound specific duty. The lawmakers granted finished dyes a 30 percent rate, plus a 5¢ per pound tax on all dyes except indigo and alizarin dyes. In addition, the law included definite time limits on the duration of the specific surtaxes. The rates, considered exceptionally steep by Kitchin in his committee report to the full House, would exist for five years as listed. If, at the end of five years, the domestic industry supplied 60 percent of U.S. consumption, then the specific rate could continue, declining by 20 percent each year to nothing over a second five-year period. On the other hand, if the industry failed to meet the 60 percent threshold, the 5¢ specific rate would end immediately. Pharmaceuticals also received the 30 percent ad valorem but with no specific duty. The act gave Bakelite and other phenolic resins a 30 percent ad valorem plus a 5¢ per pound duty.[19]

Senator Oscar Underwood, Democrat of Alabama and chief sponsor of the lower rates in 1913, quite simply hated the dyes tariff. With the inclusion of both ad valorem and specific rates, plus the generally high level of rates, the Republicans could hardly have produced a more offensive bill than the Democratically authored version, he thought. No matter what the Democratic caucus said, he for one would not "bow to the desecration of Democratic principles." Underwood, bitterly complaining of "special interests," and six other senators opposed the legislation unsuccessfully. Meanwhile, at the other extreme, Republican Senator Boies Penrose of Pennsylvania demanded protection for more industries, suggesting that because the Democrats had admitted the necessity of tariffs on dyes, they may as well admit the necessity to protect American industry more generally.[20] Both Republicans and Democrats saw the dyes tariff as a breach that could lead to further erosion of the Democrats' 1913 tariff reforms.

In the chemical community, the tariff received a favorable response from some quarters. The *Weekly Drug Markets* in July had considered the Kitchin version an acceptable compromise that gave the dyes manufacturers "practically all that they have asked for." Also in July, J. F. Schoellkopf Jr. had confided to Herty that he could accept the Kitchin version of the bill, although he preferred the rates to include indigo and the alizarin dyes, and Beckers felt the same way. Dyes manufacturer E. C. Klipstein declared the new tariff law "the most scientific" in his experience, one that finally worked to the "benefit of Americans" and "braved the anger of the great German dyemakers."[21]

The overwhelming response from the chemical community, however, was hostility and disappointment. From Monsanto came criticism for the "discrimination" against synthetic pharmaceuticals for denying a specific rate; "a puzzle in an otherwise most harmonious composition." Bernhard Hesse, who calculated a dyes industry a losing equation in economic terms, criticized the changes from the original legislation, the result of "quibbling, haggling and cheese-paring," as the worst solution. The Hill bill rates, based on the recommendations of the ACS, endorsed Hesse's own statistical analysis, and anything less, in his mind, was simply inaccurate and insufficient to build an industry. The rates in the law would only raise prices without establishing the industry. The bulk of the criticism focused on depriving indigo and alizarin dyes of the 5¢ specific per pound duty levied on the other synthetic dyes. Americans consumed indigo in larger quantities than any other dye; it was the only dye for which sales exceeded $1 million. Plus, the alizarin dyes made up an increasingly important share of dyes. As the manufacturers calculated it, excluding indigo and alizarin dyes from the 5¢ rate meant "leaving the dike wide open over 27 per cent of its length." Despite the 30 percent ad valorem rate, such an exclusion from the specific rate would hamper the industry's quest to make 60 percent of the nation's needs within five years.[22]

The five-year, 60 percent feature of the law, coupled with the exceptions to indigo and alizarin dyes, made the law far from satisfactory in the eyes of domestic producers. Within three weeks of the passage of the new tariff law, Herty and the leading dyes manufacturers met for a brainstorming session at the Chemists' Club in New York City to express their dismay at the law and to find ways to amend it. Despite their testimony before Congress, all agreed that the law's timetable failed to allow enough time for the industry to develop, consequently dissuading capital investment. More specifically, because alizarin dyes and indigo composed about 30 percent of all dyes consumed in the United States, leaving them without the specific per pound duty made the 60 percent goal that much more elusive and put the domestic production of all synthetic dyes in jeopardy. As a result of this meeting, the participants drew up a resolution to "condemn the removal" of the surtax from alizarin dyes and indigo. They pronounced the exceptions "detrimental to the establishment and development of the American Dyestuff Industry and [a] subversion of the best interests of the American people." Representative Hill, who had voted against the final form of the bill he had initiated, continued unsuccessfully to introduce legislation to include indigo and alizarin dyes. Herty sent out rounds of letters imploring correspondents to lobby Congress, especially Representative Kitchin, to change the legislation, which he argued bore the imprint of "the German influence."[23]

The Trading with the Enemy Act

On October 6, 1917, Congress passed "one of the most drastic pieces of legislation ever," declared the *New York Times*, capping six months of legislative and executive work intended "to cut off absolutely the last vestige of hope of

German agents to get aid to their nation."[24] Shortly after Americans joined the war, a working committee of representatives from the Departments of State, Commerce, Justice, and Treasury gathered to discuss trade with Germany and German-owned property in the United States. They wanted to shut down commerce "that benefits Germany and German subjects," and their proposal encompassed not only transatlantic trade with the enemy countries and subsidiaries of German firms, but also trade through neutral countries, international correspondence, and German stockholders and intellectual property. In May, the House of Representatives began hearings on the executive branch proposals, and the Senate followed in the summer. Six days after Congress passed the law in October, President Wilson issued an executive order putting into motion the new Trading with the Enemy Act (TWEA).

Enforcing the law took an array of executive branch departments and agencies. Some provisions generally had little impact on the chemical industry: the Department of Treasury gained jurisdiction and the power to license foreign insurance companies and continued its supervision of transnational financial flows, the Postmaster General monitored the foreign language press, and the new Censorship Board censored "communications by mail, cable, radio, or other means of transmission." But three provisions carried particular consequences for the future of German synthetic organic chemicals in the United States. First, the president established the War Trade Board, which superseded the export control agency set up in the Department of Commerce under the June 15 Espionage Act and also included import controls (discussed in chapter 6). Second, the Federal Trade Commission gained the authority to license foreign-owned patents, copyrights, and trademarks for use during the war. Finally, the new Office of Alien Property, established as an independent agency, would prevent the use of German property in the United States from aiding Germany.[25]

When Secretary of State Robert Lansing and Secretary of Commerce William Redfield introduced the working committee's proposals to the House of Representatives in late May, they emphasized their attempt to draft the proposal as liberally as possible. They specifically rejected more draconian models available from Civil War or contemporary European precedents that confiscated outright. "It is quite different," noted Lansing, because the others were "based on the person and not on the locus." That is, German citizens living in the United States and their property would no more be targets than American citizens, as long as they ceased trade with Germany. Additionally, as Redfield explained, the committee wanted "to provide the necessary authority to protect ourselves against aiding the enemy in a way which would provide as little interruption of our commerce as possible." Not only was a strong American economy necessary for waging war, but the United States had also become "the world's purse" and foreign trade depended on American success as a creditor. "The nations of the earth are looking to us as the financial power." Assistant Attorney General Charles Warren, who chaired the working committee and drafted the proposal, reiterated: "In my mind it is in line with the modern development of leniency toward private property in time of war. It is not in accordance with the older barbaric methods, but it is a modern, up-to-date method." In protecting private property, the authors of TWEA embraced as principle the notion that nations waged war against other nations, not against individuals and private property.[26]

The practical mixed with principle, ideology, and idealism as Americans confronted the way the international economy linked American and German property and its owners. If not confiscation, then what? The goal, offered Representative William C. Adamson of Georgia, who chaired the House hearings, was "that there will be no final injury to anybody." TWEA would "protect" enemy property from "the passions of war." "In the hands of the government," agreed Secretary of State Lansing, the enemy property "will avoid any lawless acts against it." The alternatives were to let neglected German property lose value or to leave German property in private American hands, where some Americans might rack up undeserved profits, perhaps unscrupulously. The government, on the other hand, could use any revenue from German properties to buy Liberty Bonds and help fund the war. In addition, the government might need the German property for bargaining during peace negotiations. Holding property, rather than confiscating it or leaving it in private hands, would give Americans postwar financial leverage to settle on American property in Germany (smaller holdings than German property in the United States), but also moral leverage—"an act of good faith even toward an enemy," as Secretary of Commerce Redfield suggested. Redfield saw the Office of Alien Property in a similar light. The new agency would demonstrate "that same effort toward equity which impels us to . . . to show to the people with whom, unfortunately, we are engaged in war that here is the opposite of confiscation and here is the opposite of requisition." Policymakers believed the United States could model better moral behavior for the European belligerents.[27]

Despite the liberal guidelines adopted by the law's authors, TWEA contained at least two ways it could transform quickly into a much more punitive law without amendment. First, throughout the hearings in both houses of Congress, the policymakers made it clear they had no expectation that law-abiding citizens from Germany and allied nations in the United States would become "enemies," but they explicitly reserved the right of the president to expand the definition of "enemy" simply by proclamation. Second, TWEA gave Congress the right to dispose of the property after the war. Again, the policymakers framed that provision with the expectation that Congress would deal more responsibly with enemy property than private Americans would and in most cases would return it to its prewar owners after the war and peace negotiations ended, but the law put few restraints on future congressional disposition of property.

The Federal Trade Commission and the Salvarsan Patents

The first of the provisions that directly involved synthetic organic chemicals, Section 10 of TWEA, addressed patents and evolved directly out of controversy over Salvarsan, one of the most important synthetic organic pharmaceuticals. Herman Metz controlled Hoechst's intellectual property in the United States, including Paul Ehrlich's famous chemotherapeutic treatment for syphilis, under U.S. patent since 1911. The shortages since 1914 had been acute, and as German supplies dwindled, Metz carefully doled out supplies of Salvarsan directly to physicians, avoiding middlemen altogether to limit both speculation and re-export, the latter to appease his Hoechst partners, under pressure from the German government. The British periodically allowed German Salvarsan through the blockade, but unpredictably and with considerable delay. With the incidence of syphilis rising, particularly in the military, the United States faced a growing public health crisis, and Metz and the Salvarsan patents became the primary target for physicians' frustrations. As the United States mobilized for war, synthetic organic chemicals sat at the heart of legislating the use of enemy-owned patents.

The most direct and immediate challenge to the Salvarsan patents came from Philadelphia, where a non-profit research institute, the Dermatological Research Laboratories (DRL), began to research Salvarsan in 1915. Dr. Jay Frank Schamberg and two partners founded the research institute in 1912 with funding from Philadelphia philanthropist Peter A. B. Widener, who had a particular interest in psoriasis research. Schamberg modeled his institute after the much larger Rockefeller Institute. Anticipating increasingly severe shortages in Salvarsan, Schamberg expanded the DRL's research into Salvarsan and announced at a professional meeting in May 1915 that he and his partners had developed an alternate and affordable process to make it. Partly because Widener died in November 1915, Schamberg wanted to begin manufacturing and selling in volume to fund his research institute, as well as to alleviate the shortages. Schamberg's activities brought him into conflict with Metz over patent infringement, and the two men skirmished for years, an antagonism driven by different ethical worldviews and large financial stakes.[28]

Metz continued to rely on the sporadic imports through 1915, but he also took steps to launch his own manufacturing, getting help from Hoechst as the war wore on. Metz owned the U.S. trademarks for Salvarsan and Neosalvarsan; he held power of attorney for the patents, and Hoechst was in the process of transferring ownership to Metz when the German government forbade any such written instrument from leaving the country (TWEA included a similar provision). In January 1916, Metz's brother Gustave traveled to Frankfurt, initially intending to carry the patent transfer documents back. Although thwarted in that goal, Gustave accomplished the other purpose of his trip: learning to manufacture Salvarsan and Neosalvarsan, and he spent nearly six months in the Hoechst plant, studying and memorizing every facet of production, including testing methods. When Gustave returned, he and Herman began to set up manufacturing in Brooklyn, but backlogs in machinery orders and finding substitutes for equipment unavailable in the United States slowed construction. In the fall of 1917, H. A. Metz Laboratories turned out Salvarsan commercially.

Until October 1917, Schamberg risked a lawsuit for infringing on Hoechst's patents, and that brought the two parties unhappily together in a contractual arrangement on December 5, 1915. Metz held the upper hand legally, but throughout most of 1916 and 1917 he remained dependent on German imports, a supply no one could guarantee. In the agreement, Metz allowed Schamberg's DRL to manufacture Salvarsan, but DRL needed to sell it to Metz for distribution. The agreement also gave Metz the right to demand that DRL cease or restart production. Indeed, when a large shipment arrived from Germany in May 1916, Metz ordered DRL to halt production except for local research and charity. When Metz began to run short in April 1917, he asked DRL to start up again, and he continued to buy from DRL even after he made his own and right up until TWEA passed.[29]

Despite coming to terms with Metz, Schamberg appealed to the medical profession in 1917 to help terminate the Salvarsan patents, tapping into longstanding professional misgivings about patented medicines and medical devices. Doctors only grudgingly accepted patents in the decade before the war, pushed in part by the impressive patented products of German science. Telling them that he could make Salvarsan affordably, Schamberg asked a responsive American Medical Association (AMA) leadership to help lobby Congress to let him infringe legally. Several AMA members, including physicians from leading medical institutions, sent letters to the Senate patent committee on Schamberg's behalf. The Mayo Clinic doctors protested to Congress that syphilis posed "a more serious menace to the manhood of a nation than many of the devices of an enemy," and "abundant and cheap" supplies of Salvarsan were required for "the protection of the Army and Navy." The Mayo doctors were among those who used the crisis to return to an earlier stance of entirely opposing intellectual property rights for drugs. "We urge that such abrogation shall . . . be permanent, and . . . [that] a radical revision of our patent law [prevent] private monopoly of remedial agents indispensible to the public health." Schamberg shared the concern: "Any man who commercializes his medical work degrades his profession and loses his proper medical status," as he put it in 1922. He never viewed his manufacturing as commercial work because the revenues supported the non-profit DRL.[30]

The medical profession's attack on the Salvarsan patents led to Senate hearings on June 4, 1917, where Metz defended himself with words and fisticuffs. Metz made an impressive witness at congressional hearings (and he had many opportunities to hone his skills in World War I and the 1920s). The written record demonstrates not only a thorough understanding of his business and the larger industry but also an ability to explain it to a lay audience of lawmakers, an ability no doubt aided by his own term in the House of Representatives—and he knew how to use the record. A speedy tongue complemented a quick mind (noted one senator, "If you will talk a little slower you would not ruin the stenographer"), but Metz was in a difficult spot and tensions ran high. To counter accusations of price gouging, Metz insisted on explaining the cost structure of the imported Salvarsan, which he sold for $3.50 per dose before the war and $4.50 after April 1915. Each ampule cost $1.25 in Frankfurt, plus shipping, duty, and packaging, and then war risk insurance after 1914. American-made Salvarsan brought the costs and prices down considerably; both Metz and DRL could sell to the government a dose for $1.50 and to physicians for $2.50 by the end of 1917. Insinuations of profiteering and disloyalty hurt Metz, and when Dr. George Walker of Johns Hopkins University and other doctors suggested at the Salvarsan hearings that Metz inflated prices and the patent should be invalidated, Walker and Metz scuffled in the hallway. Metz "struck" Walker, and Walker "landed an upper cut on Mr. Metz's chin," although others soon had them separated and apologizing to one another.[31]

American manufacturers hardly came running to Metz's defense, given his ties to Germany, but they agreed with him and not the medical profession on the subject of patents. Like Metz, pharmaceutical and chemical manufacturers generally favored licensing enemy-owned patents during the war, yet they saw the demand for nullification (more commonly discussed as "abrogation") as a threat to their own futures. One Eli Lilly executive wrote privately to Charles Herty in 1917 that pharmaceutical manufacturers walked carefully to avoid AMA opprobrium, but he worried that the current AMA leadership was too "radically opposed" to protection of intellectual property related to medicine. Manufacturers expected their longer-term interests lay with the German model of research and development and recouping expenses through patent protection. They also (successfully) opposed other proposed legislation that would have limited American patents strictly to processes and not products.[32]

Over the summer, Congress folded the Salvarsan problem into the larger discussion about TWEA, and TWEA addressed German-owned patents. When TWEA passed in October 1917, Section 10 addressed intellectual property rights in several parts. Enemies could still file for U.S. patents, and Americans could still file for patents in enemy countries after vetting for security (provisions eliminated in early 1918). More importantly, Americans could license enemy-owned patents; the president delegated to the FTC the right to set the terms and number of licenses per patent. In general, the FTC required licensees to submit a royalty of 5 percent of "gross sums" collected on sales using the license, and the royalty went into a trust account. After the war, under Section 10(f) of TWEA, the owner of the patent had one year to file in a federal district court to collect royalties from the trust account; the courts determined the amount paid to the owner, with any remainder in the trust account going back to the licensee. Congress endorsed Charles Warren's vision of a war policy that protected private property of enemies, including intellectual property.[33]

For licensing pharmaceutical patents, the FTC officials established an advisory committee, which determined both the number of licenses issued and assessed the ability of applicants to make drugs safely. The FTC licensed four manufacturers for Salvarsan and Neosalvarsan by early 1918: Metz and DRL primarily, plus two others with relatively small outputs. Once DRL had an FTC license, all contracts and strained politeness ceased between Metz and Schamberg, a situation that required diplomacy by the advisory committee, headed up by University of Chicago chemist Julius Stieglitz. The toxicity of Salvarsan and Neosalvarsan required particular care, and the British experience influenced American policy; more than 200 people died in Great Britain from bad Salvarsan made by inexperienced manufacturers. The drugs contained arsenic, which posed its own hazards, but patients also received the drugs by injection directly into the bloodstream, and as

Metz indelicately put it, "If any dirt gets in there you are going to have a dead man on your hands." Manufacturing required a spotless environment, careful attention to temperature, and maintaining the products in a vacuum to avoid oxidation. The dangers of Salvarsan and Neosalvarsan led to close cooperation between the manufacturers and Public Health Service (PHS). The PHS tested every batch for safety, but their chemists also engaged in joint research to improve the firms' processes, yields, and standards, which went up over 1917–1918, surpassing prewar levels of purity substantially.[34]

Alien Property Custodian

TWEA's most dramatic departure from past precedent, believed its authors, was Section 6, the provision that created the Alien Property Custodian (APC). Until just before TWEA passed in October 1917, legislators had planned to put the APC in Redfield's Department of Commerce (within its Bureau of Foreign and Domestic Commerce), but their final wording left it to the president's discretion. Wilson instead placed the APC at the head of a new executive agency, the Office of Alien Property, which would report directly to him. The APC had as his chief responsibility the execution of the new, more "civilized" policy that would conserve enemy property for the duration of the war. But in selecting his first APC, the president could hardly have picked a person more incompatible with the original spirit of TWEA. By the end of the war, the first custodian helped to transform TWEA into the punitive and confiscatory act that Lansing, Redfield, and Warren had spoken so loftily of avoiding. The confiscations had enormous consequences for the synthetic organic chemicals industry: the German chemical firms were the APC's top target.

A. Mitchell Palmer accepted President Wilson's offer to be the first APC, a post he held from October 22, 1917, to March 4, 1919. By 1917, the thirty-nine-year-old politician had won accolades in Pennsylvania as a young and effective Democratic reformer, running against the state's party machine, and he served in the House of Representatives for three terms following the 1908 election. In 1912, he worked for Woodrow Wilson's election. In return, the newly elected president offered Palmer the post of secretary of war, which the latter's Quaker convictions led him to decline. After the United States entered the war, Palmer instead accepted Wilson's offer to serve as the APC, followed by two years as the nation's attorney general, in which he contributed to the first Red Scare with the "Palmer Raids"—indiscriminate raids on organizations to find alien radicals to deport—and other challenges to civil liberties. In 1920, he unsuccessfully sought the Democratic nomination for president. His biographer traces his "lifelong distrust" of immigrants to his roots; although Pennsylvania as a whole had attracted waves of immigrants, Palmer's home county in the state's east central region remained overwhelmingly "homogenous, old-American stock." In its obituary for Palmer in 1936, the *New York Times* called his political career "one of the most meteoric in the history of American democratic politics," with its pinnacle, however controversial, being his tenure as APC and attorney general. In those posts, Palmer pushed the boundaries of his power, generating charges of abuse, corruption, and violations of civil liberties (although lawsuits in subsequent years never pinned him to explicitly illegal behavior).[35] As APC, Palmer mixed his Quaker beliefs, a political reformer's impulse, and anti-immigrant fervor into a brew that made the Office of Alien Property into a strong and aggressive agency.

First, however, Palmer had to build an organization, which he did quickly and relatively effectively, and to begin collecting reports on enemy property, a more challenging task. To carry out TWEA, Palmer established several bureaus that kept track of the accounts held in trust, addressed legal questions, supervised audits, and undertook investigations, among other things. At its peak, the office employed several hundred people, primarily in Washington and New York. By law, all enemy-owned property needed to be reported to the APC no later than December 1917, resulting in about 11,000 reports reaching the Office of Alien Property. Convinced that most such property remained unreported, however, Palmer continued to pursue his quarry through other channels, including press releases, letters to lawyers and bankers, and appeals to all 56,000 postmasters throughout the country. Reports came in on bank accounts, insurance policies, estates, and stock ownership, but by the end of the war the APC also held property ranging from cotton, beer barrels, and tungsten ore to jewelry, theater and opera royalties, and even underwear. In addition, the APC had files on more than 250 corporations; it placed representatives on boards of directors in proportion to its enemy ownership. As a result, another 22,000 reports arrived in 1918. "This organization," reported the *New York Times* in January 1918, "is at once the biggest trust institution in the world, a director of vast business enterprises of varied nature, a detective agency, and a court of equity."[36] The German-affiliated chemical firms would come to know the "detective agency" and "court" functions very well.

Palmer quickly became disenchanted with the APC mission to hold and conserve German-owned property. He pushed President Wilson and Congress to amend TWEA and expand the APC's powers; he wanted to confiscate German property and sell it to American citizens. By March 1918,

Palmer openly and vigorously advocated using TWEA to "Americanize" enemy property. Testifying before the Senate, he explained that some of the German property in trust brought in significant income. "I am today operating factories and mills and industries all over the United States . . . , most of which are making enormous profits by reason of the very conditions for which the enemy is responsible, namely, the war conditions." If Congress returned "both principal and profits to the German owners at the end of the war, [it is] a tremendous favor to the German Empire, our enemy." Instead, Americans should derive the economic benefit of the property and not only for the duration of the war. "The time has come when the ownership of some of these great German properties should be permanently separated from German capital." From Palmer's point of view, taking property from Germans was an entirely legitimate "weapon," a justifiable form of economic warfare. At least two senators questioned Palmer's intentions for the postwar world, expecting that "we will want to do business with Germany" after the war, but they had become a minority by March. While only the fire-breathing senator from South Carolina, Benjamin ("Pitchfork Ben") Tillman, put it so bluntly—condemning "making money for those whelps over there"—Palmer found an increasingly receptive congressional audience as he discussed holding millions of dollars for enemies.[37]

Beyond the numbers and profits, however, Palmer painted an insidious picture of German economic holdings in the United States. From his post, he argued, "we have seen how the German Empire, through its financial operations, has put an industrial and commercial chain all the way across this country." During the war and for years afterward, Palmer insisted that German industry and the German government were nearly one and the same. German corporate interest in the United States "constituted Germany's great industrial army on American soil. They were the far-flung lines of advance for her kultur." Under TWEA's original provisions, the APC "protect[ed] for the enemy the great industrial and commercial army which Germany had planted here with hostile intent." While he still excepted average individuals, Palmer believed Germany's "great industries . . . supported by the junker class" needed to be "rooted out." By the end of the war, Palmer characterized German-owned U.S. chemical firms as "spy centers filled with the agents of Germany long plotting against the safety of the United States. They were depositaries [sic] of secret information gleaned by the ubiquitous spies in the German employ, and without them these spies would have been almost harmless."[38]

Congress amended TWEA on March 28, 1918, with a provision attached to a war appropriations bill, and the APC immediately went after particular

German industries. Initial targets included the Hamburg-American and Northern Lloyd shipping lines, which quickly lost their piers on New York's harbor to confiscation. German insurance firms were under market and government pressure even before the amendment because of dependence on the transnational flow of funds, and most closed down by summer 1918. Also seized quickly was the Bosch Magneto Company with plants in Massachusetts and New Jersey, partly because the military used the magnetos—electrical generators used widely for engine ignition—particularly for airplanes. In July, Bosch became the first major firm sold to Americans, and it was the second largest sale under TWEA, coming in just behind the Bayer Company. The German chemical industry became the APC's most notorious target.[39]

Throughout his tenure at the Office of Alien Property, Palmer's right-hand man was Francis P. Garvan, whom Palmer had selected to run the office's Bureau of Investigation. As the assistant district attorney in New York City (1902–1910), he worked on four sensational murder trials, including the high-profile murder of architect Stanford White (by an insane millionaire in 1906 over a love interest), and appeared in the news regularly. In later congressional hearings, Palmer asserted he brought Garvan into the Office of Alien Property because he was "without a superior in the business of detection of crime in a great, broad way." Garvan's APC job included investigating companies in the United States for German ownership, and his specialty became the German chemical companies and their American importers. When Palmer became attorney general in the spring of 1919, Garvan inherited his boss's position as APC, a post he held until the change of presidential administrations in March 1921. Garvan's years in the Office of Alien Property gave focus to the rest of his life. To the day of his death in 1937, Garvan dedicated himself to promoting the American chemical industry at the expense of its German competitors. His bullying and often rude personality could cause friction even among friends and allies, but he earned the respect of the chemical manufacturers and chemists by the sheer force of will and energy he brought to their cause. Born and married into wealthy families, Garvan seemed motivated not by money but by a patriotic conviction that drove him into increasingly isolationist, autarkic policy positions and into questionable legal territory. His anti-German zeal and harsh tactics contributed to the abuses of civil liberties during the war while simultaneously advancing the cause of the domestic synthetic organic chemicals industry.[40]

###

The first six months under the APC went relatively smoothly for the Bayer Company. President Herman C. A. Seebohm reported on the German ownership of Bayer and the Synthetic Patents Company, the American holding company set up in 1913 to own and license U.S. patents filed by the German firm. In both cases, the owners were Carl Duisberg, Charles Hess, and Rudolph Mann, all in Germany. The Bayer Company's long-time lawyer, Charles J. Hardy, facilitated communications with the APC, partly because he knew Bayer's business very well and partly because he was the only American citizen on the board of directors, the rest being either German or Swiss. Hardy strongly recommended that the APC retain Seebohm as president, arguing that he knew best how to run the firm. On January 23, 1918, Frederick B. Lynch became the APC supervisor of Bayer, and he monitored business transactions. In February, the board membership changed, with all Germans but Seebohm departing and six new arrivals selected by the APC.[41]

In early March, when Lynch submitted his first major accounting to Palmer's office, two large items on the balance sheet raised alarms at the APC. First, by contract, the German and American firms split profits on any exports coming out of the United States, but Bayer held a $300,000 note payable to the German firm (after the war)—for purchasing the German firm's portion of the U.S. export business. The note dated to March 31, 1917, after the United States had cut diplomatic relations with Germany and a week before declaring war. Second, Bayer held $290,000 in Liberty Bonds, bought with money from sales in Brazil and China. The money arrived shortly after March 31, 1917, and therefore, in the Bayer accounts, they owned the Liberty Bonds, not the Germans. The APC ordered an audit and investigation, particularly wondering whether anyone in the United States had the legal authority to sign the note for the German firm. In addition, Palmer appointed to the Bayer Company a new lawyer, J. Harry Covington, a former politician from Maryland and a recently retired judge, to take a closer look. The APC lost confidence in Bayer's Hardy, fearing he was too close to the Bayer officers to protect the APC's interests; Covington also suspected Lynch could "give but superficial supervision" and lacked the expertise to sort through Bayer's affairs effectively. In July 1918, everything fell apart for the Bayer Company leadership. The APC quickly took over the Liberty Bonds and looked for ways to declare the March 31, 1917, purchase illegal. And there appeared to be a third liability: Williams & Crowell, the dyes start-up in Connecticut that Bayer had worked with since 1915.[42]

The Bayer officers' biggest problem lay in explaining the nature of William & Crowell's relationship to the Bayer Company and to the individual top managers at Bayer. Bayer executives portrayed their actions as businessmen hedging their risks in a very uncertain political and economic climate, and Hardy believed their steps observed TWEA. The APC, on the other hand, saw a deceitful conspiracy and a serious threat to the domestic synthetic organic chemicals industry.[43]

Shortly after Seebohm reported German ownership of the Bayer Company to the APC in December 1917, he and several others in the Bayer leadership purchased Williams & Crowell as individuals. Already by mid-1917, the founders of the start-up wanted to sell out. Hardy knew that the Bayer men in the United States, those who were German citizens, were not defined as enemies under TWEA or Wilson's executive order and therefore a purchase of Williams & Crowell by the individuals would not constitute trading with the enemy. Between the December 1917 APC report and the January appointment of Lynch, Hardy and the others worked out the purchase agreement; payment happened under Lynch's watch. Hardy first set up a new Williams & Crowell corporation, this one in New York rather than Connecticut. He lined up American citizens to be the founding stockholders; given the hostility toward German Americans, German names posed a commercial liability, and he sought to avoid them. Seebohm paid for part of his stock out of savings, but he also borrowed, as an individual, from the Bayer Company to pay for the remainder. The Bayer Company helped solidify the new firm by signing a contract to continue buying Williams & Crowell's dyes. The organizers expected the income of Williams & Crowell would allow Seebohm to pay back the company loan quickly. Hardy and Seebohm swore to the end that every move was legal and had nothing to do with Germany. The arrangement protected the operations and assets of the Bayer Company in addition to giving the individuals a backup plan in the event Bayer changed hands. Plus, Bayer's continuing relationship with Williams & Crowell's dyes helped to complete government contracts and benefited Americans and not Germans.[44]

Covington, Garvan, and Palmer interpreted the Bayer moves very differently. They immediately saw the plan as a permanent diversion of assets from the Bayer Company to Williams & Crowell, hiding the assets from the APC for the benefit of Germans. As Garvan accused, "They were ready to have us turn the real Bayer Co., in the form of Williams & Crowell Co., back to the Germans, leaving the old Bayer Co. nothing but a shell."[45] Garvan and others in the APC spent the remainder of 1918 sorting out the Bayer affairs with the definite aim to seize and sell the firm to Americans.

As the Bayer Company president since 1915, Herman Seebohm received the closest scrutiny from the investigators. A native and citizen of Germany, he was also Carl Duisberg's brother-in-law and the son-in-law of a lieutenant colonel in the German military. Aided by Department of Justice lawyers,

Garvan and his Bureau of Investigation questioned Seebohm and other Bayer Company leaders and employees through several depositions, some of them decidedly hostile. The lawyers baited him: "You would be a mighty poor specimen of a human being, Mr. Seebohm, if you did not have a good, big feeling for your family, for your home, for your fatherland, would you not?" Mostly the lawyers pushed Seebohm about the firm: wasn't Williams & Crowell really German-owned? Wasn't it Bayer Company money purchasing the start-up? And why did Williams & Crowell buy warehouse property with a pier in New Jersey? Was it to facilitate postwar German importing? Was it on behalf of the German government, to replace the confiscated Hamburg-American and Northern Lloyd piers?[46]

Seebohm had a skeptical audience as he explained that Bayer Company advanced money for operating costs and to secure Williams & Crowell's dyes and that he and other individuals bought Williams & Crowell. Through several interrogations, Seebohm consistently explained the flow of money: it involved Bayer Company monies, foreign receipts due or not due to Germany, and German funds to pay Bayer Company employees extra commissions directly. The New Jersey warehouse was a possible location for expanding Williams & Crowell, particularly because the sulphur dyes produced by Williams & Crowell stank too badly to expand in Connecticut or New York, but they bought it for a low price and believed there was a strong probability its value would increase. Seebohm had a particularly difficult time accounting for the significant amounts of money that flowed in cash or gold: he had sent accounting records back to Germany on the last submarine in November 1916, and he destroyed his copies when he knew the sub was safely across. Hardy, also deposed, argued that Germans in business used cash regularly in circumstances Americans found odd, but Garvan interpreted the missing paper trail as purposefully covering up money going into pro-German propaganda.

The Bayer Company received scrutiny not only as a large German chemical subsidiary, but also because Hugo Schweitzer, the head of pharmaceuticals until his death December 23, 1917, was very active in pro-German propaganda before the United States entered the war. A native of Germany and an American citizen, Schweitzer reportedly said in 1915 that "my allegiance is to the United States, but I do all I can to help Germany without hurting the United States." Schweitzer developed a working relationship with Heinrich Albert, the financial chief of the German diplomatic corps. Albert's task, while the United States remained neutral, lay in purchasing supplies (food, cotton, war matériel) for Germany and then trying to get it through the British blockade on neutral ships, a plan that met little success. He also attempted to buy matériel for use in the United States to keep it from aiding the British and French. This latter strategy brought him to Schweitzer, and they arranged to buy phenol to make aspirin, partly to keep it out of explosives. Schweitzer also joined efforts to raise money among German Americans, particularly brewers, to buy a major newspaper, another unsuccessful plan. There was nothing especially secret about Schweitzer's propaganda work, all done during the period of neutrality, but it drew attention to Bayer. Several Bayer officials later testified that Schweitzer and Seebohm had a strained relationship at best, in part over Schweitzer's extra-curricular activities (of Schweitzer, Seebohm said, "I was very glad if I did not have a fight with him every day.") But Garvan asserted that some of Bayer Company's money found its way into pro-German activities; neither Garvan nor Seebohm, who denied the charge, possessed documentary evidence to prove his respective case. Shortly after the war, Garvan testified to Congress that Schweitzer was the "head of the German Secret Service in America," providing his alleged and unwieldy German "secret service number": 963192637.[47] Garvan also regularly reprinted letters of Schweitzer's that British and American intelligence agencies had intercepted. The correspondence, assessments of the American chemical industry and market, proved Schweitzer understood his business well but hardly justified a secret service number. Schweitzer's death from pneumonia in December 1917, however, spared him from any consequences of the APC's accusations.

When the APC completed its investigation, Bayer Company and its leadership faced harsh consequences, although even two days after the armistice a lawyer in the Office of Alien Property could only say of the company's executives that he suspected they were "pro-German at heart but against whom nothing definite has as yet been obtained." The accountants discovered no sign of intentionally hiding assets, but they found that Bayer Company underpaid taxes on profit-sharing arrangements with Germans, primarily through the Synthetic Patents Company. Garvan determined that Williams & Crowell had effectively been purchased by Bayer Company, and the APC confiscated Bayer Company, Synthetic Patents Company, and Williams & Crowell and put them all up for sale. Seebohm ended up interned as an enemy at Fort Oglethorpe in Georgia, one of 6,000 aliens interned during the war; he joined 2,300 others designated as "alien enemy suspects" and considered more dangerous. Four other Bayer employees also spent time interned. At Seebohm's arrest in August, the APC press release announced it was because the Bayer executives had a "scheme to retain a bridgehead for an attack of German commercial interests on the chemical industry in this country after the war."[48]

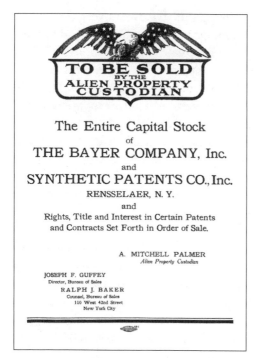

FIGURE 5.1. The Alien Property Custodian sells Bayer Company, 1918. (Courtesy of the National Archives and Records Administration, RG 131, entry 155, box 35.)

Legal complications held up the APC's plans to sell Bayer Company and Synthetic Patents Company. First, the March amendment permitted the confiscation of real property, but patents remained covered in TWEA only by the original provision for FTC licenses. Second, Bayer was only one of many German firms that made substantial legal changes in ownership in the period between the break in diplomatic relations between Germany and the United States (February 3) and the declaration of war (April 6). None of the transactions were necessarily illegal, and several of them involved ownership documents stranded in Germany. In the fall, Palmer and others in the agency pushed Congress to amend TWEA once again to permit confiscation of patents and to claim a clear title to property without the normal documents.

In justifying the second amendment to senators, the APC's lawyer Lee Bradley repeated accusations that German subsidiaries were "outposts of . . . propaganda and sabotage and general deviltry." The APC's push to sell and Americanize the firms had stalled because so many of the firms depended on patents which could not be seized and sold with the firm. "That is peculiarly true in the chemical industry, which, I believe, in the mind of the general public, is perhaps the most important of all to be cleaned up it if can be done." On November 4, 1918, one week before the armistice, the APC got the second major amendment to TWEA, receiving the right to confiscate patents and to gain clear title. In addition, corporations were obligated to cancel shares of stock owned by enemies, another complete departure from the intentions of the original law's authors. Finally, the law reiterated that people whose property was seized could file an appeal through TWEA's Section 9, but the amendment stated that the "sole relief and remedy . . . shall be limited to and enforced against the net proceeds."[49] In other words, even if owners had a legitimate claim to a factory or stock certificate or fur coat taken by the APC, they could only get money back. This last provision suited the APC goal to Americanize German-owned industry.

In the end, none of Bayer's activities were clearly illegal; even the unpaid taxes, the strongest claim against the firm, required the accountants to appeal to the Department of the Treasury to make a ruling in an ambiguous case (without specifying, the APC report also noted, curiously, "a number of less legitimate evasions of tax laws"). Palmer, Garvan, and the APC lawyers argued that TWEA and its amendments made the seizure and sale of the company perfectly legal regardless of the Bayer executives' behavior, legal or otherwise.[50]

At the conclusion of its investigation, the APC organized the sale of Bayer's plant, 1,200 patents, and trademarks on December 12, 1918—a month after the armistice. The consulting engineers hired by the APC estimated the worth of Bayer's physical plant at $2.7 million, which excluded stocks of raw materials and finished goods, working capital, and patents and trademarks. Out of a field of five bidders, Sterling Products Company of Wheeling, West Virginia, offered the winning bid of $5.3 million and immediately resold the dyestuffs end of the business to the Grasselli Chemical Company of Cleveland, Ohio, for $1.5 million. Founded in 1901, Sterling's product line consisted primarily of patent medicines, which made the purchase of Bayer a step toward scientific respectability as well as a promising investment. Grasselli Chemical, a well-known manufacturer of sulphuric acid and other heavy chemicals, began production during the war of organic intermediates for explosives and of a handful of simpler dyestuffs. Its share of the Bayer plant would allow Grasselli to expand dramatically its depth in synthetic organic chemicals.[51]

The Bayer Company employees were in no position to return fire with questions about conduct by the APC, but others, including Congress, asked questions in later years, particularly questioning the APC-appointed leadership of the firm. Who was Frederick B. Lynch, the administrator appointed to lead Bayer? He was a Democratic politician from Minnesota who served in a leadership position with the Democratic National Committee and who suggested Palmer as a presidential candidate in the 1912 convention and again in 1920. Lynch also became the APC administrator of the Botany Worsted textile mills in New Jersey. The Bayer Company's APC-appointed lawyer, J. Harry Covington, was a former member of Congress from Maryland, had served on the Democratic National Committee, and earned $46,000 in legal fees from the APC, including $17,500 from Bayer. One of the new APC-appointed board members was Nicholas Brady, Francis Garvan's brother-in-law. In the years afterward, accusations of political patronage dogged Palmer—in the case of Bayer, the patronage was real and the actions aggressive, but not definitively illegal, as was the case for other APC officials.[52]

As with Bayer, the executives of other German-affiliated chemical firms in the United States generally aimed to respect the interests both of U.S. national security and of their German partners. The original design of TWEA seemed to offer them a way to make it through the war by balancing the two demands while walking a difficult but viable legal tightrope, but few made it past a hostile APC without a misstep. Palmer and Garvan no doubt assumed correctly a sympathy or loyalty to the German partners, particularly in firms bound by family ties or long, successful partnerships. But because the APC goal by early 1918 lay in "Americanizing" the German chemical industry, Palmer and Garvan attacked any international bonds that might survive the war and revive in peace. Despite creative lawyers who worked overtime for the German-affiliated firms and executives, the APC rode the hostile political climate and detected treasonous motives at every turn. Once the March amendment to TWEA passed, the APC lawyers permitted no innocent interpretations of behavior as they went after the chemical firms and their leadership. The personal costs for some in the industry could be very high, with non-citizens particularly at risk.

In the case of Heyden, a U.S. court in May 1921 ordered the German president of the American firm, George Simon, to pay a $20,000 fine and sentenced his American lawyer, T. Ellet Hodgskin, to prison for two years for their actions during the war. Although caught up in Schweitzer's "phenol plot," Heyden's vulnerability in APC investigations lay elsewhere. The jury found Simon and Hodgskin guilty on two counts: defrauding the U.S. government by declaring an unauthorized dividend in December 1917 and failing to report German ownership of the firm. The key issue centered on whether ownership of the American firm legitimately transferred to Americans in 1916. Simon and Hodgskin believed they had, in good faith, separated the American firm from German ownership long before the United

States entered the war. Garvan, however, declared they had created only "an apparent American ownership" that was merely "an attempt to camouflage the real ownership." The "beneficial interest," he and the other lawyers argued, still defaulted to the Germans. This argument appeared regularly in APC charges.

George Simon came to the United States in 1900, responsible for establishing Heyden's American subsidiary at the relatively young age of twenty-eight, and he maintained a close and profitable working relationship with the German parent firm. In a 1906 agreement, still in effect in 1914, the German firm provided financing and technical processes in exchange for all profits left after issuing 8 percent dividends—and they owned 745 of 750 shares of the stock. At the beginning of August 1914, Simon was vacationing with his family in Germany. He and the German parent firm's management discussed the potential dangers to the American enterprise, and Simon left Germany with discretion to make decisions he deemed in the best interests of the two companies. As with Bayer, trust between the two sides had been built over years of business, and the Heyden partners also relied on informal, verbal understandings, which became a liability under close, hostile APC scrutiny. Simon had applied for American citizenship in 1910, but he erred in filing papers in Manhattan rather than Brooklyn, and he remained a German citizen throughout the war. His wife and two children were American citizens, as was his naturalized father-in-law, Richard Kny, a wealthy man who also became caught up in Heyden's fate.

Simon decided that the discretion granted to him in 1914 included permission to sell the Germans' stock to an American, his lawyer Hodgskin, in May 1916. Kny loaned $150,000 to Hodgskin to purchase the stock, and Hodgskin sent the money to Germany. The money reached Germany, but the Germans could see no safe way to send the stock ownership certificates to the United States, and Simon therefore nullified those and issued new certificates to Hodgskin, the legality of which the APC challenged in 1918. To have "an arrangement that was absolutely fair to both parties," Simon asked that the sales agreement include an option for the German firm to buy back the stock within eighteen months, the "beneficial interest" that the APC subsequently targeted, although the eighteen months expired before the armistice arrived. After TWEA passed in October 1917, the Heyden company reported two shares of stock as being German-owned but not the 745 held by Hodgskin. Operating under the assumption his was an independent, American firm, Simon—after receiving legal advice other than Hodgskin's—declared a 10 percent dividend in December 1917. The APC investigators, however, found Hodgskin's ownership merely a "camouflage" for the former German owners, and the dividend therefore a dispersal of assets that rightly belonged in an APC trust. The APC thus confiscated the company and sold it to the highest bidder, Allen A. Ryan, the son of a successful typewriter manufacturer.[53]

Simon and Hodgskin appealed their criminal convictions, arguing that "what they did, when they did it, did not amount to a crime," and the appellate court acknowledged their actions could be "capable of an innocent interpretation" but that the eighteen-month claim on the stock demonstrated that the "determined purpose was the preservation and protection" of the German firm's interest "from all enemies of Germany including the United States." The court also dismissed as minor the prosecutor's repeated characterization of Simon as a Prussian: "I say he is a Prussian because he looks like one; he is in evidence before this jury. . . . Loyalty is his middle name. Loyalty to the Chemische Fabrik von Heyden, loyalty to Germany, loyalty to the Fatherland, loyalty to the Hohenzollern." The appeals court upheld their convictions.[54]

After jousting with Dr. Jay Schamberg over Salvarsan, Herman Metz also faced a Francis Garvan determined to confiscate his firms—Farbwerke Hoechst (primarily cash assets by 1917), H. A. Metz Laboratories (Salvarsan), and H. A. Metz & Company (dyes)—as part of the larger effort to Americanize the industry. After the 1913 contractual readjustments, Metz held clear title to his firms, but Garvan argued that the "beneficial interest" still accrued to the Germans because Metz carried sizable debt due the German firm. Metz agreed to pay for an APC-approved financial audit until the accountants' bills reached $20,000 on the way to a five-month final tally of $30,000—"highway robbery was an amateur game compared to this," Metz later complained to Congress, convinced that Garvan kept digging for German ownership where none existed to "ruin" him. The APC-appointed board of directors for his firm intended to authorize payment to the APC-appointed accounting firm, until Metz intervened. The expensive audit showed that Metz owed the German firm over $200,000 for purchasing stock back from Hoechst after the antitrust suit in 1913. The collateral for the bank note was Metz's stock—if he failed to pay the loan, the German manufacturer would again own the stock. The debt, the APC insisted, placed Metz at the mercy of the German firm and its claim on ownership. Metz fought back with more force and resources than any of the other targeted chemical firms, filing suit under TWEA's Section 9 and getting court injunctions to prevent APC actions. "I ventured to suggest to [Garvan] that the Alien Property Custodian under the statutes was not authorized to found an American dyestuff industry, [and] . . . my counsel frequently called his attention to the

Table 5.1. APC Actions toward Selected German-affiliated U.S. Firms

Firm	Fate as of April 1922	$ in APC account
Badische Company (inactive)	held in trust	$554,455.02
Bayer Company	sold Dec. 12, 1918	$5,310,000.00
Berlin Aniline Works (Agfa)	liquidated & held in trust	$581,308.59
Cassella Color Co. (inactive)	held in trust while litigated	$85,000.00
Farbwerke Hoechst (inactive)	held, litigated, returned	$199,000.00
Heyden Chemical Works	sold Mar. 27, 1919	$1,500,000.00
Kalle Color & Chemical Co.	liquidated & held in trust	$261,623.64
Merck & Company	sold May 9, 1919	$3,750,000.00
Rohm & Haas	sold Mar. 4, 1920	$300,000.00

Note: Most trust deposits remained undispersed as of April 1922. The list does not include firms such as H. A. Metz & Company or Kuttroff & Pickhardt, which were incorporated in the United States during the war to replace the German-sounding Farbwerke Hoechst and Badische Company. Metz won his lawsuit on November 28, 1921, and the APC returned his $199,000.00.
Source: *Report of the Alien Property Custodian*, 67th Cong., 2nd sess., April 10, 1922, S. Doc. 181 (Washington: GPO, 1922); for Cassella and Farbwerke Hoechst, see also *Annual Report of the Alien Property Custodian for the Year 1922*. 67th Cong., 4th sess., January 24, 1923, H. Doc. 525 (Washington, D.C.: GPO, 1923).

fact that the trading with the enemy act was a sequestatory statute and not a confiscatory law." Metz avoided the sale of the firms in part by arranging them to go into an escrow account rather than an APC trust account, and the lawsuits remained unsettled until well after the armistice. For Metz and Garvan, 1918 was only round one of a long fight between two powerful men with extensive political and economic networks; they remained archenemies until their deaths in the 1930s.[55]

During the war, Merck's sales and profits grew dramatically. The Rahway plant expanded to three times its prewar size, and the company offered 3,500 to 4,000 products—not all synthetic organic chemicals—to the market. Early in 1918, however, George Merck surrendered to the APC the 80 percent of the company stock owned by the German Merck. The custodian put the 8,000 shares of stock up for sale after the armistice, but the company encountered a different fate than the Bayer or Heyden companies. George Merck astutely arranged financial backing from large New York bankers and orchestrated the purchase of the company back from the APC. George Merck's company would grow into one of the largest pharmaceuticals manufacturers in the twentieth century.[56] How did Merck escape the worse fate of Bayer and Heyden?

The experience of Merck & Company was remarkable in comparison to the fate of Bayer, Heyden, and even Metz. George Merck navigated Merck & Company through the APC shoals with less damage than the other German chemical subsidiaries sustained, in part because he had been an American citizen since 1902 and in part because he and his lawyers clearly thought carefully about their approach. Merck filed a formal petition with the APC on April 3, 1918, requesting that the APC take over the 8,000 shares in trust, their ownership to be determined through legal channels. Merck and his representatives made no apparent missteps, paying the $22,000 accounting bill, negotiating board membership before turning over the stock, and filing suit under TWEA's Section 9 to reclaim the assets. A Merck & Company lawyer also recalled how the APC representative told them in April 1918 (after the March TWEA amendment) that "there would be no likelihood at any time of any sale by the Alien Property Custodian," and the Merck side expected the 80 percent of stock owned by Germans to be held in trust and eventually returned. The audit found the firm's books in order, albeit with the Germans' "peculiar faculty of surrounding their business operations with a great deal of secrecy and cloaking, as to things they consider confidential." The APC patent lawyer found that neither Merck & Company nor E. Merck owned particularly significant U.S. patents and primarily licensed from others. Merck's reputation was another of the firm's assets: "There is no doubt that the name of 'Merck' enjoys an enviable position in the drug and chemical industry, owing to the quality of its products," wrote the accountants. With George Merck's conservative business strategies, the firm had accumulated a surplus of more than $4.5 million by the end of 1918. The financial assets and George Merck's track record helped when the APC notified somewhat surprised Merck & Company officials in the fall of 1918 that it would sell the 8,000 shares of German-owned stock.[57]

Merck & Company continued to negotiate with the APC as the sale approached: they agreed not to contest the sale if George Merck could bid for the 8,000 shares. Before the May 9, 1919, sale, Merck lined up the capital from major investment bankers, Goldman Sachs and Lehman Brothers, with contingency plans to ensure he could offer the highest bid. In a twenty-five-minute auction, Monsanto and American Aniline, one of the newer dyestuffs firms, both bid up the price, but Merck's group won at $3.75 million. A few months later, the bankers helped orchestrate a public stock offering, selling preferred shares broadly to raise capital but all the common shares—and therefore complete control—to George Merck. Garvan, wary of George Merck's German ties, insisted upon custodian-appointed trustees to ensure continued American ownership for ten years following the sale. Curiously, neither the *New York Times* nor the *Philadelphia Inquirer* carried any news about the sale, in contrast to splashing Bayer and Heyden across the headlines, suggesting that the APC opted not to send out their usual provocative and accusatory press release. George Merck kept ownership and control of his firm, but Palmer and Garvan had succeeded in "Americanizing" it.[58]

By "Americanizing" the three main German-owned subsidiaries during the war, the APC attacked the German industry in ways difficult to imagine without the vehemence of war; nevertheless, the APC kept up the pressure long after the halt in open warfare on November 11, 1918. Palmer became attorney general in March 1919, and Garvan, as the new APC, carried on his mission with passion. Actions of the APC created legal tangles for years to come—not only the suits provided for in TWEA's Section 9 (disputing APC ownership claims) and Section 10 (FTC patent licensing and royalties) but also lawsuits about charged illegalities and corruption. After the confiscation of the subsidiaries, what remained of German-owned chemical property were the cash deposits held in trust accounts awaiting congressional disposition—and all of the German-owned chemical patents except those sold with Bayer.

###

The armistice stopped the soldiers, but the United States and Germany legally remained in a state of war until 1921, extending a period of unsettled political, legal, and economic conditions. At the end of 1918, the promotional policies brought together two sets of industry advocates who had largely worked separately. The manufacturers and their allies lobbied hard for tariff protection; Palmer and Garvan in the APC office took on the German chemical industry as a patriotic mission, attacking the most German of industries. The concern about German-owned patents, especially, led Garvan to a closer relationship with the domestic chemical manufacturers. Despite the significant investments flowing into making dyes and related chemicals, the expectation of strong postwar German competition loomed large in the strategies of all industry advocates. They prepared to fight in an "economic war" after the world war, which included maintaining and expanding promotional policies won during the war.

6 / Surviving the Peace

Economic War, 1919–1922

The armistice on November 11, 1918, ended the fighting in Europe but initiated a turbulent postwar economic transition from a war economy to a peacetime economy. The three years following the war proved crucial to the American synthetic organic chemicals industry, particularly the dyes and pharmaceuticals manufacturers. Most industrialists felt they still lacked the ability to compete on equal footing with the German industry and continued to work closely with the federal government to encourage the development of the industry. The policymakers and American manufacturers strove to prevent the return to the prewar globalized commercial network, which they believed would spell disaster for most American manufacturers. During the war, the industry profited from the market protection provided by the war's severance of international trade and by embargoes imposed as war measures. After the armistice, the synthetic organic chemicals manufacturers and their supporters attempted through legislation and regulation to extend the beneficial aspects of the war economy.

In fact, the supporters frequently employed the metaphor of war to describe the conditions the domestic industry faced after the armistice. Blending the dangers of World War I and the expected economic battle, industry advocates urged the government in the name of national defense to protect the industry from foreign competitors. The public debate over the kind and degree of government help to the industry occurred in the context of the immediate postwar years when nationalist sentiments and fears still ran high. In the aftermath of the war, industry advocates found few opponents who denied the strategic significance of synthetic organic chemicals to the military, although some debated whether any organic chemicals except intermediates were crucial to national defense. The most extreme of the industry proponents—Francis P. Garvan, in the Office of Alien Property, especially—conflated the German military opponent and the German economic opponent, insinuating that international trade with Germany bordered on treason.

This chapter covers the domestic industry's strategy to win the postwar economic contest for the American synthetic organic chemicals market. Throughout the immediate postwar years, the manufacturers benefited from creative public-private cooperation designed specifically to aid the industry, which proved effective but also raised questions about decision-making processes without sufficient transparency to Congress and other interested parties. The industry also reaped rewards from the continuing resentment toward Germany and German Americans, which helped win additional support for protective measures, especially because the industry advocates urged promotional policies in the name of national defense. By mobilizing political support in the wake of the war, the manufacturers received exceptional help from Congress and a variety of agencies in the federal government. At the end of 1922, a collection of federal policies were in place that shielded the domestic industry from most outside competition and promised to foster this relatively underdeveloped industrial sector. In chapter 7, we see that the policies of the war and immediate postwar years required ongoing defense by the industry advocates.

The Chemical Foundation, Inc.

Europeans had established the rationale for patents by the time the new United States embedded a patent provision in the U.S. Constitution. Nations granted patents in the expectation that they would facilitate economic development by encouraging and diffusing knowledge of new technology. Inventors publicized their inventions in exchange for monopoly powers for a limited period of time. The writers of the U.S. Constitution, by granting Congress the power to issue patents "to promote the progress of science and useful arts," accepted the justification that a patent system would increase the amount of knowledge in the public domain. By the late 1800s, Americans wondered about the monopoly power of patents and whether the rise of corporations with research and development (R&D) facilities had fundamentally altered the limits that traditionally constrained the power of inventors. The ability of corporations to obtain a group of related patents, or to pursue excessive infringement suits, to bolster their market power posed a challenge to American antitrust policies.[1] When Germans wielded the monopoly power of U.S. patents, the fear of the corporate power of patents grew among the generation of World War I. Supporters of an American domestic synthetic organic chemicals industry saw the German chemical patents as an intolerable obstacle in the way of the industry's success. And invalidating German patents, as the British had done, would not be enough help for the American industry, its advocates believed. In a public-private collaboration, Americans formulated a unique policy solution to the problem of German-owned U.S. chemical patents.

Francis P. Garvan pursued his mission as the industry's most aggressive advocate in the federal government after the armistice. Until the U.S. Congress declared the war officially over in 1921, most of the war agencies continued to function under wartime regulations. The Office of Alien Property actively pursued additional investigations of enemy-owned property well beyond the armistice. A. Mitchell Palmer, the first Alien Property Custodian (APC), became attorney general in the spring of 1919, and Garvan assumed the top post in the Office of Alien Property until the end of the Wilson administration two years later. As APCs, Palmer and Garvan believed that depriving Germans of property in the United States was a justifiable form of economic warfare, even after the armistice. In the anti-German hysteria that continued after the fighting ceased, few opposed their actions.[2]

After the APC's office sold the Bayer Company in December 1918 and the Heyden and Merck companies in 1919, it held little additional real property of the German dyes and pharmaceuticals firms. However, the APC still retained the Germans' intellectual property: the patents and trademarks filed in the United States through 1917. Other than the 1,200 Bayer patents sold to Sterling and Grasselli, the patents under the APC's control remained largely untouched. But the Bayer sale raised alarms within the APC's office, as well as among American chemical industrialists. The sale of so many patents in a single lump to one firm generated fears in the Office of Alien Property that an American monopoly would simply replace the so-called German monopoly. The antitrust suit against Du Pont just prior to the war kept the industry under the constant suspicion that it could become a monopoly, and the APC's office wanted to avoid contributing to concentration of the patents in a few hands. Naturally, the sale also raised an outcry from other domestic manufacturers who worried that the patents would continue to restrict the ability of the majority of American manufacturers to produce important dyes and pharmaceuticals that remained under patent. Du Pont executives had explored the possibility of purchasing the Bayer patents, but they had difficulty evaluating the value of the patents and chose not to bid on the Bayer property. However, the executives recognized that the patents could pose a problem for Du Pont if they fell into the hands of their chief domestic rival, National Aniline.[3]

Garvan and two other lawyers in the Office of Alien Property worked to find an arrangement that seemed to satisfy the desires of the chemical industry and the legal and political parameters allowed to the APC. Neither

Garvan nor the American industrialists wanted the patents to revert to the German owners. Consulting with members of the American Dyes Institute, the trade association for domestic synthetic dyes manufacturers, Garvan and his patent lawyers devised the organization of the Chemical Foundation, Inc. The APC insisted the stock for the non-profit corporation should be owned by as many different chemical companies as possible to prevent even the hint of monopoly. More importantly, the Chemical Foundation, as the owner of the patents, would be allowed to issue only non-exclusive licenses for patents. Any firm that desired to manufacture under a particular patent need only request a license, in addition to showing 75 percent American ownership.[4] While the British simply nullified their German-owned patents, the U.S. government conserved some of the value of the patents and, in fact, created an institution to redirect formerly private royalties into the collective defense of the domestic industry. Managed as a group, the patents provided another obstacle in the way of German attempts to regain their market share. In theory, the description in the patents also released vital technical knowledge to American manufacturers, but the reality would be otherwise.

As Garvan planned the Chemical Foundation in January and February of 1919, the APC had not yet seized any of the chemical patents beyond the ones sold with the Bayer Company. The rest remained in trust. The Office of Alien Property, with the help of the largest chemical manufacturers, raced to implement their plan before a peace treaty technically ended the state of war and limited the powers of the APC, which they expected could occur within weeks. Under the auspices of the American Dyes Institute, Du Pont and National Aniline supplied money and patent lawyers to develop a list of patents that they wished the APC to seize and convey to the Chemical Foundation. The APC's staff worked around the clock to process the seizure but essentially rubber-stamped the selection made by the American Dyes Institute lawyers, who possessed a broad working definition of the "chemical industry"—the patents included those for stainless steel, for example. The first and most important sale to the Chemical Foundation sent about 4,500 patents to the foundation in exchange for $250,000. By 1921, when the Senate officially ended the state of war with Germany, the total sale amounted to $271,850.[5]

The top personnel of the Chemical Foundation initially came almost wholly from the Office of the Alien Property. Garvan became the Foundation's first president, because, according to Garvan's public pronouncements, he had become the country's foremost expert at exposing the trickery of the German chemical firms. More importantly, the creators of the

Chemical Foundation envisaged a quasi-governmental institution that would be almost an extension of the Office of Alien Property. Several other employees of the APC besides Garvan also held positions in the Chemical Foundation's hierarchy, including the high-ranking "dollar-a-year" patent lawyers who were close advisors to Garvan and Palmer.[6]

In its first four years of existence, the Chemical Foundation's leadership developed its organizational form and operating procedures. In 1921 the foundation possessed 4,764 patents, 874 trademarks, and 492 copyrights. From 1919 to 1922 it licensed 103 manufacturers to use its patents; about half of the licensees were manufacturers of synthetic organic chemicals. The foundation collected an income of $700,000, primarily from royalties. It collected $270,000 in royalties from domestic manufacturers, and, in an additional insult to the former German patent owners, the Chemical Foundation acquired another $365,000 in royalties from imported dyes and pharmaceuticals covered by its patents. On average, the Chemical Foundation executives charged domestic manufacturers 4 percent of sales for royalties; they charged importers 10 percent.[7] When faced with criticism about the Chemical Foundation, Garvan liked to compare the income figures of the foundation to the total sales of dyes by domestic manufacturers, including those not covered by foundation patents. For the four-year period, the total sales of American-made dyes almost reached $250 million—the foundation patents covered just under $7 million of dyes sales, or about 3.3 percent of the total. Garvan emphasized the small income relative to the total sales to dismiss suggestions that the Chemical Foundation, when it paid $250,000 to purchase the initial collection of patents and trademarks, had committed a fraud of extraordinary proportions.[8]

Despite the substantial income, expenses of $850,000 left the nonprofit corporation with a net loss after four years. The largest single expense of the Chemical Foundation from 1919 to 1922 came from the distribution of pamphlets, books, and articles that promoted the chemical industry and the study of chemistry. During the four years, the foundation spent $340,000 to send over fifty different items to an array of groups and individuals, including businesses, trade associations, civic groups, newspaper editors, educators, agricultural agents, and the Boy Scouts, not to mention politicians. For instance, in August 1919, the foundation sent 144,372 copies of Garvan and Palmer's *Aims and Purposes of the Chemical Foundation* to twenty-nine categories of recipients. Popular science books, such as 73,867 copies of Edwin Slosson's *Creative Chemistry*, frequently received mass distribution by the foundation. In March 1921, over 100,000 pamphlets entitled "Shall America Remain the Only Important Country at the Mercy of the German

Chemists?" went out. Garvan delivered a speech to a joint session of the American Chemical Society and the Society of Chemical Industry in September 1921. The Chemical Foundation sent the speech, full of Garvan's typical accusations of German infiltration and condemnations of pacifists, to 648,429 people, including doctors, clergymen, and women's clubs.[9]

A second set of expenses included general operating costs of the Chemical Foundation's office in New York City. The office staff consisted of the usual accountants, telephone operators, and stenographers. But the foundation also required a complement of lawyers and six chemists to protect its patents from infringement and to evaluate prospective licensees for proper American ownership. In 1921 and 1922, the years that the Chemical Foundation distributed the largest quantity of literature, about thirty female clerks handled the vast mailing program. Finally, the foundation maintained an office in Washington, staffed by William Keohan. Keohan had been a part of the APC's public relations staff, which had been so effective during the war that President Wilson asked A. Mitchell Palmer to avoid encroaching on the propaganda turf of George Creel and the Committee on Public Information. Keohan continued to carry on public relations duties for the Chemical Foundation and also lobbied and monitored members of Congress.[10]

The corporation's charter outlined a procedure for issuing non-exclusive licenses to all applicants that passed the requirements for American ownership and "competence," as judged by the foundation. The creators of the Chemical Foundation pointed to the non-exclusive provision as evidence that the Chemical Foundation would serve the public good rather than particular private interests. In trying to implement the policy, however, the Chemical Foundation administrators ran into two problems. First, the lack of exclusivity removed incentives for domestic manufacturers to invest in developing dyes under the Chemical Foundation patents because other firms could exploit the patent before research costs could be recovered through sales. Second, precedent existed for a different management of the German patents. When held in trust, the Federal Trade Commission appointed an advisory committee to determine which firms and how many firms might use the pharmaceuticals patents. During the war, the FTC committee had limited the number of licenses to what the committee members believed the market could bear, while still rendering the licensed manufacturers a profit to compensate for the risk and research expenses. Garvan appointed an advisory committee, too, adopting the FTC committee members, who initially continued FTC policies. In May 1919, Dow Chemical Company requested a license for procaine (the Americanized name of Novocaine) because the firm produced all the intermediate ingredients necessary for the drug and expected it could undersell the existing manufacturers, all of whom bought these ingredients from Dow. The existing manufacturers, Abbott Laboratories, Rector Chemical Company, and H. A. Metz Laboratories, felt threatened by Dow's competition and protested to the Chemical Foundation. All three firms, which operated under the original FTC licenses granted during the war, argued that they had first taken the risk to manufacture procaine at the Public Health Service's request and claimed they still operated under huge debts incurred in the course of this patriotic duty. The advisory committee sided with the existing manufacturers, but Dow justifiably accused the Chemical Foundation of violating its charter and playing favorites. Dow won that battle, and the Chemical Foundation henceforth observed its non-exclusive provision.[11]

By 1922, the Chemical Foundation had become the foremost proponent and defender of the U.S. synthetic organic chemicals industry. Its promotional and public relations functions resembled those of trade associations, which flourished in the 1920s with Secretary of Commerce Herbert Hoover's encouragement.[12] Unlike most trade associations, however, World War I profoundly colored the goals and activities of the Chemical Foundation, whose administrators actively engaged in an "economic war" that followed the military war. The battlefields of the economic war included the peace treaty negotiations, during which the APC's representatives worked to protect the interests of the domestic industry. And Garvan threw all of the resources of the Chemical Foundation behind winning extended and expanded tariff protection from Congress. By depositing German patents in the Chemical Foundation, the United States simultaneously deprived the German firms of the monopoly benefits conveyed by patents and gained a tool to augment other market barriers against German imports.

The Treaty of Versailles and Chemical Reparations

The end of the war left the transatlantic world in political and economic disarray, and advocates for the U.S. synthetic organic chemicals industry nervously followed the negotiations and decisions in the peace process. The war had changed the international political economy of synthetic organic chemicals, and American manufacturers desperately wanted to hold on to the gains they had made during the war. They feared the peace process could lead to the return of confiscated German property and very quickly to the resurgence of powerful German competition, which they did not yet feel ready to meet on equal terms. As the Allies sat down to work out the shape of the postwar world, advocates of the U.S. synthetic organic chemicals industry sought to prevent the resumption of the prewar international dyes and pharmaceuticals trade. They lobbied the peace commission and often worked closely with their British counterparts to install measures that would inhibit the German firms from reclaiming their prewar market domination. That included attempts to prevent the German firms from working again with their prewar American importers. The seriousness with which the peacemakers viewed the German chemicals industry was reflected in the Treaty of Versailles, which contained specific provisions devoted to German dyes and pharmaceuticals.[13]

When the Allies gathered in Paris to determine the conditions of peace, they wanted Germany to pay for the costs of war, not least because most of their countries suffered more damage from war than Germany had and were also deeply in debt, largely to the United States. Each nation's peace commission, however, arrived not only with a complement of staff but also with the conflicting goals and agendas of multiple interest groups. Once the commission undertook discussions on reparations, the diverse goals of economic interests became starkly apparent. As historian Bruce Kent has described, the political situations within Great Britain, France, Belgium, and the United States made the idea of reparations difficult to abandon; the leaders worried about the radicalism of domestic taxpayers who would have to bear the costs of the war if Germany did not. On the other hand, nearly every Allied country had raised protective tariffs during the war to encourage domestic industries, and the industrialists from those countries resented competition from cheap reparations products. Reparations payments relied on the ability of Germany to export successfully to accumulate the funds to pay the Allies, but protective barriers hindered German exports, thus creating a situation that promised to accomplish little toward settling war debts. Although some German industries such as coal and iron resisted the imposition of reparations, the German dyes manufacturers found reparations a way to keep their dyes on the world market in the heavily protectionist postwar years.[14]

Advocates for the U.S. synthetic organic chemicals industry arrived in Paris in force, hoping to influence the outcome of negotiations and protect their interests. Both Du Pont and National Aniline sent over representatives to report back information gleaned from observations and conversations with various officials. Garvan continued to support the industry from his post in the Office of Alien Property, sending over representatives to make it clear the United States should not return any confiscated property. The Department of Commerce was present, squabbling with the Department of State over jurisdiction related to foreign commerce. The show, however,

AND—GAS WILL BE A DECIDING FACTOR IN THE NEXT WAR

FIGURE 6.1. U.S. Chemical Warfare Association cartoon, "Disarming Germany." (Courtesy of the Manuscript, Archive, and Rare Book Library, Emory University, Atlanta, Georgia. Charles H. Herty Papers, box 98, folder 21.)

belonged to the Department of State, and supporters of the industry did not entirely trust the diplomats to look after their interests, nor did the diplomats trust the industry's representatives. In the context of intense politicking, the industry and the diplomats crossed paths on many issues, both in cooperation and conflict. Two issues particularly loomed large for the industry's representatives in Paris: the presence of technical experts to advise the diplomats and the administration of the reparations provisions of the treaty.

The leaders of the American industry found the U.S. officials in the peace mission cordially cooperative but far less accommodating than they desired. Although the American peace mission traveled to Europe with a full complement of staff, nearly 1,300 people at its peak, the Department of State shied away from enlisting technical advisors from the industry. The other Allies, however, appeared in Paris with top executives of their dyes or textile industries, and the American industry feared that the Department of State would fail to safeguard American interests against future competition from the Allies as well as the Germans. The U.S. diplomats, on the other hand, felt that designating a representative from the chemical industry would only prompt other American industries to demand a similar voice.[15]

The first struggle over the role of technical advisors came when the American manufacturers expected to join Allied visits to German chemical plants. The U.S. peace delegation, however, yielded to German howls of protest about the industrial espionage, and the diplomats forbade representatives of the American chemical companies to tour German plants to inspect production techniques. Instead, non-technical U.S. Army personnel accompanied British and French dyes experts on Inter-Allied Commission excursions to dyes and pharmaceutical plants in the occupied territories. One U.S. visitor, Major Theodore W. Sill of the Chemical Warfare Service, toured the German plants, noting their size, cleanliness, and the employers' tendency to retain unnecessary labor to inhibit "Bolshevikism." Sill's report consisted of general observations rather than detailed technical notes, commenting, for example, that in "equipment" and "layout" American plants could rival the German factories. He noted a greater reliance on labor-intensive practices than in American plants, which made more extensive use of mechanical conveyances. However, the Germans employed several methods he thought Americans might profitably adopt, such as types of pipe reinforcements and fittings, location of water towers, and a more effective filter press. Most importantly, he concluded that the cumulative experience of labor and chemists remained a significant advantage for the Germans.[16]

Contrasted with Sill's general observations were the detailed notes on factory processes taken by the British and French. Deprived of the opportunity to make their own notes, the American industry, aided by a few sympathetic American officials, sought to persuade the Allies to share their wealth of information. Not surprisingly, the British and French placed limits on their generosity and generally refused to help American competitors. In August 1919, however, the Department of Commerce acquired through diplomatic channels a copy of British and French notes on plants for explosives, powder works, and poison gas but not dyes plants. The department contacted the American Dyes Institute to invite interested manufacturers to review the notes, but it is unclear whether any accepted the invitation.[17]

The dispute over experts drew the Department of Commerce into the fray. On the day after the armistice, the president of National Aniline, William J. Matheson, wrote Secretary of Commerce William C. Redfield to request help in sending a "thoroughly reliable American . . . through the lines in the aniline dye district to see what the Germans are now doing." He suggested that two of his most experienced chemists, R. C. Taggesell and Bernhard C. Hesse, should travel immediately to Germany, either under the auspices of the company or the Department of Commerce. The department officials turned him down, partly because they said the French "promised" to share the results of their survey. In the meantime, Matheson dispatched a top executive, Henry Wigglesworth, to Paris to look after the firm's interests. For reasons that are not entirely clear, by February 1919 the Department of Commerce designated Wigglesworth as an official representative charged with investigating the chemical industries in France, Belgium, Switzerland, Italy, and Scandinavia. Upon his arrival, Wigglesworth presented himself to the American peace mission and demanded, as the official Department of Commerce representative, to see the German plants. He annoyed the American officials, who refused him permission to tour the plants. Wigglesworth's own arrogant personality caused many of his problems, but he also fell victim to a jurisdictional struggle between the Department of Commerce and the War Trade Board, which worked closely with Department of State and became a part of it in the summer of 1919. Years before the war, officials in the Departments of Commerce and State sparred frequently over which department should supervise foreign commerce. The feud continued through years of the Wilson administration as Secretary Redfield aggressively sought to expand his department's role in boosting foreign commerce.[18] Wigglesworth's ties to the Department of Commerce probably won him few confidants among the diplomatic corps.

Having lost the battle to inspect German plants, the representatives of the industry turned their attention to the provisions of the peace treaty, which also had them concerned about the lack of technical expertise in the American peace mission. Initially, the American manufacturers rejected all attempts to bring German dyes into the United States under the reparations provisions of the treaty. They resented the competition not just of German dyes but of apparently free German dyes. They acquiesced to reparations dyes for two reasons. First, they worried about extra reparations dyes in the hands of their competitors in Allied countries. On the issue of dyes, the textile industries held sway over the delegations from France, Belgium, and Italy, which were determined to obtain German dyes as soon as possible for their ailing industries. Great Britain and the United States, which intended to foster domestic dyes industries, held the minority position and generally lost out to the other Allies' demands. When France, Belgium, and Italy insisted on a supply of German dyes, the British and American negotiators and industrialists urged their governments also to claim reparations dyes to have some control over the amount on the world market. As Wigglesworth argued, any dyes that the United States failed to claim would compete against either American dyes or American textiles through European suppliers.[19]

Second, the American textile industry wanted to have the dyes. Through the war, the textile industrialists had become relatively staunch supporters of a domestic dyes industry and maintained that policy at the armistice. However, the textile industrialists increasingly agitated for a supply of vat dyes and other high quality specialty dyes the American manufacturers could not produce. Like the American manufacturers of dyes, they worried about competition from other textile producing countries that had access to inexpensive reparations specialty dyes.[20]

Through the spring of 1919, the peace treaty took a more definite form. By March, however, the negotiating parties recognized that the implementation of the final peace treaty lay months in the future and began discussions for short-term arrangements to alleviate the economic privations in Germany, as well as to satisfy Allied shortages. The Brussels Agreement of March 14 allowed Germans to export a wide variety of commodities to gain capital for purchases of food. The French, Belgian, and Italian textile industries aggressively sought to acquire supplies of German dyes, although negotiations for implementing the Brussels Agreement moved slowly.[21] A flurry of cable correspondence flew between the American mission in Paris and the Department of State in Washington while officials tried to determine whether and how the United States should take advantage of the agreement with respect to dyes. In the end, the three Allies acquired a small amount of dyes under the Brussels Agreement, but the United States chose not to acquire any dyes.

While the United States passed on the dyes under the Brussels Agreement, the opportunity prompted the Department of State to consider the mechanics involved in claiming and distributing reparations dyes, and it set off a stark power struggle for control of the process. Several federal agencies jockeyed for jurisdiction over the goods, including the Department of Commerce, the War Trade Board, and the APC, where Garvan used his post to argue that the Chemical Foundation should control reparations dyes. At first, officials in the Department of State turned to the Department of Commerce to survey the textile industry to find the demand for German dyes. The Department of Commerce quickly assembled the names of four textile manufacturers and four representatives from the domestic dyes industry to form an advisory committee.[22]

Meanwhile, Garvan exercised influence from his position at the Office of Alien Property. The agency maintained a representative in Paris to advise peace negotiators on issues of alien property, and the representative, among other things, made it clear the United States would return none of the chemical plants or patents. Perhaps Bernard Baruch, the War Industries

Board chair and an American delegate to the peace conference, received from the Paris representative of the APC the idea that the Chemical Foundation should supervise the importation of German dyes until the peace treaty came into effect. At any rate, near the end of March 1919, he sent a telegram to the Department of State suggesting Garvan be contacted. The telegram's message only found its way to Garvan two weeks later, probably held up by the War Trade Board officials in Washington who protested the idea that the Chemical Foundation should assume powers over importing and exporting, functions explicitly granted to the War Trade Board.[23]

Garvan was irate that Baruch's proposal had not reached him promptly, and he immediately took steps to accept Baruch's offer. He rejected cooperation planned earlier with the Department of Commerce, although he promised to keep the advisory committee. The Chemical Foundation designated Wigglesworth and Leonard Yerkes of Du Pont as official representatives in Paris. The only obstacle that remained was the War Trade Board's resistance. The officials at the War Trade Board, however, maintained their objections on jurisdictional grounds and prevented the Chemical Foundation from acquiring power over the pre-treaty phase of dyes imports. In addition, the Paris officials, including Bernard Baruch, had just received additional complaints from Wigglesworth and Yerkes about the treatment they, as representatives of the American industry, received from the American mission. The persistence of the two chemical industry men rubbed the government officials the wrong way, and one wrote back to Washington revealing "misgivings" toward "certain chemical interests" that clearly behaved out of "self interest."[24]

Although its leaders agreed to allow the Chemical Foundation to arrange the actual shipment and distribution of the dyes, the War Trade Board emerged as the final authority on the imports and exports of dyes by mid-May 1919. Within a few weeks, the Chemical Foundation was no longer considered for this more limited role. The War Trade Board appointed the advisory committee originally selected by the Department of Commerce to help the board determine which dyes the United States should allow into the American market. Customers would need an import license, and the War Trade Board would grant them only for dyes and pharmaceuticals without competing American-made products.[25] By asserting their claim to control the reparations dyes, the officials of the War Trade Board became a significant counterbalance to the powers of the Chemical Foundation. The War Trade Board possessed none of the close ties to a single industry or firm and, therefore, retained an alternate view of U.S. interests that included the American textile industry, as well as diplomatic concerns with the Allies and Germany.

In June, the Germans signed the Versailles Treaty, in which the dyes section comprised two main features. First, the Allies claimed 50 percent of all German dyes stocks extant at the armistice, an amount totaling roughly 20,000 tons. Second, for five years the Allies could claim up to 25 percent of Germany's new production of dyes. Although the United States never signed the Treaty of Versailles, it participated in treaty clauses related to reparations and placed unofficial representatives on the Reparations Commission, the organization the Allies established to supervise reparations.[26]

The treaty took effect in January 1920 after the majority of Allies ratified it. However, in the summer of 1919, the Germans and European Allies alike were anxious to resume economic production and began negotiations to exercise, prior to ratification, the treaty provision granting half of German dyes stocks to the Allies. Again, as under the temporary Brussels Agreement, the French, Belgian, and Italian representatives sought dyes for their domestic textile industries and pressed the Interim Reparations Committee to speed the process of claiming and distributing dyes. Against initial resistance from the British, who sought to protect their domestic dyes industry, Allied dyes experts began negotiating with the Germans to claim a share of the 10,200 tons of German stocks due to the Allies. The Department of State, warned American diplomats in Paris, should immediately send over its own dyes expert to participate in the negotiations.[27]

The selection and appointment of a dyes expert brought productive, if somewhat strained, cooperation between the American dyes industry and the Department of State, primarily the War Trade Board section, which joined the department in the summer of 1919. The diplomats asked Garvan to suggest a representative with sufficient technical and commercial knowledge to defend effectively the interests of the American industry in negotiations with the Allies and Germans. It accepted his recommendation that Charles H. Herty, still the editor of the *Journal of Industrial and Engineering Chemistry* and staunch advocate for maintaining a domestic industry, should travel to Paris. Funded by the Chemical Foundation but with the official sanction of the Department of State, Herty spent the fall of 1919 in Paris and London negotiating the release of German dyes to the United States.[28]

By the time Herty arrived, the European parties had already decided to claim and distribute 5,200 tons of the German stocks subject to the 50 percent provision of the peace treaty. Using consumption and production figures from 1913–1914, the Allies divided the allotment proportionally, with the United States and Great Britain receiving rights to the majority of the dyes. Herty's primary task centered on selecting the desired dyes for the U.S.

market. The War Trade Board cabled to Herty orders for dyes collected from the American textile industry, and Herty advised the U.S. officials which of the dyes in the allotment should be claimed for the home market. In this complicated task, Herty had the assistance of Eric Kunz, one of the Du Pont employees in Paris. Initially, the United States claimed just 20 percent of its 1,500-ton share of the Allies' 5,200 tons.[29]

Herty spent long hours correlating the orders to the reparations dyes allotted to the United States and discovered that the supply and demand matched up poorly. Consequently, Herty talked directly with the German manufacturers to obtain a supply of the crucial dyes above and beyond the U.S. reparations allotment. Named the "Herty Option," this secondary agreement allowed the Germans to sell as much of their half of the dyes stocks as they wished, at prices they determined.[30] Ultimately, the poundage and value of dyes sold under the Herty Option exceeded that of dyes fulfilling the reparations provision.

While the War Trade Board had asserted its authority to control the flow of reparations chemicals to the United States, the War Trade Board delegated the daily management of the dyes to the Textile Alliance. The Textile Alliance had been organized early in 1914 to combat fraud in the textile industry but, with the advent of war, became involved in wartime commerce with Britain. Led by textile manufacturer Albert M. Patterson, who spent the spring of 1919 in Paris as an economic advisor to the American mission, the Textile Alliance managed the shipments of reparations and Herty Option dyes from Germany for distribution in the United States. Because the United States held only unofficial status on the Reparations Commission, the result of the U.S. Congress's refusal to sign the Treaty of Versailles, the Textile Alliance paid the commission for the U.S. allotment of dyes, unlike the other Allies, whose accounts against Germany were credited. Consumers paid in advance for orders placed with the Textile Alliance, which sold the dyes at the cost paid to the Reparations Commission plus the expenses of shipping, overhead, and patent royalties to the Chemical Foundation, when applicable. Sometimes the Textile Alliance purchased dyes in the U.S. allotment that had not been specifically ordered by a consumer. In these cases, underwriters bore the cost of the advance purchases, and the Alliance sought to sell the dyes afterward. The Textile Alliance maintained a technical expert as a liaison to the Reparations Commission, to prevent against fraud, and to supervise dyes shipments. In Antwerp, the Alliance rented a warehouse to store dyes before shipping importable dyes. Dyes headed for the United States required an import license from the War Trade Board, which granted licenses only for dyes not manufactured in the United

States, dubbed "importables." "Nonimportable" dyes stayed in the warehouse until the Textile Alliance could exchange them for importable dyes or, in the case of indigo, sell them to China. As the variety of dyes manufactured in the United States increased, the number of importable dyes types decreased, which led to a considerable overstock in the Antwerp warehouse. In the United States, the Textile Alliance maintained a warehouse in Hoboken, New Jersey, from which dyes were shipped to the ultimate consumer. (Secret service agents recovered dyes stolen from the warehouse in July 1920, reportedly with the help of a local cat with an unnaturally orange tail, a sign the cat had found the dyes.)[31]

The advocates of the U.S. synthetic organic chemicals industry approved of the Textile Alliance as the central importing and distribution agency. This kept the dyes out of the hands of the traditional importers for the German firms. Indeed, the Herty Option was designed partly to prevent importers from handling German dyes as Herty endeavored to route all German dyes—whether reparations or not—through the official importing agency. Some War Trade Board officials noted that Herty exceeded his authority in negotiating directly with the Germans for purchases of dyes, but the agreement remained in force. The domestic manufacturers tried to discourage purchase of any reparations dyes, even from the Textile Alliance, by seeking to publish the names of customers, but the War Trade Board forbade the release of such information. The domestic manufacturers and their supporters were therefore upset when the War Trade Board officials decided they lacked the authority to grant exclusive importing rights for all dyes imports to one organization. Consequently, in late September 1919, the government officials announced that any importer could receive an importing license from the War Trade Board.[32] Supporters of the American dyes industry found this unwelcome news and expressed displeasure that the government would allow the old importers for the German companies to resume prewar ties, thereby threatening the nascent domestic industry.[33]

By 1920, the system for delivering reparations dyes revealed a fundamental problem from the Allies' perspective. The Allies grew increasingly frustrated at the inefficiency of the system that failed to correlate German supply with Allied demand. In 1920, the Germans satisfied the first provision of the treaty related to 50 percent of dyes stocks and began fulfilling the second provision, which required them to turn over 25 percent of new production for five years. Under both provisions, the Allies tried strategies to improve the relationship between supply and demand. Under the second and final distribution of dyes from the 50 percent of German dyes stocks, the Allies divided the allotment up not by total tonnage but by paying more

Table 6.1. Reparations Dyes Received through the Textile Alliance

Purchases from Reparations Commission	4,501,807 lbs.
Dyes returned to the factories (-)	2,236,084 lbs.
Dyes received in exchange (+)	487,806 lbs.
Dyes received and sold (total)	2,733,013 lbs.

Note: This table is for the Textile Alliance's Account No. 2, the main account that handled the reparations dyes under the Department of State. Account No. 1 covered the 300 tons of dyes taken earlier, and Account No. 3 covered dyes purchased by the Textile Alliance after it lost the official sanction from the Department of State. The high percentage of dyes returned to Germany for exchange highlights the imbalance between supply and demand. The lower number of dyes received in exchange reflects the American demand for more expensive, higher quality dyes.
Source: William C. Dickson, Special Accountant, Department of Justice, "Report of Investigation of the Textile Alliance Incorporated of New York City in Re. Reparation Dyes," p. 47, National Archives and Records Administration, RG 60, entry 233, box 1.

careful attention to individual dyes types. For instance, the United States received only 5 percent of the indigo in the allotment, but 55 percent of an indanthrene blue dye. The Allies applied the same division of spoils to the dyes collected under the second provision of the treaty covering the new production. A second strategy for rectifying supply and demand included voluntarily swapping nonimportable dyes for importable dyes with the Allies and Germans.[34]

The imbalance between supply and demand persisted, however, and concerned the American dyes industry. Near the end of 1920, the American Dyes Institute sent Joseph H. Choate Jr. to Paris to press the American reparations representatives into demanding that the Germans alter production to suit the desires of American dyestuff producers and manufacture more specialized dyes. Choate, a lawyer, had been Garvan's right-hand man in the APC's investigations of German chemical interests in the United States and also played a significant role in forming the Chemical Foundation. Choate met with representatives of the British Dyestuffs Corporation to encourage them to prevail upon the top British member of the Reparations Commission—who had more influence than the unofficial American representative—to force the Germans to produce the desired dyes. Although his trip provided American manufacturers with up-to-date information about conditions in Paris, Choate felt the U.S. diplomatic officials were uncooperative and left Europe uncertain of any reform in reparations dyes policy. By the middle of 1921, however, a change in reparations procedures appeared desirable to all Allies. The European Allies discovered that they had been dumping unwanted reparations dyes into each other's national markets. As a result, the Allies forced the Germans to change the procedure and accept orders in advance for the dyes the Allies wanted. The Germans accepted the new arrangement out of necessity; the manufacturers needed the orders to stay in business, and the German government needed the revenue to pay the reparations debt. In addition, the Allies pledged not to resell reparations dyes in the markets of the other Allies.[35]

Even as the Textile Alliance more efficiently delivered the reparations dyes, opposition to the entire arrangement grew in the United States. Members of Congress questioned the legality of the relationship and worried that the arrangement with the Textile Alliance encouraged a monopoly in the dyes importing business, even after the War Trade Board granted licenses to other importers. In December 1921, representatives of the Department of State severed their relations with the Textile Alliance. Loss of official sanction from the U.S. government caused problems for the Textile Alliance in Paris; the Reparations Commission hesitated to grant the lower reparations prices to an agency unaffiliated with a government. The Textile Alliance's lost prestige coincided with increased resistance from the German government toward reparations payments and refusal by German manufacturers to fill some orders from the Alliance. As a temporary solution, the Commission allowed the Textile Alliance to continue receiving the lower reparation prices until June 30, 1922, when all sales to non-government entities had to be at market prices. This marked the end for the Textile Alliance. When it had to pay market prices, its managers developed plans to discontinue operations and dispose of existing dyes stocks. In November 1922, the Textile Alliance—and the United States—relinquished its share of reparations dyes, which the other Allies subsequently split four ways.[36]

The Textile Alliance's control of reparations dyes never translated into a complete monopoly on dyes imports. Beginning in mid-1920, as the German companies increased output, the old importers handled more of the German dyes trade. Although the Textile Alliance hired experts to help in the application of dyes, customers recognized the importers as more efficient and knowledgeable, and the importers gradually cut into the market share of the Textile Alliance. In 1921, the War Trade Board issued import licenses for 4.7 million pounds of dyes from all sources. Of this amount, the licenses to the Textile Alliance accounted for 730,000 pounds from Germany and 30,000 pounds reexported from France. As of February 1922, the Textile Alliance had purchased 3.7 million pounds of dyes and sold 3.2 million pounds for total sales of $7.2 million.[37]

The U.S. collection of reparations dyes left a legacy, however. Under the original agreement with the government, the profits earned by the Textile Alliance would be split between the Textile Alliance and the U.S. Treasury, with the understanding that the money would support research.

Seventy-five percent of the profits went to a special account in the U.S. Treasury, and the Alliance retained 25 percent to disperse in industry-related research as its officials chose. The Textile Alliance accumulated just under $2.1 million in net profits, of which $1.4 million went to the U.S. Treasury. The money sat undistributed until legally cleared for distribution in 1930, and the Textile Alliance donated its share to the Philadelphia School of Textiles ($248,507.90), Princeton University's endowment for pure science ($124,253.95), and MIT ($28,057.34). With the government money, President Herbert Hoover directed the establishment of the Textile Foundation to support scientific research and education.[38]

The treaty provisions for synthetic pharmaceuticals were the same as for synthetic dyes, but the domestic economic situation differed. That is, the United States and Great Britain both felt that their domestic production of synthetic organic pharmaceuticals more than satisfactorily met domestic demand. In the spring of 1920, the United States announced that it would not participate in either the distribution of existing German stocks or in future German production, although the War Trade Board diplomats reserved the right to change policy if some shortage appeared in the American market. The policy did not change, and the United States accepted no reparations pharmaceuticals.[39]

The United States rejected the Treaty of Versailles and, not having borne the brunt of war costs, had little need for reparations. That the United States therefore accepted synthetic dyes as a form of reparations from Germany signifies the degree to which the American government had committed to support the domestic synthetic organic chemicals industry.[40] In the politicking of peace, the efforts of the domestic industry and its supporters showed clearly their desire to establish the global commerce in synthetic organic chemicals along an entirely different trajectory than the prewar patterns. The Americans determined not to return confiscated German chemical property, participated in reparations dyes shipments, and at least temporarily prevented the old importers from resuming business.

Tariff and Monopoly

While representatives of the U.S. synthetic organic chemicals industry lobbied the Department of State to wage "economic war" at the Paris peace conference, they turned their attention to Congress for increased tariff protection. The rhetoric of the lobbyists remained intensely anti-German, and they pressed vigorously the relationship of the industry to national defense. Determined to revise the tariff of 1916, the industry supporters brought a coordinated line-up

WHY THE DYE INDUSTRY NEEDS PROTECTION
It will Surely Be "Torpedoed" Unless This German Submarine Can Be Intercepted

FIGURE 6.2. "Why the Dye Industry Needs Protection." (*Drug & Chemical Markets* 8 [January 5, 1921]: 8. Courtesy Othmer Library of Chemical History, Chemical Heritage Foundation, Philadelphia.)

of witnesses to explain the need for extended and expanded protection for an industry that seemingly had grown so large during the war. The lobbyists particularly included the Chemical Foundation and American Dyes Institute, which generally spoke in unison. In a wonderful irony, the industry advocates looked internationally—to Great Britain—to obtain additional strategies to build the protectionist and isolationist walls higher. As the industry pushed Congress in hearing after hearing, however, many politicians began to question the activities of the industry, the APC, and the Chemical Foundation, which led to separate Senate hearings in 1922 to consider the possibility that the United States had fostered a monopolistic domestic industry. The hearings, full of intense political jockeying and accusations, revealed both real and perceived threats the industry felt with the arrival of peace.

The tariff of 1916, with its specific duties scheduled to decrease in 1921, seemed inadequate to industry backers, especially given the uncertain rates of exchange following the war. Already in spring 1919, Nicholas Longworth, a Republican congressman from Ohio (and future Speaker of the House), introduced a bill into the House of Representatives that recommended tariff rates considerably steeper than in the tariff of 1916. His amended bill led to hearings in both the House and Senate in 1919 and early 1920 but failed to become law. As the tariff of 1916 expired in 1921, the Senate held another hearing, and Congress passed a temporary emergency extension to the existing tariff. In 1922, after the Senate investigated charges of monopoly with yet a fourth hearing, Congress legislated more definitively on the tariff question.

The list of witnesses in the various hearings included many of the usual suspects: Francis Garvan, Charles Herty, Herman Metz, but also Irénée du Pont, the president of Du Pont, military officers, textile industrialists, and an assortment of many others. After serving as Garvan's assistant in the Office of Alien Property, lawyer Joseph H. Choate Jr. became the chief lobbyist for the American Dyes Institute in April 1919. His task was to get the Longworth bill passed. Choate orchestrated a highly organized campaign, coordinating testimony from dyes and pharmaceuticals manufacturers of varying sizes and also from supportive textile manufacturers. In the Longworth bill, the industry sought rates of 50 percent ad valorem and 10¢ per pound on dyes, pharmaceuticals, and other finished products; 35 percent ad valorem and 6¢ per pound on intermediates; and they removed indigo and other dyes from the free list, giving them the same rates as other dyes.[41] Choate embraced the high rates, but also pushed for an additional form of protection tried by the British.

By 1914, Great Britain had a decades-long attachment to free trade, and although their economic empire began to crumble in World War I, many British refused to join the rush to protectionism in the industrial world after the war. Like the United States, however, the British wanted to protect their dyes industry from German competition. Indeed, the British government had directly invested in the British Dyestuffs Corporation. To protect the industry without using the politically unpopular tariffs, the governmental Board of Trade supervised an embargo and licensing system. No dyes could arrive in Britain without a license, which meant that the board could exclude dyes for which there was a domestic manufacturer. Adverse court rulings suspended the system for most of 1920, which allowed the British textile industry to stockpile foreign dyes, but early in 1921 Britain passed a law to institute a licensing system to last for ten years. The government subsequently extended the system through World War II.[42]

Choate studied the British system and convinced Representative Longworth to amend his own bill to include a similar embargo and licensing provision. Unlike the British system, however, the Americans would combine a tariff with the licensing for twice the protection. Americans also believed they had their own working model in the War Trade Board, the war agency that supervised imports and exports and continued to grant the import licenses until the official peace. Supporters of the licensing system argued that the Germans could undersell any tariff, no matter how high the rate, particularly as the value of the German currency headed downward. They appreciated the discrimination a licensing board could exercise to permit entry only to those dyes that the U.S. industry could not make. The textile industry could still receive the high-end dyes only the Germans or Swiss made, but the American dyes industry would not be challenged on American-made dyes. Opponents of the licensing system thought it an unwarranted intrusion of government into an economy at peace and wondered about the fairness of any licensing body. Others accused the Democrats of embracing the license system to avoid the higher tariffs their constituency disliked while still showing support for the industry. Even more than the proposed high rates, the embargo and licensing system became the center of debate in the hearings.

Choate and his line-up of witnesses pursued several arguments in their quest to win votes for a higher tariff. Fresh from the war, the witnesses hammered on the relationship between the domestic synthetic organic chemicals industry and national defense. Before the Ways and Means Committee, a Chemical Foundation lawyer argued that "the coal tar chemical industry is so inseparably interwoven with the industrial independence of a nation, and its program for national defense, that it must at all hazards exist in a country which is to be in reality one of the great independent nations of the world." Choate brought in military officers from the navy and army, particularly the Chemical Warfare Service, to testify to the indispensable role of the industry. Showing a conspicuous coordination in their testimony, the military men reiterated three central points before all three congressional hearings on the tariff. First, they and other witnesses argued, the chemical relationships and overlap between war chemicals and dyes and pharmaceuticals was real, and the development of a domestic industry meant the United States could maintain the capacity to produce war gases and high explosives. In their specific comments before congressional hearings, the officers cited several examples to support their arguments. An army ordnance officer discussed the ability of the Calco Chemical Company to bring its dyes expertise to bear on the development of TNA; another cited the use of chlorine in synthetic organic chemicals for both peace and war; and a third noted the sale of surplus phosgene to the dyes industry. If the United States had had a dyes industry before the war, testified a rear admiral, the country would have had a ready supply of toluene and "we would not have been forced to build plants during the war at tremendous expense to the Government and at great loss of time." Given the great consumption of toluene for explosives compared to even a large dyes industry, he exaggerated; but the shared intermediates made a convincing argument. "The dyestuff industry," argued William Sibert, the head of the Chemical Warfare Service in 1918, "is the one peace-time enterprise which will ... furnish us with the plants and equipment which can be hurriedly converted to essential uses in time of war." His successor, Amos Fries, quoted a British report about a Bayer dyes plant the Germans converted to TNT production in six weeks.[43]

Second, the officers focused on the value of a trained pool of chemists and other technical personnel as a source of expertise the military could deploy in a crisis. Fries believed this the most significant contribution a thriving domestic industry would make to national defense. "You can not train a chemist in six weeks, six months, or a year. We had to have men who were experts," and the American chemical warfare program never had enough chemically trained personnel for the entire duration of the war. "We could not get enough trained men over there to do what we should have done, although we combed all the ground under general orders to get every man who even claimed he was a chemist." A supply of trained chemists was more important than the supply of raw materials and available factories to national defense, Fries argued.[44]

Finally, the officers expected that future warfare would inevitably involve chemical gases. Indeed, argued General Sibert, "it was the unexpected use of gas on a large scale that caused the Germans to nearly win the war" in the spring of 1918. With gas warfare a continuing threat, the United States would need to stay abreast with the latest developments in chemistry as a matter of national defense. Through their research laboratories, a domestic dyes and pharmaceuticals industry would play a central role in the nation's ability to meet future threats.[45] These three arguments appeared time and again as military representatives endorsed protectionist measures for the industry as necessary for national defense.

A second major theme of the proponents of the Longworth bill focused on the danger that the German chemical industry still posed to the U.S. industry. In some cases, their argument overlapped with the one for national defense. For instance, Herty called on Longworth and the other members of Congress to seek destruction of the German dyes plants because the plants provided Germany with potential resources for the next war. In more extreme statements, Garvan told the House committee that "the war never would have started except for the German dye factories," which had given Germany a sense of superiority. E. C. Klipstein, a manufacturer and importer of Swiss dyes, suggested that the German government began fostering its dyes industry in 1870 as part of long-term war planning.[46]

More frequently, proponents of the industry fretted about the economic war they perceived at hand. The American industry needed the exceptional protection granted by the licensing provision because the German dyes industry posed such an enormous commercial threat, they argued. Allied observers of the German plants in the occupied zone noted that the factories suffered little war damage and sat prepared to resume commercial production. Others believed that Germany would attempt to regain lost markets, especially the American one because other European countries had already instituted protectionist measures. Furthermore, they expected the Germans to spare no effort, legal or illegal, to regain their prewar dominance. "The records show the ruthlessness of the German trust, and they purpose to attack us in every way known to man, legal and illegal." The German firms would be helped by the two remaining American importers, Metz and Kuttroff, Pickhardt & Company, argued Garvan, who submitted to the tariff committees the APC's reports on each. At one point before the House committee, Garvan asserted that Kuttroff, Pickhardt & Company "stand for everything that is evil in the whole Hun system."[47]

The manufacturers testified that tariffs alone would be inadequate to protect their industry for the next several years. They knew that the German manufacturers still surpassed the Americans in chemical knowledge, research, and experience, and they had to make the case that Americans, while not yet equal to the Germans, could catch up in a reasonable timeframe. The German superiority was especially evident in the American producers' inability to produce vat dyes other than synthetic indigo, and a Dow executive testified that his firm produced indigo at a cost of 65¢ per pound. Before the war, the Germans sold it for 15–18¢. The manufacturers warned that much remained for them to learn, even about the dyes that Americans made in bulk. Several manufacturers, including Irénée du Pont, noted that their companies had not yet reached yields similar to the German makers—that American dyes manufacturers still consumed more raw materials per pound of finished dye than did the Germans. As he put it, no tariff "will protect a baker who must spoil 19 out of 20 of his biscuits," and du Pont requested ten years of embargo protection to overcome the higher costs of production that resulted from American manufacturing inexperience relative to the German.[48]

Opposition to the tariff bills, either parts of them or in their entirety, came from several quarters, including textile manufacturers, importers, and politicians concerned about political philosophy and the tactics of the industry proponents. The textile industrialists, as the largest consumers of synthetic organic chemicals, potentially wielded the economic clout to make or break U.S. dyes manufacturers. In the aftermath of the war, however, the textile manufacturers disagreed among themselves about the tariff legislation. Nearly all textile manufacturers favored the development of a domestic dyes industry, but they differed on the level of sacrifice they should make to support the dyes manufacturers. Textile industrialists who had worked closely with dyes manufacturers on war boards generally sympathized with them. For instance, Frank D. Cheney, a silk manufacturer and a member of the War Trade Board Advisory Committee, solidly backed the legislative agenda of the dyes industry. Rufus Wilson, the secretary of the National Association of Cotton Manufacturers, articulated the official viewpoint of many in the textile industry when he stated that his industry had learned a lesson in the war about relying on a foreign supplier for an essential ingredient in their manufacturing process. He argued that "enlightened self-interest" motivated textile manufacturers to support the Longworth bill.[49]

On the other hand, several textile manufacturers opposed a licensing provision, which resembled the system conducted under the War Trade Board. Although some textile producers viewed the War Trade Board licenses as a minor, tolerable, and worthy inconvenience, others resented the system as an unnecessary encroachment on their business. First, the system created delays in receiving dyes. If approved immediately, as were 90 percent of the cases, the consumer's application for an import license required up to a week to process. If the War Trade Board administrators believed an American company manufactured the desired dye, the consumer needed to contact the company and receive a negative reply before the War Trade Board would issue a license. Occasionally, the consumer found the American dye a technically inadequate substitute for the requested foreign dye and the burden of proof lay on the consumer to demonstrate the inferior quality, a time-consuming and frustrating process. After the licensing, the consumer then waited for the dyes to cross the Atlantic, making the entire procedure compare very unfavorably with the prewar years when the importers maintained a stock of dyes and could deliver within two or three days. The textile industrialists argued that such delays were detrimental to their business, which required swift changes to keep up with fashion seasons.[50]

Some members of the textile industry also opposed the licensing system because they feared it betrayed trade secrets. The information given to the licensing agency, whether the War Trade Board or some other body

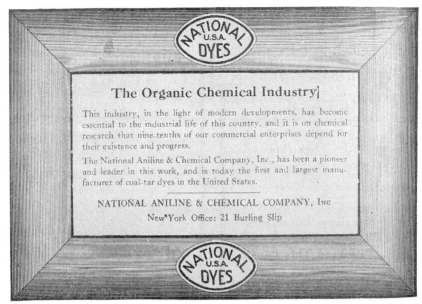

FIGURE 6.3. National Aniline & Chemical Company advertisement, 1922. (*Drug & Chemical Markets* 10 [March 15, 1922]: 664. Courtesy Othmer Library of Chemical History, Chemical Heritage Foundation, Philadelphia.)

proposed in legislation, could reveal a textile manufacturer's production strategy for the upcoming season. The manufacturers feared that such information could be easily discovered by competitors, especially when the competitors sat on an advisory board like that of the War Trade Board. No evidence suggests that the advisory board became a conduit for leaks, but, as a matter of policy, the officials handling the licenses informed domestic manufacturers of consumers applying for a dye made by those companies. Some opponents to the bill insinuated that textile manufacturers who sat on the War Trade Board advisory committee received special attention from the license administrators.[51]

Like some textile industrialists, many importers and dealers of dyes opposed the tariff bills because it ran counter to their own economic interests. Of the importing agencies for the six major German dyes companies, only Kuttroff, Pickhardt & Company and Herman Metz remained in business, and Metz had become primarily a manufacturer rather than an importer immediately after the war. The importers of dyes from countries other than Germany presented the importers' case, although Metz and Paul Pickhardt were also prominent witnesses. Metz, in fact, earned the committees' respect as one of the most knowledgeable witnesses in technical and sales matters related to dyes and pharmaceuticals. Importers such as Walter F. Sykes, who imported French dyes and represented the U.S. Dyestuff and Chemical Importers' Association, found the Longworth bill unacceptable for its "excessively high" tariff rates and its "radical" licensing plan. The bill would severely curtail the extent of prewar business the importers might recover. Not only did it forbid importation of any foreign-made dyes in direct competition with the kinds produced in the United States, but it also put the importer in a financially risky position if he tried to order early a supply of an importable dye to anticipate orders, because he would be unable to sell them in the United States should a domestic manufacturer subsequently begin production of the dye prior to a sale. In an attitude less vehement than that of Sykes, P. R. MacKinney, an importer of Swiss dyes, suggested that the licensing system would be tolerable if importers had representation, but he also remained unconvinced that the high tariff rates in the bill were justified.[52]

Political ideology and partisan jousting also shaped the hearings. Republicans, the "party of protection," gained a majority of seats in the House in the election of 1918, which placed Joseph W. Fordney at the head of the Ways and Means Committee. Fordney, a Michigan acquaintance of Herbert Dow's, had been a strong ally of the industry's in 1916 and continued to advocate protection for the industry, but he believed the licensing proposal an unnecessary and unwieldy intrusion of government into business. Protectionist Republicans knew that high tariff barriers encouraged retaliation in other countries, which hurt U.S. exports, but they believed that the American market should be the nation's first concern. Fordney argued that the United States "consumed ... 93 per cent of all the agricultural and manufactured products of this country" and likened export trade to "chasing rainbows." On the other side, George Demming, a lawyer who helped organize the textile opposition during the Senate hearings, urged Congress to think of the economy in global terms. The United States, he argued, could not afford "to view the affairs and happenings of the world from afar" but "must realize that the welfare of the world is bound together, and that which adversely affects one nation, affects all." The nation had become a creditor in the world economy, and "national policy" should change to take into consideration its new status.[53]

The initial tariff bill failed. Even stripped of the licensing provision, the bill fell victim to filibuster, and industry advocates nervously awaited a peace treaty, which would bring resumed trade with Germany. In 1921, Congress passed emergency legislation that temporarily extended the tariff of 1916 beyond its expiration and that transferred the remaining vestiges of the War

Trade Board from the Department of State to the Department of Treasury. Frederick S. Dickson, an official trained as a banker, continued as the chief administrator of the licensing system begun under the War Trade Board. Congress retained the licensing system with a series of temporary extensions.[54]

In the meantime, the industry advocates lost no time in rallying support for another attempt at legislation. Garvan and the Chemical Foundation increased the amount of money spent on their "educational" campaign to $227,600 in 1921. Often in coordination with the Chemical Foundation and the American Chemical Society, the American Dyes Institute arranged for speakers to appear before a wide variety of bodies to advocate the necessity of embargo legislation. The ADI naturally concentrated efforts on the textile industry but also made a special endeavor to persuade women's groups to support favorable legislation. William Corwine, the secretary of the ADI, believed that women remained the most skeptical that the quality of American dyes could match those made in Germany. Working with Herty, Corwine organized speakers before women's groups, including the American Home Economics Association. The American Chemical Society also targeted women's groups such as the League of Women Voters and the National Federation of Women's Clubs.[55]

Congress introduced new legislation in 1921, and this became law as the Fordney-McCumber Tariff Act in September 1922. The Republican Congress had clearly rejected the embargo and licensing plan as an unnecessary measure that could not be administered fairly, despite a last minute, temporary appearance in a conference committee report. Instead, the Republican legislation imposed exceptionally high rates and, intriguingly, introduced a feature called "American valuation." Customs officials would apply the tariff rates not on the imported good's value at the port of export but on its selling price in the United States. Particularly as the German currency depreciated, and as American costs of production stayed higher than the German, the American valuation provision would become a good friend to the domestic manufacturers, although they did not immediately recognize its value to them.

The act placed the steepest tariffs ever on dyes and intermediates derived from coal tar. Dyes received a 60 percent ad valorem and 7¢ per pound specific duty, and the rate on intermediates was set at 55 percent ad valorem with a 7¢ per pound specific duty. After two years, the ad valorem rates would drop 15 percent to 45 percent and 40 percent, respectively. The rates, on average, exceeded those originally proposed by Longworth in 1919. The act also immediately terminated the remnants of the licensing administration that had continued to function after the War Trade Board's demise.[56]

In the course of the hearings on the tariff, members of Congress increasingly questioned the activities of the APC and the Chemical Foundation. Some began to suspect a "dye monopoly" centered on the Chemical Foundation. To some extent, the criticism was partisan—the Republicans, in control of the executive and legislative branches after the election of 1920, sought to expose wartime corruption under the previous Democratic administration. But the combatants did not easily divide along party lines: Democrats included Garvan, Herty, and Metz; Republicans included Choate, Senator George H. Moses, a strong critic of the dyes industry, and the du Ponts. In 1921, Senator Robert M. LaFollette and a fellow Wisconsinite, Representative James Frear, denounced the circumstances surrounding the formation of the Chemical Foundation. LaFollette, the independent and progressive Republican, viewed the dyes industry as another example of corrupt relations between big business and government. As an opponent of American participation in the war and a defender of civil liberties during the war, LaFollette found additional reason to criticize the methods of the Office of Alien Property. Frear, another Republican, registered the bitterest condemnation of the "dyes trust" in speeches on the floor of the House during the summer of 1921, by denouncing the sale of the patents to the Chemical Foundation as a "fraud" engineered by Du Pont in an attempt to devise a legal cover for monopoly practices. He referred to a "Big Six" of the largest American companies and especially targeted the "octopus of Wilmington," noting that the capital investments of the two largest American chemical corporations were three times the size of the German firms put together. Frear planned to introduce a resolution to have the attorney general investigate the sale of the Chemical Foundation patents. "No more remarkable evidence of private greed and official miscarriage will be found in all the pages of our history," he declared, matching the rhetorical excesses of the other side.[57]

Mounting criticism of the dyes industry led to an investigation by members of the Senate Committee on the Judiciary in the spring of 1922. Led by Republican Samuel M. Shortridge of California, the subcommittee listened to testimony for four months to determine whether a monopoly existed in the dyes industry. The Shortridge hearings resulted from a Senate resolution sponsored by Senator William H. King of Utah, a western Democrat, who echoed charges made by Senator Moses of New Hampshire on the floor of the Senate.[58]

King submitted roughly three interrelated charges for the subcommittee to investigate. First, he maintained that the lobbying by the industry supporters had crossed the line into the realm of unacceptable and corrupt behavior. In 1920 alone, the American Dyes Institute spent over $70,000 on its "legislative campaign," a figure King thought enormous. The extent of the lobbying efforts by industry advocates frightened him, and he took a dim view of the expensive "nation-wide propaganda" that overwhelmed the public with one-sided polemics. Second, King charged that a monopoly already existed among American dyes manufacturers. He submitted a chart, borrowed from a speech by Moses in the Senate and a parody on the ubiquitous "dye tree" chart circulated by industry supporters, that showed Du Pont in the center with ties to the Chemical Foundation, the APC, the American Dyes Institute, the American Chemical Society, the Textile Alliance, and others. He pointed to Irénée du Pont's testimony during the Senate hearings in December 1919 in which du Pont admitted that situations existed in American business when a big company "might soak" a "little fellow sticking his head up," although du Pont denied that any company in the dyes industry faced such a situation. Prices of the most important dyes remained well above prewar levels, which King charged to fixing among members of the American Dyes Institute. Why, King asked, should Congress grant an embargo—unprecedented protection—when the industry currently supplied 90 percent of domestic consumption?[59]

Finally, King wanted the Senate committee to investigate improper and corrupt conduct on the part of government officials in the employ of the Office of Alien Property. The Senate, however, removed that section of his original resolution, placing such an investigation beyond the scope of the Shortridge hearings. Still, in his opening statement, King censured the activities of Garvan and others in the Office of Alien Property. The confiscation of property, he argued, constituted a "breach of international morality" that appeared especially reprehensible for a nation trying to "wear the crown of moral primacy in the world." He reminded his audience, "We do not make war upon individuals; we make war upon the Nation. The war is over now, and if we can begin to think in terms of world peace and world amity it will be better for the United States as well as for the world."[60]

Senator Joseph F. Frelinghuysen, a Republican from New Jersey and a strong ally of the dyes industry, added an amendment to King's resolution that included an investigation of the importers of German firms. Frelinghuysen argued that the issue before the country was "a question of national policy . . . for the maintenance of the United States and our people rather than industrial greed for any man or set of men. If we allow German domination we are unfaithful to those who served in France, we are unfaithful to our trust for the American people." An American monopoly, he said, was preferable to a German monopoly.[61]

With the task outlined before them, Senator Shortridge and his colleagues listened to thirty-eight witnesses discuss the development and current state of the U.S. dyes industry. All the manufacturers, Herman Metz included, vouched for the cutthroat competition among the American dyes manufacturers and denied the existence of a monopoly. Despite the queries about the nature of the American Dyes Institute's open pricing system, little evidence appeared to suggest any price fixing or division of markets in the industry. As in past hearings, Garvan recited all the perceived subversive activities of the importers for the German firms, but the Shortridge hearings marked a shift for Garvan. Through the war and earlier hearings, he had been on the attack, warning of the German threat. In the Shortridge hearings, Garvan felt obligated to explain in detail the activities of his office and to deny any illegality in past actions, although Senator Shortridge generally proved a friendly audience for him. Garvan was well aware that the Department of Justice, at Congress's request, was beginning an investigation of the sale of patents to the Chemical Foundation.[62] The hearings ended in June with little fanfare. Shortridge never submitted a final report, although he announced in the Senate that he had uncovered no dyes monopoly.

By a carefully orchestrated, well-funded, and persistent lobbying effort, which included mobilizing public opinion, industry advocates acquired one of the most important policies for the immediate future of the synthetic organic chemicals industry. The Fordney-McCumber Tariff Act of 1922 returned the nation to an upward climb in tariff rates and was intentionally prohibitive for most synthetic dyes. The new American valuation provision extended the level of protection, although it would require new working procedures in the customs houses, which would determine its ultimate stringency. After all of the more unusual measures during and after the war designed to promote the domestic synthetic organic chemicals industry, the industry received its most significant aid from the most traditional of policy tools, the tariff.

###

By the end of 1922, the domestic synthetic organic chemicals industry had received exceptional promotional measures to erect steep barriers to guard against the return of the German chemical firms to the American market. The Chemical Foundation spent royalties on the former German patents to lobby for additional protection, industry advocates ensured the peace negotiations would not undo promotional measures, and Congress bestowed generous tariff protection on the industry. In each case, advocates and opponents for the measures engaged in powerful political struggles, in which

Table 6.2. U.S. Tariff Rates, 1913–1922

	DYES		INTERMEDIATES	
	ad valorem	specific	ad valorem	specific
1913 act	30%	—	10%	—
1916 act	30%	5¢	15%	2½¢
Longworth bill	50%	10¢	35%	6¢
1922 act	60%	7¢	55%	7¢

Note: Per the 1916 act, if, after five years, the U.S. dyes industry supplied 60 percent of U.S. needs, the industry would be rewarded with another five years of the 5¢ specific rate, although it would decline by 20 percent or 1¢ each year, becoming zero at the end of the second five-year period. If the U.S. industry did not supply 60 percent of U.S. needs after the first five-year period, then the specific duty was to end immediately. Per the 1922 act, Congress set the ad valorem rate on dyes and intermediates to go down 15 percent after two years; the rates became 45 percent and 40 percent, respectively, on September 22, 1924.

Sources: "The Dyestuff Tariff before Congress," *Metallurgical and Chemical Engineering* 15 (July 15, 1916): 65–66; "Status of New Tariff Law on Dyestuffs," *Metallurgical and Chemical Engineering* 15 (September 15, 1916): 283–84; "Chronology," submitted in Poucher testimony in Senate Committee on the Judiciary, *Hearings, Alleged Dye Monopoly* (1922), 507–12; "Tariff Rates of New Law Compared with Old," *Oil, Paint and Drug Reporter* 102 (September 25, 1922): 19–20; U.S. Tariff Commission, *Census of Dyes and Other Synthetic Organic Chemicals*, 1925, p. 42.

the advocates claimed the moral high ground, linking their cause to national security and playing on fears of German commercial and military power. More economically divided in their interests and weaker politically, the opponents of the promotional measures modified but could not stop the momentum that the industry's advocates possessed coming out of the war.

The political fighting mostly involved American interests and American combatants, but it occurred against a European backdrop. Not only did the Germans pose a commercial threat, but they also retained the ability to resume chemical warfare in the future, domestic industry advocates believed. Americans also watched the British and French in their attempts to counteract the economic strength of the German interests. In this international context, Americans took political steps to ensure their continuing presence in the industry.

The unusual political and legal steps on behalf of the industry, most taken in the heat of war or the uncertain immediate postwar years, drew increasing attention and criticism even as the industry advocates lobbied for extended protection. With the change of administration, the taming of anti-German sentiment, and the greater awareness of the activities of the APC, the industry would face a backlash in the 1920s. While the firms concentrated on surviving a competitive market, the industry continued to be deeply enmeshed in battles to secure its political and legal gains.

7 / Customs, Courts, and Claims

The Industry and the Law, 1922–1930

With the war increasingly behind them, Americans attempted to return to a peacetime economy, but the war's political and economic fallout would consume the synthetic organic chemical manufacturers' attention through the 1920s. Isolationism permeated the United States, and Americans lost trust in the benefits of global ties. The synthetic organic chemicals manufacturers, with their hope to survive the peace and thrive in the 1920s, certainly had no desire to return to a prewar state. They worked hard to establish a new status quo in which the German firms would never regain their old footing in the American market. With the Tariff of 1922, the manufacturers had acquired an important victory, but the uncertainties of the immediate postwar years carried through the 1920s, and the manufacturers continued to call on the government for support. While the domestic industry had won the favor of the executive and legislative branches of the federal government, the courts became integral to the industry's future in the 1920s, and it was not clear the judiciary would be nearly as sympathetic as the other branches.

The Department of Commerce and U.S. Tariff Commission and related customs agencies continued to be among the manufacturers' best allies in the federal government. In particular, the manufacturers' leading trade association worked very closely with customs officials to ensure that administration of the new tariff would provide the maximum protection. Paradoxically, as the United States and its nascent synthetic organic chemical industry increasingly hunkered down in its isolationist mood, there was also a concerted effort to study world markets. The Department of Commerce and other government agencies collected a vast storehouse of statistics, analyzed economic trends, and stationed economic observers in consulates around Europe and elsewhere in the world. The manufacturers had learned well the power of market information during the war, and they supported wholeheartedly the government's expanded programs to monitor international trade. The continuing political and economic instability in

Europe, as well as the revival of the powerhouse German firms, kept Americans' attention on Europe even as they reinforced their trade barriers to the outside world.

The legal fallout from war measures also demanded continued attention to Germany and its chemical industry. The war left behind a morass of disputed property rights, and the 1920s included a major challenge to the legality of the Chemical Foundation and the settlement of other war claims. At the beginning of the decade, it was not at all certain that the confiscated German property would still be in U.S. hands once Congress and the court system sorted through the complicated claims. As the anti-German hostility of the war years subsided and cooler heads began to reassess wartime decisions, the domestic synthetic organic chemicals manufacturers feared that policymakers and judges, without the wartime sense of urgency, would turn their backs on the industry and allow German firms to regain the American market.

Tariff Administration and the Department of Commerce

Shortly after Congress enacted the Tariff of 1922, the preeminent Harvard economist Frank W. Taussig published a harsh critique of the new protective legislation. He was skeptical of the administrative and procedural elements it would require, and he was convinced the dyes provisions were the consequences of "military excitement" rather than "cool economic considerations." But the ardent (and ideologically lonely) free trade advocate found even more to dislike in the tariff. Protectionism had been the law of the land for decades, and Taussig bemoaned the terms of the debate. Rather than view differences in labor costs and natural resources as the grounds for advantageous trade among nations, theorists of protectionism ignored free trade and debated the qualities of a good tariff. Increasingly, Progressive Era protectionists argued that an ideal tariff would be one in which the customs duties perfectly offset the cheaper costs of production abroad, creating a "fair" competition between domestic and foreign goods, a view the supporters of a domestic synthetic organic chemicals industry endorsed wholeheartedly. The theory was a "sort of fetish," complained Taussig, and "fatally unsound." Taken to its logical extreme, the protectionists' logic would eliminate all international trade; it was an economic theory for an isolationist age. Protectionists were pleased with the new "flexible" provision of the legislation, which allowed the president to change the tariff without congressional approval, within certain limits. In other words, the flexible provision offered the opportunity to adjust the customs duties as costs of production changed, coming closer to the protectionists' idea of a perfect, "scientifically" determined tariff.[1]

The tariff was only one part of the ongoing relationship between the federal government and the synthetic organic chemical industry as the nation shifted from a war economy to peacetime. Herbert Hoover's Department of Commerce, as well as the U.S. Tariff Commission, formed close working relationships with the manufacturers, primarily through the industry's new trade association, the Synthetic Organic Chemicals Manufacturers' Association (SOCMA). Both the U.S. Tariff Commission and Department of Commerce supplied important market information to the industry through the collection of statistics and regular reports. Both operated chemical divisions, which had cordial relations with trade associations and the industry at large. The two government units aggressively studied the world markets, collecting information particularly about European manufacturers and markets even as they used that information to reinforce the increasingly isolationist and autarkic U.S. policies.

Although the Chemical Foundation lobbied on behalf of the industry, the manufacturers in the early 1920s felt they lacked an effective trade association to represent their interests, particularly in the congressional hearings leading to the Tariff of 1922, and to facilitate networking. The American Dyes Institute sputtered along, but it increasingly lost members when courts declared open price societies illegal; criticism during the Shortridge hearings in 1922 worsened its reputation. In addition, with membership limited to synthetic dyes manufacturers, the ADI's base of support was narrower than the range of organic chemicals manufacturers operating in the United States.[2] Consequently, the manufacturers of dyes, pharmaceuticals, perfumes, and other synthetic organic chemicals convened in October 1921 to establish SOCMA, which they hoped would be more inclusive than the ADI.[3]

SOCMA's organizers chose Charles H. Herty as the association's first president. Although Herty was unforgiving toward those he believed hindered the development of the industry, he also had a reputation as a congenial organizer, a man who could twist arms and motivate manufacturers and government officials to pursue a coordinated agenda. Herty maintained an extensive set of contacts with people in universities, government agencies, and industry, which made him an appealing choice to the new association. In addition, Herty's affiliation with academia seemingly set him above the manufacturers' fray because he had no financial ties to any of the companies in SOCMA.[4] The association offered Herty a salary ($25,000), several times the amount an academic or industrial chemist normally received, and agreed to hire Herty's assistant, Lois Woodford, as well.[5]

In the first weeks of SOCMA's existence, the trade association worked out its organizational structure and elected members to fill positions as officers and committee members. SOCMA established a board of governors to decide the broad scope of the association's activities. In addition, the manufacturers divided into four sections according to their product lines: dyes, medicinal and fine organics, crudes and intermediates, and specialty chemicals. Companies like Du Pont, which manufactured a variety of chemicals, sent representatives to more than one section. To generate income, SOCMA implemented dues and assessments. Each firm paid a flat fee of $250 for dues, but the bulk of the association's budget came from assessments, fees charged relative to the size of a firm's business in synthetic organic chemicals. As a result, Du Pont paid more than twice as much to SOCMA as the second largest member firm and placed numerous representatives in the association's hierarchy. Neither National Aniline nor its parent firm, Allied Chemical, joined the trade association, adhering to the policy of secrecy established by its president, Orlando Weber. Initially, the association attracted about seventy members, but the number declined as the result of mergers, downward trends in the business cycle, and occasional dissatisfaction of members who decided the cost yielded too few benefits. By 1926, thirty-eight members remained in SOCMA.[6]

The single largest mission of SOCMA in its early years was monitoring the implementation and administration of the Tariff of 1922. The tariff contained new features that, on paper, benefited the industry, but all of the interested parties understood that its actual impact would depend on how the U.S. Tariff Commission devised routine procedures to fulfill the law's provisions. The tariff act contained two provisions that demanded the attention of the chemical manufacturers: the "American valuation" provision and the "flexible" provision.

SOCMA invested enormous resources in shaping the procedures for the "American valuation" provision. Traditionally, the United States and most other countries assessed duties based on the price of the article in the country of export; customs officials could obtain the price from the invoices at the port of departure. This was "foreign valuation." The Tariff of 1922 dictated that synthetic organic chemicals would henceforth have duties assessed on the value of the articles once they had arrived in the United States. Duties under American valuation were determined in one of two ways. If the imported article was "substantially equal" to an article made in the United States, the customs officials identified it as "competitive" and assessed the duty based on the "American selling price," the price for which the domestically made product sold on the U.S. market. If the imported article had no domestically made rival product, the officials labeled it "non-competitive" and assessed duty based on "U.S. valuation," which was the article's selling price in the United States minus some deductions for transportation, insurance, general expenses, and profit. Among other things, American valuation kept the duties high when Germany's currency went into its hyperinflationary tailspin.[7]

The customs officials faced difficult practical problems. They needed to know which synthetic organic chemicals American manufacturers produced and their U.S. market prices. They had to interpret the ambiguous congressional directive to determine whether a domestic and an imported article were "substantially equal" and therefore "competitive." And they needed the personnel and facilities to undertake their work. The last problem they addressed by building a laboratory in New York to test chemical samples of imported and domestic chemicals to aid their work in appraising the imports. The others proved more challenging.

To determine which chemicals were competitive, the customs appraisers requested the domestic manufacturers and importers to submit lists of chemicals considered "non-competitive."[8] The manufacturers' list contained 154 items, while the importers suggested that 2,598 imported organic chemicals had no comparable product among American-made chemicals. Needless to say, the manufacturers and importers had rather different assessments, and rumors floated around that manufacturers sometimes inflated their prices to raise the duty on their foreign rival.[9] The customs appraisers' initial plan for compiling comprehensive lists of competitive and non-competitive chemicals fell by the wayside as manufacturers protested the impracticality of the method. Instead, the appraisers would decide each case individually and gradually develop a list of importable organic chemicals to guide their everyday application of the tariff law. Manufacturers and importers could challenge the appraisers' decision and ask for a reconsideration before the three-member Board of General Appraisers and then, if necessary, before the Court of Customs Appeals. During the first two years under the tariff act, the U.S. Customs Service found that appraising imported chemicals and assessing the appropriate duty was a cumbersome process, but they gradually developed a system as their information accumulated and challenges were addressed.[10]

The importers pressed a series of important test cases through the customs judicial system that set precedents for subsequent regulations on the importation of synthetic organic chemicals. One important decision, in a case over azoflavine, confirmed the constitutionality of the section of the tariff act that mandated the American valuation. Other cases established regulatory guidelines. Imported natural organic chemicals could be

Table 7.1. Competitive and Noncompetitive Dyes Imports, 1928

	Number of Products	Pounds	Percentage of Dyes Imports
Noncompetitive	813	2,799,000	52
over 3,000 lbs.	184	2,386,000	44
less than 3,000 lbs.	629	414,000	8
Competitive	312	2,627,000	48
over 3,000 lbs.	117	2,461,000	45
less than 3,000 lbs.	195	165,000	3
TOTAL	1,125	5,426,000	100

Source: "Analysis of Coal Tar Dye and Color Imports for 1928," compiled from the 1928 Department of Commerce Monthly Import Bulletins by SOCMA, c. 1929, Charles H. Herty papers, box 77. Courtesy of the Manuscript, Archive, and Rare Book Library, Emory University, Atlanta, Georgia.

considered "competitive" to domestic synthetics and therefore subject to higher duties. Second, appraisers considered the relative concentration or strength of imported dyes, charging proportionally higher duties for more concentrated products. Every few months, the U.S. Tariff Commission issued an updated list containing the dyes selected as the standard by which the strength of comparable dyes would be judged. Third, if an American manufacturer produced a specified organic chemical but had not yet sold the chemical—generating no American selling price on which to assess the duty on a comparable imported article—the appraiser charged a duty on the imported chemical based on U.S. valuation rather than the American selling price, contrary to manufacturers' wishes. Finally, the court decisions established the domestic chemical as the standard of comparison; any imported product that would replace a domestic article for a particular use was automatically deemed "competitive," whether or not it also offered a wider variety of applications than the domestic chemical.[11]

The significance of these cases lies in the influence SOCMA wielded on them through its financial, legal, and technical resources. Legally, the court cases pitted the government, as collector of customs duties, against the importers, who challenged the legality of the tariff provisions and related customs regulations. Government victories in the cases served the interests of the domestic organic chemicals manufacturers, however, and SOCMA invested a considerable portion of its budget and personnel to assist the government's lawyers, who were understaffed and overworked. Shortly after the tariff measure passed in 1922, SOCMA hired chemist Harrison F. Wilmot as its technical advisor and the New York firm of DeVries and Doherty as the association's legal counsel.[12] Wilmot and the SOCMA counsel advised the government's lawyers on all aspects of the cases, including chemical analysis, "interviewing and subpoenaing witnesses," and, in the courtroom, cross-examining witnesses put forward by the importers.[13] In 1923, SOCMA devoted 40 percent ($36,000) of its year's expenditures ($89,000) on salaries and fees related to "tariff administration." The expense declined to $22,000 in each of the following two years but remained the second largest item in SOCMA's budget (only salaries for Herty and his staff were larger).[14]

SOCMA played a crucial role in generating a favorable implementation of the tariff legislation. Whereas lobbying for the tariff itself occurred on a temporary ad hoc basis, overseeing the daily bureaucracy of the government's customs administration required constant attention. Although not helpless, the importers commanded fewer resources and won few administrative or legal battles against an allied government and domestic industry.[15] The maneuvering over the customs regulations largely occurred out of the public spotlight and received little scrutiny even from members of Congress sympathetic to the importers, yet decisions at the customs level made a significant impact on the efficacy of the tariff, which had taken several years of acrimonious debate to pass.

While the importers clearly lost nearly every battle related to the implementation of procedures, the numbers suggest the continuing gap between the American and German industries. Non-competitive dyes by far outnumbered competitive dyes; that is, Germans and other foreign makers still produced a wide variety of dyes not replicated in the United States. The measure of imports by weight reflected Americans' focus on the higher volume dyes, while most of the imported dyes were niche products of low volume.[16] When Congress began debate leading to the infamous Smoot-Hawley tariff of 1930, the importers argued that the relatively few competitive dyes created a domestic monopoly by prohibiting imports, even if only one American maker produced a dye.[17]

Establishing the procedures for American valuation was the first and largest challenge in the Tariff of 1922; the second major change in the tariff legislation was the "flexible tariff" provision. The flexible provision drew much less attention from SOCMA, because Congress, having granted the industry the extraordinary measure of American valuation, excluded from the flexible provision most products derived from coal tar. That eliminated most, but not all, of the products represented by the SOCMA firms. The provision embodied the "equalization of costs" protectionist theory that so aggravated Frank Taussig. It allowed the president to alter the duties on most products after an investigation by the U.S. Tariff Commission. The president could not move articles on or off the free list, but he could change valuation from foreign to American valuation (not relevant to dyes, as they were by statute assessed by American valuation already), and he could raise or lower ad valorem rates up to 50 percent. The writers of the provision hoped it would permit American duties to adjust as the costs of production abroad shifted.[18] Among the synthetic organic chemicals that the U.S. Tariff Commission investigated were phenol, barbital, methanol, and Bakelite.[19]

The flexible provision survived domestic challenges, including whether Congress violated the Constitution by delegating to the president its right to determine revenue issues. But even more contentious were the foreign challenges. Under the provision, an American or American firm could request the U.S. Tariff Commission to undertake an investigation of costs, which the president would consult in the final decision. To complete their mandate, the investigators needed to know the costs of production of manufacturers to assess accurately the differences between domestic and foreign makers. But how, exactly, should the U.S. tariff experts learn the costs of production of the foreign manufacturers? The procedure would not endear them to Europeans, whether recent enemies, allies, or neutrals.

The American tariff experts believed, justifiably, that the most accurate information on foreign costs of production was in the account books of the foreign manufacturers, and therefore Americans after 1922 increasingly demanded to see the manufacturers' records. The Europeans who refused access to American officials risked having their goods denied entry to the U.S. market.[20] The French, who were already upset that Prohibition had deprived them of a prime export to the United States, decried "America's customs inquisition."[21] Foreign manufacturers were appalled; as one observer put it, "the new influx of young men demanding details which a French manufacturer would not give even to his own Government has caused a wave of sentiment extremely prejudicial to the United States."[22] The British suggested that even if "the United States has not made enemies in the true sense of the word, . . . it certainly has made no friends" in its quixotic attempt to pinpoint continuously changing costs of production.[23] Many of the Europeans suspected information gathered by the American officials, who had diplomatic immunity, made it back to their American rivals.[24] In the Reichstag, Germans "denounced [the officials] . . . as industrial spies"; by 1926, even the editorial page of *Drug and Chemical Markets*, a vocal supporter of high tariffs on synthetic organic chemicals, charged that the U.S. Treasury inspections in Europe had "degenerated" into an "industrial spy system" and called for its end.[25] Most protests did not involve synthetic organic chemicals, but some other chemical products featured prominently. The Norwegian Nitrogen Products Company, an American importer, protested the revelation of secret trade information all the way to the Supreme Court.[26]

In practice, American officials quickly faced a rising tide of European opposition and therefore sought alternative sources of information, such as local trade journals and other publicly available material. The significance for the story of synthetic organic chemicals lies not so much in the specific cases under the flexible provision but in the international scope of the efforts to implement a tight American tariff policy. Perhaps Europeans and other trading partners could be excused for not celebrating Americans' quest for an ideal tariff in the 1920s. Americans sought to glean significant trade information from the Europeans largely to frustrate the competition of Europeans; the American economic isolationism was an informed one.

The U.S. Tariff Commission was not the only government agency helping synthetic organic chemicals manufacturers learn trade information around the world. To an even greater extent, the Department of Commerce under Herbert Hoover expanded its role in collecting and publishing market information, and SOCMA also developed close links to Hoover's department during those years.[27]

The synthetic organic chemicals industry had benefited from the department's investigations and statistics since early in the war, but Hoover's arrival as secretary of commerce brought more systematic cooperation between the industry and the department. Hoover, in fact, spoke to the chemicals manufacturers at their initial meeting about the virtues of trade associations, particularly the benefits of eliminating waste through cooperation and the potential role of the association in promoting university research.[28] Hoover devoted no more attention to SOCMA than to any other trade group, but the department's reorganization under Hoover worked to the benefit of the industry. The Department of Commerce organized a Chemical Advisory Committee comprising important representatives from different branches of the chemical industry. The committee met annually to offer suggestions to the department and also to learn about the relevant services provided by the Department of Commerce.[29] Within the Bureau of Foreign and Domestic Commerce, Hoover organized divisions along product lines, including a new Chemical Division, which served as the industry's primary contact to the department.[30] Historian Ellis Hawley has written at length about Hoover's vision of an "associative" relationship between business and government, which, among other things, included a desire to convert the Department of Commerce "into a vast research and consulting service for economic decision makers."[31] This was certainly true in the case of synthetic organic chemicals. The manufacturers received an enormous amount of market and trade information that was especially important to an industry with so little experience and such limited knowledge of past trends in the field.

The department's Chemical Division, established in August 1922, quickly became one of SOCMA's most reliable sources of information about the development of the industry. Herty, acting on behalf of the association, was in frequent contact with Charles C. Concannon, the chief of the division beginning in November 1922. Concannon headed the Chemical Division for nearly three decades and thus provided an element of continuity for both the Department of Commerce and the chemical industry through his accumulated knowledge and expertise.[32] Early in 1923, Herty complimented Concannon on his work in the Chemical Division, noting "a kind of zip and go atmosphere there that spells accomplishment."[33]

Concannon and his Chemical Division offered two basic kinds of information to the synthetic organic chemicals industry. First, it regularly collected and distributed statistics on chemicals imports and domestic production, which served the new industry well as chemical manufacturers sought to learn the demand for organic chemicals. The Chemical Division often collaborated with its counterpart in the U.S. Tariff Commission, which continued to collect import statistics for its annual censuses. In addition, however, the manufacturers encouraged the Department of Commerce to tabulate the statistics more frequently in a monthly report.[34] As a result, the Chemical Division negotiated with customs officials to place a Department of Commerce representative in the customs office in New York.[35] Over the next few months and years, the monthly report expanded to include a greater number of chemicals, including those synthetic organics not derived from coal tar.[36]

Second, Hoover's drive toward expanding foreign trade led to the collection of relevant trade information that proved useful to members of SOCMA intent on selling in foreign markets. The foreign trade information from the Department of Commerce was extraordinarily rich and diversified, from sweeping overviews of a nation's chemicals industry to specific price information on a particular product. The department's foreign commercial attachés, stationed in fifty locations around the world, investigated market conditions and sent back reports to the appropriate commodity division in the Bureau of Foreign and Domestic Commerce. Sometimes the attachés conducted surveys of foreign consumers, which would guide American manufacturers' decisions about export business.[37] In 1923 and 1924, the department appointed Frederick E. Breithut as a special trade commissioner, and he traveled extensively in Europe and wrote a number of thorough reports, including ones on Germany, Italy, and Britain. Other authors produced similar reports on Switzerland, France, and Japan and unpublished reports on Poland and Austria, among others.[38] Succinct overviews of the U.S. industry's primary rivals, these reports discussed the number and size of manufacturers, production figures, imports and exports, domestic consumers, and legislation when the information was available.[39]

The Chemical Division also had informal lines of communication and shared unpublished information with the industry. Manufacturers could visit the division's offices to consult files, and Du Pont and National Aniline regularly sought information from Concannon and his colleagues.[40] When Germans protested that the unofficial U.S. delegation to the Reparations Commission gave the Department of Commerce German trade secrets, American diplomats advised Department of Commerce officials to dispense the statistics "confidentially" rather than through open publication.[41] Because German manufacturers would quote prices only to a potential customer, the division's Berlin office used a "subterranean channel" to obtain price information when they could.[42] The department also tried to keep up with the almost continual rumors circulating among American and European manufacturers, whether about dyes seized by the French in the Ruhr occupation or about impending cartel agreements among European chemical industries.[43]

During the 1920s, nowhere was the relationship between government and the industry any closer than with the Department of Commerce and the tariff administration. The mission of these governmental agencies included a mandate to aid American manufacturers, and the industry viewed the agencies as important allies—absolutely vital allies in the case of tariff administration—against the outside world. The relationship in the early 1920s still carried an urgency flowing from the war years, but, with the industry securely behind exceptional tariff walls, peacetime concerns and habits increasingly defined the contacts between industry and government. Protectionism and an easy cooperation were the norm. Some of the more extreme promotional measures during the war, however, generated a backlash that pitted the government and industry against one another.

USA v. The Chemical Foundation, Inc.

While 1922 brought reassurance to the domestic synthetic organic chemicals manufacturers through the Fordney-McCumber tariff, they were less certain about the survival of the Chemical Foundation as the nation moved beyond the war. The confiscation of German chemical patents and their administration by the Chemical Foundation presented a striking departure from the traditional American defense of private property and was made possible only by the extreme political climate of World War I. As the intense anti-German sentiment faded and after the Congress and presidency switched from Democratic control in World War I to Republican dominance in the 1920s, the actions of the Alien Property Custodian (APC) and the Chemical Foundation faced more intense scrutiny. Through the several tariff hearings and the Shortridge hearings, critics raised questions about the propriety of the Chemical Foundation's creation, and enough criticism emerged by 1922 to stimulate a Department of Justice investigation into the legality of the foundation. Advocates for the industry believed the Chemical Foundation provided an additional bulwark against the German competition, but confiscation pushed the boundaries of acceptable promotional policies.

Consequently, when *USA v. The Chemical Foundation, Inc.* opened in 1923, several issues were at stake. At the most basic level, the court case would determine whether the formation of the Chemical Foundation and the sale of patents to it violated U.S. law. Second, the trial raised questions about German chemical patents and patenting strategies. During the war, and again during *USA v. The Chemical Foundation, Inc.*, the concept of patent monopoly acquired sinister overtones, especially because chemical patents seemed to promote the progress only of German science and useful arts. In their domestic rivalries, the Germans had developed overarching strategies that harnessed the power of groups of patents in addition to the power of individual patents. The Chemical Foundation's designers hoped it would become a tool to harness groups of patents to benefit American industry. Third, the witnesses' testimony in Wilmington suggested the complexity and difficulties of the wartime strategies designed to encourage development of the domestic synthetic organic chemicals industry and to reach a competitive level of technical competence. The case illustrated how much—and how little—technical and economic expertise American chemical companies had gained in the intense years since the outbreak of war in Europe.[44]

To some extent, the Department of Justice inquiry derived from partisan politics. The new Republican administration of Warren Harding sought to uncover mismanagement in the wartime governance of President Woodrow Wilson's Democrats. Attorney General Henry M. Daugherty, pushed and prodded by Congress, spearheaded attempts to uncover suspected wartime frauds, particularly profiteering from war contracts. In addition, an audit of the Office of Alien Property in 1922 raised questions about the wartime stewardship of property. Besides the general quest for malfeasance, criticism specific to the Chemical Foundation attracted more attention with the change of party. In April 1922 Daugherty warned President Harding that "hardly a more important duty" faced him than a thorough investigation of the Chemical Foundation.[45] The *Washington Post* suggested the case was "in some respects one of the most important suits ever filed by the government. . . . The government's action is regarded as an indication of its determination to live up to the best American traditions in upholding international law and the high standards of conscience of the country in dealing with the property of enemy nationals."[46]

In addition to the pursuit of real and perceived wartime fraud, a second force also pushed Department of Justice officials to bring suit against the Chemical Foundation. As noted earlier, the Trading with the Enemy Act (TWEA) contained a provision, Section 10(f), that allowed patent owners whose patents had been used under government license to claim royalties. To collect, the patent owners needed to file a federal lawsuit within one year of the formal declaration of peace, which occurred on July 2, 1921. By the deadline in mid-1922, about 125 lawsuits had been filed against the Treasury and FTC licensees in various U.S. district courts. Among the claimants were the original German owners of alien property and, in about half the cases, the Chemical Foundation.[47]

As the new owners of the German property, the Chemical Foundation claimed royalties from the FTC licenses for patents in its possession. If successful, its suit would enrich the corporation's coffers and simultaneously prevent any of the royalties from returning to the Germans. If unsuccessful, the legal status of the Chemical Foundation could be called into question. The Section 10(f) lawsuits made it imperative for the Department of Justice to act quickly on its case against the Chemical Foundation. Direct challenge to the Chemical Foundation's ownership of alien property would be far more difficult if one of the district courts with a Section 10(f) suit permitted the Chemical Foundation to collect FTC royalties from the Treasury, thereby acknowledging its rightful ownership of the patents. Consequently, the department forged ahead on its case, but it also coordinated the U.S. attorneys in the Section 10(f) cases to delay them until the primary suit could be filed in October. After a series of short-term postponements, the government arranged for all of the Section 10(f) suits to be tried after the settlement of *USA v. The Chemical Foundation, Inc.*[48]

The Department of Justice and the Chemical Foundation filed their briefs in the fall of 1922. Through the winter and the following spring, the two parties prepared their cases for the trial, which opened on June 4, 1923, before Judge Hugh M. Morris in the U.S. District Court of Delaware (the Chemical Foundation had been incorporated in Delaware). A native Delawarean and a Democrat, Morris had received his appointment in 1918 from President Wilson at the behest of Delaware's two Democratic senators. Leading the government's case was Colonel Henry W. Anderson, a Republican corporate lawyer from Richmond, Virginia, who specialized in railroad issues. During the war he had directed the American Red Cross in parts of Eastern Europe and afterward returned to private practice in Richmond before becoming Special Assistant to the Attorney General in 1921. Isidor J. Kresel argued the case for the Chemical Foundation. Born in Austria, Kresel immigrated to the United States as a child, began a law practice in New York City in 1900, and later joined the district attorney's office of New York County. During the war Kresel had been hired by the APC to investigate at least one alien-owned firm. At the end of the Wilson administration, Kresel served as Special Assistant to the Attorney General in a case against the meatpackers. Each side descended on Wilmington with a complement of lawyers and witnesses in June 1923, and the long, dramatic case began. Over forty witnesses took the stand, and the final published record contained some twelve volumes of testimony and exhibits running close to ten thousand pages.[49]

The government pursued three lines of attack in its case against the Chemical Foundation. First, Anderson argued that the APC's sale of the patents to the Chemical Foundation was illegal simply because it violated TWEA. Second, he contended, the sale occurred as a result of a conspiracy among the industry to monopolize trade. Finally, and most extensively, the government viewed the $271,850 purchase price as completely inadequate, if not entirely fraudulent.

TWEA, as originally passed in October 1917, ordered the APC to act as the trustee for the owners of alien property. The seizure of the patents was legal and appropriate for a state of war, Anderson contended, but their sale violated the APC's role as trustee, even if it was executed with presidential sanction. Only Congress could legally dispose of the property. In a secondary argument, Anderson reasoned that even if the custodian had conducted the sale legally, his role as trustee obligated him to obtain full value for the patents. Anderson argued that the reputation of the United States in world opinion would suffer considerably if it so flagrantly tossed aside expected rules of conduct regarding property. Drawing on Woodrow Wilson's hope that the United States could serve as an example of justice and democracy to the rest of the war-weary world, Anderson pleaded for the court to hold the nation's principles above financial gain. He admitted that precedents existed in American legal history to permit Congress to confiscate alien property during war, but only to benefit the public good. The sale of patents to the Chemical Foundation, he argued, was for private benefit, not public good, and was therefore illegal.[50]

To make his argument compelling, Anderson needed to explain why the amendments to TWEA that seemingly provided congressional authority for the patent sale were, in fact, inadequate justification. His explanation related to the government's second charge, namely, that the sale of patents had been part of a conspiracy by domestic industrialists who had manipulated officials in the Office of Alien Property. The amendments were passed through Congress as riders on revenue bills, which, he said, prevented a full debate that would have generated more protest in Congress. Anderson pointed to the industry representatives who served simultaneously on the board of the ADI and a government advisory board, and he noted the ubiquity of Du Pont's Morris Poucher. He cited the "open price society" of the ADI, in which members shared pricing information, as an example of monopolistic tendencies among the dyes makers. Finally, Anderson asserted, the fact that Du Pont's and National Aniline's lawyers had selected the patents to be seized provided the clearest evidence that an unholy alliance existed between the manufacturers and the APC. Compounding the effects of the high tariff and other policies, the Chemical Foundation restricted trade by making international trade prohibitive, a violation of American antitrust laws.[51]

Anderson's accusations implicated the two key figures in Wilson's Office of Alien Property, A. Mitchell Palmer and Francis P. Garvan. Both men testified in *USA v. The Chemical Foundation, Inc.* about their activities in the Office of Alien Property. Garvan's testimony to the court made abundantly clear the zeal he brought to his investigative work. Although his appearance in *USA v. The Chemical Foundation, Inc.* was relatively subdued compared to his earlier appearances before congressional committees and to his various public speeches and writings, Garvan repeated again his usual charges: that the German firms and their importers engaged in cartel behavior, that the importers' firms became centers of spying and propaganda, and even that the German chemical industry had influenced American tariff legislation to prevent the establishment of domestic firms.[52] He supported his testimony by submitting into the record evidence such as intelligence reports and postwar books critical of legal cartels in the German economic system and of the German chemical firms more specifically.[53] To the detriment of the government's case, the prosecuting lawyers never adequately rebutted or neutralized Garvan's accusations of misbehavior among the importers and the German chemical firms, despite the fact that his allegations contained only kernels of truth amid much larger portions of exaggeration.

Turning the focus of the case on the activities of the German firms and their importers became an important strategy for the Chemical Foundation defense lawyers, but Isidor Kresel also had to address the primary government arguments more directly. First, he rebutted the government's contention that the amendments to TWEA occurred under conspiratorial conditions. The defense insisted that members of Congress and President Wilson were fully alert to the seizure and sale of the chemical patents. Kresel included transcripts from congressional debate over the 1918 amendments to demonstrate that members understood the issues and implications for the original act and the chemical industry. Moreover, former Undersecretary of State Frank L. Polk, who had issued executive orders while President Wilson participated in peace negotiations in Europe, testified that although he signed the relevant orders on Palmer's advice and without knowing the details about the proposed Chemical Foundation, both he and Wilson were in accord with Palmer.[54]

To rebut the charge of monopoly within the newly founded American dyes industry, Kresel called to the stand a wide range of dyes manufacturers and chemists. A representative of the ADI open price society declared that the attempt to exchange pricing information had been an abysmal failure from the start. Kresel made Anderson's accusations seem like a wild conspiracy theory lodged against respectable industrialists and patriotic government officials. He suggested that Anderson's application of the Sherman antitrust act to the Chemical Foundation was a most creative—too creative—use of the law. The government's case, Kresel asserted, was driven by partisan politics, and he accused officials in the Harding administration of attacking Wilson for cheap political gain now that the former president was in ill health and could not defend himself, a claim Anderson vehemently denied.[55]

Although the two sides sparred vigorously over the issues of improper construction of TWEA and monopoly conspiracies, by far the most important debate of the case centered on the value of the patents. These included not only thousands of patents for dyes and pharmaceuticals but also a collection of about fifty "Haber patents" for nitrogen fixation and a variety of miscellaneous patents unrelated to the chemical industry. The government saw the price as the strongest evidence that the patent sale had occurred under wrongful conditions. Selling the patents at a fraction of their worth amounted to fraud or outright "confiscation," defined by Anderson as seizing and selling the property without adequate recompense, which, he reiterated, the act did not condone. But how much were the Chemical Foundation patents worth? The answer to such a question was open to widely varying interpretations, and each side explored the assumptions underlying patents and the American patent system to establish its case. Under the peculiar rigor of a legal inquiry, a patent's inability to represent technical knowledge with precision became painfully obvious and opened the door to challenging the value of the German chemical patents.

Each side brought in expert witnesses to estimate the patents' value. Anderson started his case by calling Karl Holdermann, a forty-year-old chemist who had worked for the German firm BASF since 1906, to testify to the value of the patents. Because Holdermann was in the United States only in the spring of 1923, the court came together for one day on May 8, a month before the opening of the case, to hear his testimony. Holdermann believed that the Haber patents alone would have carried a market value of at least $17 million in 1919. In 1913 his firm had collected 1.3 million gold marks ($325,000) in profit on products manufactured under BASF's dye patents and sold to the United States, figures that made the $271,850 purchase price seem woefully low. Just the expense for filing a patent averaged $85, and the Chemical Foundation purchased the initial 4,500 patents for an average of $55 each.[56]

In another measure of patent value, the government lawyers investigated the royalties from FTC licenses issued on the patents that were eventually sold to the Chemical Foundation. From 1918 through 1922 the U.S. Treasury collected close to $500,000 on the relevant FTC licenses. When added to the royalties collected by the Chemical Foundation, the sum exceeded $1 million. Anderson pointed to the royalty figures as proof that the Chemical Foundation patents were far more valuable than the sale price would indicate. In addition, the APC, in determining price, had ignored the amount of FTC royalties, which would have suggested a far higher value, according to Anderson.[57]

Another important witness for the government was Herman Metz. The government attorneys believed that the patents for Salvarsan and Neosalvarsan were among the most valuable purchased by the Chemical Foundation and that Metz, as the prewar supplier of Hoechst's antisyphilitic drugs, would be in the best position to testify to their worth. Under his FTC licenses on Salvarsan and Neosalvarsan, Metz paid $140,000 to the Treasury from 1919 to 1922. On the two products in 1913, his importing firm had sales of $380,000 and a net profit of $150,000, figures that peaked in 1916 with sales of $970,000 and profits of $430,000. Metz estimated that the market price of the Salvarsan and Neosalvarsan patents would have been about

$1 million in 1919. The patent for the local anaesthetic Novocaine, another Hoechst product, would have been worth another $200,000, at a minimum. In addition, the trademarks for Salvarsan, Neosalvarsan, and Novocaine, in Metz's opinion, were worth as much as the patents.[58]

Had Anderson and his Department of Justice colleagues understood more fully the nature of the battle Garvan had waged against the importers of German chemicals, they might have reconsidered their decision to call Metz to the stand. In his cross-examination, Kresel portrayed the importer as a mere puppet of his German suppliers, suggesting that Metz had voted as the Germans wished while a member of Congress early in the war. Kresel grilled Metz about his acquaintance with Ambassador Johann H. G. von Bernstorff and other officials in the German embassy, again implying that Metz answered to Berlin. In 1918 soldiers being treated for syphilis died after receiving Salvarsan; Kresel suggested that the defective Salvarsan came from Metz, who, because of German sympathies, intentionally supplied a poor product. He accused Metz of profiteering by charging the government excessive prices for the antisyphilitic drugs necessary for the army and navy. Finally, Kresel blamed Metz for instigating the government suit against the Chemical Foundation at the behest of the Germans.[59]

Near the end of Metz's testimony, Anderson could only express dismay at the way Kresel "attacked this witness and his credibility and his character and loyalty with a severity" that he had never seen before.[60] But Metz and Garvan had tangled previously, and their mutual antagonism ran deep. Kresel, of course, also denounced Holdermann's testimony. The choice of Holdermann and Metz as the two leading government witnesses affiliated with the industry played into Kresel's strategy to portray the lawsuit as benefiting only the Germans. He spoke of the German chemical industry as the "invisible plaintiff" behind the government case.[61]

Unfortunately for the government, Anderson most likely had little choice in expert witnesses. Nearly all American synthetic organic chemicals manufacturers held stock in the Chemical Foundation, and many benefited from the licensed patents. If the Chemical Foundation lost the lawsuit, the manufacturers faced an uncertain fate that could cost them dearly in back royalties and future profits. American chemists, whether through economic or other motivations, also supported the Chemical Foundation, and more than once a Department of Justice investigator found a technically qualified chemist who had to be dropped from consideration as a witness because he was sympathetic to the defendant.

The Chemical Foundation, on the other hand, drew expert witnesses from the cream of the American chemical community.[62] Using witnesses

FIGURE 7.1. U.S. Chemical Warfare Association, "National Safety." (Courtesy of the Manuscript, Archive, and Rare Book Library, Emory University, Atlanta, Georgia, Charles H. Herty Papers, box 98, folder 21.)

from industry and academia, Kresel developed an alternate set of criteria for measuring the value of the Chemical Foundation patents by focusing on the workability of those patents in the United States. Prominent chemists from universities, Du Pont, Calco, and Abbott Labs testified that the German patents excluded vital information, preventing American chemists and manufacturers from producing dyes and chemicals solely from the description in them. Instead, industrial chemists were required to invest an extraordinary amount of time conducting additional research to obtain commercial products supposedly covered by the patents. In no case did the manufacturers make a commercially viable product from the patents without gaining supplementary information, they testified. When asked to place a dollar figure on the patents, one witness after another agreed with Abbott's chief chemist, Ernest H. Volwiler, that the value of the patents was "purely nominal." Some credited the patents chiefly with a certain "nuisance value," functioning primarily as a means to keep German competitors out of the American market.[63]

Volwiler explained that Abbott Labs had to invest $17,000 and two and a half years of research to supplement information from the German patent in order to produce cinchophen. Novocaine had cost even more in time and money in cooperative research between Abbott and the University of Illinois before Abbott could bring it to market. Elmer K. Bolton of Du Pont recounted some of the difficulties his company encountered in attempting to employ the German patents. One of the biggest problems was simply identifying which patents were related to which dyes. Beyond that, because Du Pont chemists did not have appropriate methods to measure the progress of a reaction, they regularly achieved only low yields of poor quality. They eventually discovered that the patented processes required modifications, such as including additional ingredients, altering the temperature, or changing the concentration of solutions, before the researchers could achieve commercial products. In short, Bolton stated that the patents "do not have any practical value without the know-how."[64] Such a judgment could presumably apply to nearly all patents at all times, but in the years immediately following World War I and in the court case, this general characteristic could be used to register complaints specifically against the German chemical patents.

Anderson had difficulty contesting the chemists' testimony that they had all been forced to alter substantially the processes described in the German patents to achieve the desired products. All through the hearings, Judge Morris had adopted a liberal policy on the kinds of evidence permitted by each side in the case. However, as Anderson pushed the industrial chemists to describe in detail how they had departed from the procedures outlined in the patents, Morris implemented his first restrictions on evidence and curtailed Anderson's probe. The chemists said they feared divulging trade secrets, and Morris felt strongly that participation in the trial should not jeopardize the companies' investments and make manufacturers reluctant to appear in court in the future. The judge's decision prevented Anderson from questioning the validity of the chemists' statements, and the lawyer could only observe obliquely that it was in the chemists' self-interest to condemn the patents.[65]

In addition to the industry chemists, Julius Stieglitz, one of the most respected organic chemists in the country, appeared for the Chemical Foundation to criticize the patent descriptions. Stieglitz had received his Ph.D. from the University of Berlin and had trained under August W. Hofmann, the famed German organic chemist, before joining the chemistry department at the University of Chicago. During the war Stieglitz conducted research on Salvarsan, and he found the patent description defective on more than one count. The patent failed to state that the starting material needed to be absolutely pure. Normally manufacturers expected to settle for some lesser degree of purity, he said. The patent also neglected to mention the point at which moisture had to be excluded. For five months, Stieglitz and his staff of graduate students experimented on Salvarsan until they achieved a safe product.[66]

Anderson made little attempt to dispute the fact that American chemists and chemical companies ran into significant difficulties in manufacturing the synthetic organics. He agreed with Bolton and rest of the chemical community that "know-how" played a key role in the production of chemicals. In fact, Anderson turned the industry's argument around and argued that the lack of chemical "know-how" in the United States—rather than any inadequacies in the German patents—caused the manufacturing problems. The U.S. laws governing patents mandated that the description be clear enough for "any person skilled in the art or science" to make the product. The descriptions in the German patents, Anderson argued, would have been satisfactory for skilled organic chemists experienced in dyes and pharmaceuticals. Neither the Germans nor their patents could be faulted simply because the American chemical industry and its scientists were not adequately "skilled in the art." Under cross-examination Bolton, the Du Pont chemist, admitted that his company's chemists were generally young and unpracticed in dyes, especially vat dyes. Not surprisingly, other industry chemists similarly dated their first exposure to synthetic organics to the war.[67]

In the government's rebuttal to the initial defense argument, Anderson produced organic chemists to testify that they had obtained satisfactory results following the descriptions in the German patents. Anderson's witnesses worked for a medical school, a government agency, or Herman Metz, but none was employed by one of the major firms backing the Chemical Foundation. Walter G. Christianson, a young chemist with only a bachelor's degree from Harvard (1918), gained brief exposure to industrial practices at Du Pont and Kodak before returning to Harvard to conduct research on arsenicals at the medical school in 1919. He testified that by following the patent for Salvarsan, he had, on his first attempt, produced the drug at a quality that passed the Public Health Service's standard. Christianson said that he subsequently performed some 1,200 experiments related to various stages of the drug's production and consistently acquired a yield of 90 percent.[68]

In his cross-examination, Kresel walked Christianson step by step through the Salvarsan patent and compared it to Christianson's laboratory notes. At several points Kresel noted the chemist's departures from the patent's description: for example, he began the experiment with different amounts of constituent chemicals, employed a higher temperature, and stirred the solution at one point. Christianson defended himself, however, by stating that he merely reduced the amount of chemicals proportionately and raised the temperature to speed the reaction; neither change affected the validity of the patent. He admitted that stirring was not necessarily something a chemist should have known to do, although repeated experiments quickly showed its significance. Christianson's lab notebook and testimony gave the government case valuable evidence that the German patent was sufficiently descriptive to one "skilled in the art."[69]

Harry D. Gibbs, a chemist at the Public Health Service who had worked for Du Pont from 1919 to 1922, also testified that he produced Salvarsan by following the steps in the patent when he worked in the Bureau of Chemistry of the Department of Agriculture. In fact, when Garvan had asked Gibbs to denounce the patents prior to the formation of the Chemical Foundation, Gibbs recalled that he told Garvan that his experience with the patents would not allow him to do so. Gibbs believed that a skilled chemist would also know to begin making Salvarsan with pure intermediates, especially since the drug was taken internally, although the patent said nothing of the purity of the stock.[70]

Anderson had one last chemist to call to the stand: Louis Freedman, whose testimony provided a dramatic finale to the trial in Wilmington. Anderson saved one of his strongest witnesses for last, but he surely knew by this time that Kresel was capable of discrediting even a solid witness. Although Freedman found himself uncomfortably in the spotlight, he was not unequipped to confront the challenge. He had received his B.S. in chemistry from Yale in 1915 and a Ph.D. from Columbia in 1922, making him one of an increasing number of American-trained Ph.D. chemists. Between his two degrees, he had obtained industrial experience at the Calco Chemical Company. At Calco, Freedman had experimented extensively with cinchophen, a synthetic organic chemical developed and manufactured by the Bayer Company in Germany and employed medicinally to treat gout and rheumatism. According to Freedman, his first experiment on cinchophen, which Calco named Atophan, produced a product with a 60 percent yield using the patent. Freedman's work contributed to Calco's success in placing cinchophen on the market in September 1917.[71]

In the middle of Freedman's testimony, just after the young chemist described his success at Calco using the German patent for cinchophen, Judge Morris asked Freedman to replicate his cinchophen experiment for the court. Freedman demurred. Five years had passed since he last performed the experiment; he would be working in an unfamiliar laboratory with someone else's chemical stock; and then there was his wedding in three days. He volunteered to return after his wedding, but the court case was nearly complete and Morris wanted to finish as soon as possible.[72] The stakes were high as each side hoped to prove its point about the viability of German patents. As if Freedman's task were not already challenging, there was one additional reason for the defense lawyer to attack his testimony: Freedman worked for Herman Metz. As a Metz employee, he was open to Kresel's insinuations that his testimony against the Chemical Foundation proved his faulty patriotism.

Arrangements went forward to set Freedman up in Dr. Gellert Alleman's chemical laboratory at Swarthmore College, in the nearby Philadelphia outskirts, under the watchful eyes of chemists representing each side in the case. At 10:00 Saturday night Freedman began lab preparations for his cinchophen synthesis. Some of the stock, or constituent, chemicals arrived by messenger, and Freedman pronounced himself reasonably satisfied with the equipment in the Swarthmore laboratory. He and the witnesses retired for the evening at about midnight and returned to the lab at 7:00 on Sunday morning to begin the chemical work in earnest. Freedman ran three separate experiments at the same time—two to make cinchophen and one to make pyruvic acid (pyroacemic acid), one of cinchophen's intermediates. The chemists spent all of Sunday and most of Sunday night in the laboratory, breaking briefly at 5:30 A.M. on Monday. They returned three hours later to examine the results of the experiments and, exhausted, reported back to the courtroom in Wilmington on Monday afternoon.[73]

Before Freedman took the stand, the official witnesses—Gibbs for the government, Bolton of Du Pont and M. L. Crossley of Calco for the defense, and Gellert Alleman for Judge Morris—testified as to whether Freedman had accurately followed the patent description. Gibbs said that Freedman had followed the patent in all significant respects. Bolton and Crossley found two deviations, the first when Freedman added two ingredients quickly one after the other rather than simultaneously in the first cinchophen experiment and the second when he added more pyruvic acid than the patent listed. Freedman denied that the deviations had any relevance. In the first case, he believed that the extra time was meaningless, but he corrected the discrepancy on the second cinchophen experiment. As for the pyruvic acid, he tested its specific gravity to evaluate its purity and increased the amount to compensate for impurities. Alleman, the court's official witness, admitted that he simply did not know if the deviations were important, although he suspected that they were.[74]

What were the results of Freedman's experiments? The young chemist obtained cinchophen, the observers agreed, but it was of poor quality. The first experiment produced a yield of 26.9 percent and the second a yield of 14.3 percent. The yields were entirely unsatisfactory for commercial manufacture. Freedman argued that he had been allowed insufficient time to complete the final purification process that would have improved the product considerably. He also explained that the second experiment's lower yield resulted from a higher percentage of mechanical loss because he had to halve all the ingredients. The defense's chemists, on the other hand, attributed the lower yield to Freedman's having followed the patent's specifications more closely by adding the ingredients simultaneously.[75]

The experiment provided a dramatic finale to the hearings, but its significance was not so clear-cut. The defense claimed a small victory—humbling a "Metz man"—and asserted that the exercise proved the inadequacy of the German patent. Anderson, on the other hand, noted that cinchophen had indeed been the end result of the experiment, which was conducted under less than ideal circumstances. He argued that patents were frequently filed and issued on the basis of laboratory work and not held to standards of commercial viability. Drawing on advice from a patent lawyer in attendance, he advanced the concept of a "pioneer patent," in which the early patents—such as that for the telephone—are somewhat crude but are not in any way voided when subsequent improvements occur.[76] The creation of cinchophen from the German patent description validated the patent, he argued.

The testimony closed in late July 1923, and the lawyers completed their cases with closing arguments later in the autumn. In his oral statement Kresel attacked the government suit for playing into the hands of the Germans. He highlighted the German connections of the witnesses appearing for the plaintiff, especially berating Karl Holdermann and Herman Metz ("the German tool") and the chemists who worked for Metz, including Freedman. In a written brief, the Chemical Foundation's defense systematically refuted more than twenty separate charges of illegal activity. The government summarized its case in a brief filed in November, reemphasizing the APC's wrongful interpretation of TWEA, the inadequate price of the patents, and the abuse of public office by Garvan. Whether the Germans had indeed committed bribery and the other offenses prior to the war was "entirely immaterial" to the case. The point, Anderson and his colleagues wrote, was that the government wrongly lost control of the property held in its trust and had a legal claim to it.[77]

With the case argued, the two sides waited for Judge Morris's decision, which would arrive in January 1924. Neither side felt completely sure of the judge's sympathies, although the Department of Justice had certainly lost in the news media.[78] Throughout the summer, the suit had attracted representatives from the Associated Press and the major New York newspapers. The Chemical Foundation actively sought to shape the coverage of the trial to its benefit, as became evident in an incident embarrassing to the press. During

Metz's testimony, the reporter for the *Wilmington Morning News*, who was the local Associated Press representative, was not present in the courtroom. He filed his article on time, however, because William Keohan, Garvan's public relations man, provided him with a story. In Keohan's version, Metz admitted to paying a German spy to aid investigations of the U.S. dyes industry. The story was so completely inaccurate that Judge Morris called reporters into his chambers to reprimand them, and the following day papers printed retractions, admitting the "statements were false in every particular." Duly concerned, attorneys for the Department of Justice paid more attention to publicity and considered issuing daily "memoranda" to present their side of the case.[79]

On January 3, 1924, Judge Hugh Morris reported his decision, finding in favor of the Chemical Foundation. Morris dismissed out of hand the government charge of conspiracy among industrialists and government officials, which Anderson "failed utterly" to prove. Ample precedent, dating to the Revolutionary War, existed for the government to confiscate alien property, and Morris found nothing in error with TWEA or its implementation. The act specifically granted the powers of an "absolute owner" to the APC, which meant that Palmer and Garvan had acted legally in selling the patents. Although the judge agreed that the patents had been sold at less than market value, he ruled the sale legal and added that the significance of the patents and the industry to national security compensated the "public interest" for any strict financial loss suffered by the U.S. Treasury. Clearly impressed by the "amazing" list of German holdings prior to the war, Morris accepted the defense's contentions that Germany, through illegal activities, had "prevented" the establishment of a domestic industry. The Department of Justice appealed the case to the Court of Appeals and then to the Supreme Court. In each case, the higher court ruled with Judge Morris in the District Court, agreeing that the "construction" of TWEA had been appropriate and that the formation of the Chemical Foundation was legal. The final ruling from the Supreme Court arrived in October 1926, ending four years of litigation.[80]

On the eve of the next world war, at least one critical legal observer would look back and castigate the behavior of the APC in World War I and the courts' decisions as flagrant violations of international law. Substantial agreements and customs existed to protect individual property in the event of national conflict, James Gathings argued. He had little doubt that Palmer and Garvan had exceeded their powers, and the introduction to his book referred to the Chemical Foundation, on page 1, as the "most notorious of the cases" conducted by the APC. But Gathings also recognized that, although the legislative and executive branches of the American government felt obligated to abide by international law, the American court system had a long record of upholding the right of the United States to confiscate enemy property during war.[81] Perhaps Anderson had been aware of the generally ineffective role that international law played in the national courts. He never relied on that body of law to bolster his case, although he wanted the judge to understand the implications for the reputation of the United States in the eyes of the international community should he legitimate the confiscation of German property.

With the verdicts establishing the legality of the Chemical Foundation, Garvan and the foundation concentrated on promoting American chemistry and the chemical industry for the remainder of the nonprofit corporation's existence. Through 1936, the Chemical Foundation had issued 973 licenses to users involving 871 patents. Of the 973 licenses, 729 were for chemical patents, and 512 brought no royalties. Du Pont became the most frequent licensee and dyes the most frequently licensed set of products. Interestingly, the two products that generated the most income for the Chemical Foundation were not synthetic organic chemicals at all but stainless steel and ammonia, patents caught up in the expansive sweep through the patent office in February 1919.[82] The last of the former German patents expired in 1936, marking an end to the foundation's royalty income. As of 1941, the Chemical Foundation had collected about $7.5 million on license royalties and another $1 million from infringement settlement claims, fees, and interest. The foundation had spent the $8.5 million on chemistry research, particularly related to medicine and agriculture, on promoting chemistry education in schools, and on chemical journals, as well as on legal expenses, overhead, and other matters.[83]

The experience of the American synthetic organic chemicals industry reinforced the idea that the legal instrument, the patent, only imperfectly captured the scientific and technological knowledge it represented. From the perspective of the domestic manufacturers, the Chemical Foundation failed as a mechanism to convey the patents' technical knowledge to the new American industry. The testimony in *USA v. The Chemical Foundation, Inc.* suggests the difficult challenge the manufacturers faced in deriving benefit from the patent descriptions. If the Chemical Foundation failed in diffusing the technical knowledge in patents to the young American industry, it was more successful in using those patents against the German firms, either through the collection of royalties on imported German products or through the foundation's lobbying functions on behalf of the American industry. The United States aided its infant synthetic organic chemicals industry by temporarily weakening a segment of the national patent system.[84]

After the Chemical Foundation court case, the German firms could not expect the return of their confiscated patents. The possibility still existed, however, that they would be reimbursed for the value of their lost assets, either through congressional legislation or through Section 10(f) lawsuits. Because the Section 10(f) lawsuits were held up until the final appeals in *USA v. The Chemical Foundation, Inc.* had been exhausted, the cases reached trial only in 1928. In the meantime, Congress faced the dilemma of millions of dollars of unsettled war claims, and for much of the 1920s the German firms could hope for compensation through Congress if the court cases failed.

War Claims

In 1926, eight years after the armistice that ended World War I, members of Congress debated once again what to do with the German and Austrian property the U.S. government had seized during the war. In the 1920s, Americans returned to "normalcy," the anti-German animosity declined, and many more Americans began to question the wisdom of confiscating private property. *USA v. The Chemical Foundation, Inc.* attracted significant attention in the mid-1920s, not least because the court decisions ran almost completely opposite to the general trend in Congress. Reflecting growing American doubts that Germany alone was responsible for starting the war—as the Treaty of Versailles had stipulated—Congress's proposals to return the property gathered momentum. In 1923, Congress passed the Winslow Act, which returned up to $10,000 to individuals whose property had been seized, but that still left alien property amounting to $265 million in 1928.[85] Throughout the 1920s, the Americans sent mixed signals to Germany about the ultimate disposition of the property and assets of the German chemical firms, including those subject to treaty reparation clauses and the Chemical Foundation patents. Powerful interest groups, federal courts, international diplomacy, and political partisanship all influenced the unpredictable decisions that by 1931 established a fairly complete if not altogether consistent set of policies affecting the assets of the German chemical firms held by the United States.

War claims were part of a broader problem—a transatlantic political economy still deeply unsettled in the war's aftermath. In the executive branch, President Warren Harding and Secretary of State Charles Evans Hughes sought diplomatically to reduce the more punitive aspects of the

Treaty of Versailles. The Republican administrations in the 1920s worked to stabilize the international economy, partly by encouraging the reconstruction of the German economy. These efforts reached fruition in 1924 in the Dawes Plan, which restructured German reparations payments and provided millions of dollars in loans to help rebuild Germany's economy. This American-sponsored initiative contributed significantly to a German recovery in the late 1920s, before the Depression. In addition, as part of the top-level diplomatic negotiations, Hughes indicated to the Germans that he would encourage steps toward resolving the question of German assets seized by the United States. In 1922, Germany and the United States had established the Mixed Claims Commission to determine American property claims in Germany, which included losses from ships sunk by Germans but rarely included property related to chemicals. With the amount of the American claims fixed, the issue of German claims in the United States came before Congress.[86]

Members of Congress understood that their impulse to return property ran against American court decisions as well as against actions by European allies, but many Americans increasingly interpreted the wartime confiscation of property as a step too far, a violation of private property rights that had sullied American principles and traditions. Americans, said Representative J. W. Collier of Mississippi, were in no way obligated to return any property to Germany, but the United States needed to be guided "by the American standard of justice and equity."[87] Similarly, Rep. Henry T. Rainey of Illinois pronounced, "We are not following precedents. . . . We are refusing to follow the decision of the Supreme Court in the Chemical Foundation case. We are refusing to follow the provisions of the treaty of Versailles. . . . We are refusing to follow the provisions of our own treaty with Germany, the treaty of Berlin, because that treaty gives us absolutely this property. We are refusing to follow our allies in the World War—they all confiscated German property and kept it." Rainey concluded by saying that the legislation would, in fact, set a precedent "for all the nations to follow through all the centuries."[88] Rainey's hyperbole found wide resonance in Congress and the wider public.

After two years of negotiating, Congress emerged with the War Claims Act of 1928.[89] In accordance with the act, the United States returned close to 80 percent of the German property held in the U.S. Treasury; the other 20 percent went into a new German Special Deposit Account, which Congress used to pay private American claims against Germany. When Germany paid its obligations to the United States over subsequent decades, the other 20 percent due Germans could be passed on to the claimants.[90]

To determine the amounts of the German claims against the United States, Congress placed authority over claims in the position of the War Claims Arbiter, the referee or judge who listened to the claims and passed judgment about their validity and value. Although the legislation provided general guidelines, Congress left the details of administering the law to the arbiter. The first arbiter was Edwin B. Parker, who had been a respected member of the Mixed Claims Commission, and that experience influenced the way he organized the process of deciding claims. Before deciding any single claim, he established a set of rules that would determine eligibility and procedures and speed along the decisions about individual claims.[91]

Perhaps no claims before the arbiter were potentially as complicated as those from the German chemical companies; the task before the arbiter lay in charting a course that fulfilled the desire of Congress to return property while also accommodating the decision of the Supreme Court in the Chemical Foundation case. In 1925, the major German firms, including Bayer, BASF, and Hoechst, merged into the giant I.G. Farben, which pursued the claims of its predecessor firms. The Germans and their American allies hoped to receive compensation for royalties on patents used by the American military and civilian contractors during the war and for confiscated property, including the patents owned by the Chemical Foundation. In the hearings before the arbiter, government lawyers challenged the scope and validity of claims, although such challenges were to conform to the spirit and purpose of the War Claims Act, which was to return German property.[92]

Lawyers for I.G. Farben and for the government produced written statements, upon which they elaborated in oral arguments before the arbiter. The arbiter had proposed that each side address certain questions. First, he asked the lawyers to consider whether or not the War Claims Act covered the patents sold to the Chemical Foundation. In other words, should Congress provide restitution to I.G. Farben for the patents and patent royalties in the hands of the Chemical Foundation, despite the U.S. Supreme Court ruling that established the sale's legality? The lawyer for I.G. Farben argued that the War Claims Act covered the Chemical Foundation property because, as a creation of the Office of Alien Property, the Chemical Foundation was essentially a "quasi-government" agency. Therefore, the U.S. government had directly benefited from German chemical property, which was ample justification for compensation under the new act.[93] The government lawyer, on the other hand, argued that the Supreme Court decision clearly defined the Chemical Foundation as a private corporation, completely independent of the federal government. As a result, the government should not be liable under the War Claims Act for any claims after the sale in 1919.[94] After consideration of each argument, the arbiter ruled that the War Claims Act did not include the Chemical Foundation patents. However, such a ruling did not preclude I.G. Farben from claiming patent royalties prior to the date of the sale to the Chemical Foundation in April 1919.[95]

The second major question facing the arbiter centered on how to interpret the congressional mandate to pay for patents "used by or for the United States." For example, would the act cover patents used by a contractor who supplied products to the government? Or would it cover only those patents used directly by a government agency, such as the Department of War? Again, the War Claims Arbiter chose the most conservative, narrow interpretation, limiting the clause only to those patents directly used by a government agency.[96] Unfortunately for I.G. Farben, the arbiter's procedural decisions greatly reduced the scope of the claim that the German firm could make.

After the arbiter's general procedural decisions, he and his office began to process the individual claims. Perhaps the significance of the claim by I.G. Farben can be measured in part by noting that it was "Docket No. 1" from among 1,000 claims. I.G. Farben's lawyers could still file for patent royalties for those patents the Department of War and, to a lesser extent, the Department of the Navy had directly infringed upon from August 1, 1914, to April 5, 1917, and from November 12, 1918, to the date the APC sold the patent, if such a sale occurred.[97]

The two military departments appointed officers and several assistants who investigated patents and worked closely with the War Claims Arbiter to establish the value.[98] I.G. Farben had no way of knowing to what extent the patents of their constituent firms had been used, which meant the reports from the military officers became almost entirely the sole source for estimating the value of patent use by the United States. In the case of I.G. Farben, these military reports were also protective of military secrets—the Department of War was extremely reluctant to indicate that it had used a certain set of patents. The military officers divided the German patents into three categories: Group A, which consisted of pharmaceuticals, including Salvarsan; Group B, the use of which the military officers deemed secret;[99] and Group C, a catch-all category that included miscellaneous patents related to photography, metallurgy, and other uses. Of these groups, the value of Group B far outweighed the others.[100]

Lawyers from the government and I.G. Farben agreed to negotiate a settlement rather than litigate the case before the arbiter, partly because of the expense, but also because the I.G. Farben lawyers felt disadvantaged when nearly all the evidence was in the hands of the government.

Unfortunately for them, they were also at a disadvantage when attempting to negotiate compensation. Early in 1928, the government suggested a lump sum settlement of $350,000, which the lawyers for I.G. Farben negotiated up to $500,000. The government lawyers agreed that $500,000 was "fair" based on the estimates they received from the military.[101] Only later did the I.G. Farben lawyers learn that the military officials had estimated the patent value at $620,000, and their anger at the disclosure resulted in another meeting before the arbiter in which they asked for a settlement at the higher figure. The government lawyers refused to adjust the settlement without a full-blown litigation.[102]

Although the I.G. Farben lawyers stopped short of litigating their case, they filed a motion, memoranda, and working notes arguing for a larger compensation. They tried several lines of argument. They noted that the value as calculated by the military officers amounted to a royalty rate of 2.5 percent, whereas the original TWEA had suggested up to 5 percent, and many unrelated court decisions had imposed patent royalty rates at 10 percent and even 25 percent.[103] They argued again that *USA v. The Chemical Foundation, Inc.* had clearly demonstrated a government right to license any patent from the Chemical Foundation and that the courts recognized a public interest in the formation of the Chemical Foundation. Therefore, fair compensation to the German patent owners would include consideration at least of Chemical Foundation patents used by the government.[104] On January 23, 1931, the arbiter issued the final decision, granting I.G. Farben an award of $537,555.69, which, with interest since 1921, amounted to $739,175.89. In May, I.G. Farben received payment for about one-quarter of the award and likely received 80 percent of it in the early 1930s. The United States continued to hold the other 20 percent of all German assets while waiting for Germany to pay American claims—a matter taken up again in the midst of World War II war claims negotiations in the 1950s.[105]

The year 1931 not only saw the final judgment under the War Claims Act, but it also marked the end of litigation of the Section 10(f) suits for patent royalties that the German chemical firms had filed back in 1921 and 1922 and that had been suspended for the duration of *USA v. The Chemical Foundation, Inc.* Once the Chemical Foundation case was settled, district courts around the country resumed proceedings on the suspended Section 10(f) lawsuits, which determined the amount of royalties due owners whose patents had been infringed during the war. Many of the German firms, as well as the Chemical Foundation, claimed the royalties collected under the FTC licenses. In 1924, the litigating participants agreed to choose four test cases from among the dozens of suits filed in district courts.[106] In each of the four cases, the defendant (or licensee) was the Du Pont Company, while in three of the cases, the plaintiffs were German chemical companies. The Chemical Foundation became the plaintiff in the fourth test suit. During the suit, the number of cases doubled when the Department of Justice joined the litigation, arguing that the APC was the effective owner of the patents in 1921 and 1922. Like the Chemical Foundation case, the Section 10(f) suits started in Judge Hugh Morris's District Court in Delaware and proceeded to the Supreme Court. For the German plaintiffs, Judge Morris rendered the worst possible judgment, ruling that the German firms had no claim to the patent royalties because the Germans had lost ownership of the patents when the APC seized them. Under Section 10(f), only the legal owner could file for royalties, and in 1921 and 1922 when the claims had to be filed, ownership resided in the Chemical Foundation. Therefore, only the Chemical Foundation could claim the patent royalties collected during and immediately after the war.[107] In the Department of Justice, where feelings still ran high against the Chemical Foundation, the district court verdict was viewed as "disastrous" because all the royalties would end up in the hands of the Chemical Foundation or American chemical companies, rather than the original German owners. As a result, the officials joined with the lawyers for the German firms in appealing the district court decision. As in *USA v. The Chemical Foundation, Inc.* the appeals reached the U.S. Supreme Court, which in 1931 upheld the decision of the district court. The Supreme Court ruling effectively ended any chance Section 10(f) of the TWEA would serve its original purpose as a guarantee of patent property rights. The Germans would receive no compensation from Section 10(f), the first and last major opportunity the Americans had presented to return German assets.[108]

###

In the 1920s, even as the government demobilized and dramatically reduced its scale and scope, the synthetic organic chemicals industry continued to obtain support from the federal government, despite some opposition and backlash. Americans implemented aggressive promotional policies, which gathered the backing of the courts in the 1920s and provided extraordinary protection as the American industry benefited from the renewed isolationism. World War I led to an extreme nationalism that created the political and public support for the more dramatic policies. Political scientist David Hart has noted that science and technology policy is too often analyzed as separate and different from American politics, but the story of the U.S. synthetic organic chemicals industry shows how deeply embedded the policies were in domestic and international politics.[109] Viewed both as strategic to national security and a bulwark against the risks of international trade, the domestic industry benefited from the government's ability to provide trade information, to increase the costs of rival German products, and even to confiscate German property.

The policies also reflected Americans' growing faith in the power of the federal government to shape the economy and to aid business in the early twentieth century. Many Progressive Era reforms tamed the excesses of the Gilded Age, but Americans also viewed the federal government as an ally of business in new ways, particularly coming out of the war. Even if government support did not yet resemble the Cold War years, the policies still shaped the decision making of entrepreneurs.[110] With policies ranging from tariffs and economic studies to confiscations and war reparations, the federal government gave the infant industry a chance to grow, but the manufacturers still needed to invest heavily in their own capacities. The policies gained manufacturers the time to build expertise and organizational capacity and to develop the ability to innovate and take the industry in new directions.

8 / An "American" Industry, 1919–1930

By 1930, the American dyes and pharmaceuticals industry had reached a stage of development sufficient to prevent another deep crisis if nations again disrupted international commerce. The industry satisfied policymakers' basic expectations for its role in the national defense, and, after a decade of learning and consolidation, the leading firms usually turned a profit. Americans, however, depended on the exceptional advantages of the steep tariffs and confiscated set of patents to hold their position in the U.S. market, and German firms recovered many of the international markets where Americans had ventured during the war. At the end of the 1920s, the representatives of the Synthetic Organic Chemicals Manufacturers' Association (SOCMA) reported to a congressional committee that the industry made a relatively small percentage of the total number of dyes consumed in the United States. The American industry, they said, produced "90 per cent of the poundage . . . 80 per cent in value . . . [and] 25 per cent of all of the different varieties."[1] Using synthetic dyes as a frame of reference, the extraordinary efforts to build an American industry had yielded satisfactory if not stellar results by 1930.

Expanding the narrative beyond synthetic dyes and pharmaceuticals to the larger sector of synthetic organic chemicals, however, provides a story with a different arc. The synthetic organic chemicals industry underwent a profound transition in the 1920s, and the German ruler by which Americans had measured their progress during the war became increasingly irrelevant. While dyes remained a notable segment of the industry, the sector expanded and changed, and entirely new product categories and new raw materials redefined the boundaries of the industry. Decades of work had created an enormous technical advantage for the German firms in dyes and pharmaceuticals, but in the newer developments, the Germans had little or no head start on American rivals, and the two sides competed with relative parity.

On the most elemental level, the chemistry shifted from the aromatic, ring-molecule, benzene-based organic chemistry to the aliphatic, chain-molecule, ethylene-based organic chemistry. Petroleum and natural gas became the new source material for synthetic organic chemicals, beginning to supplant coal, the traditional "German" raw material. The product set expanded from relatively small batch quantities of synthetic dyes and pharmaceuticals to bulk, or commodity, production of lubricants, solvents, antifreeze, lacquers, and plastics that particularly sold well in the American consumer society of the 1920s. Automobiles, especially, helped to "Americanize" the synthetic organic chemicals industry. Instead of looking backward, where Americans saw only German progress and American failure, Americans in the industry looked forward to a future with rapid expansion in plastics, synthetic alcohols, petrochemicals, and thousands of products with industrial and consumer uses.

Significantly, in 1928, the value of the new synthetic organic products exceeded the value of products derived from coal tar. "Continued progress in this [aliphatic] field of chemistry," wrote the chemists of the U.S. Tariff Commission, "will probably raise the United States to a position comparable to that held by Germany before the war in the manufacture of coal tar chemicals." From 1921 to 1927, they noted, the coal tar part of the U.S. industry had tripled in size; the aliphatic part had grown thirteen times over. Synthetic dyes and pharmaceuticals filled niche markets. Even high-volume products like indigo or aspirin were dwarfed compared to commodity organics. Many of the new products found a market in the American automobile industry, an advantage the Germans were less able to exploit, partly due to a smaller European market for automobiles.[2]

The shift to commodity aliphatic chemicals coincided with other important developments in the chemical industry, including new technologies on both sides of the Atlantic and the expansion of chemical engineering in the United States. Technologies that allowed high-pressure and high-temperature processing permitted a wider variety of reactions. The Germans' Haber-Bosch process for synthetic ammonia (1913) set the stage for other high-temperature, high-pressure syntheses. In 1923, BASF followed up their Haber-Bosch success with their first high-pressure, high-temperature synthesis of an organic commodity product, synthetic methanol. In the United States, the new technologies and the birth of petrochemicals spurred further development of chemical engineering, as firms faced the challenge of manufacturing and handling with precision enormous quantities of chemicals, often at those high temperatures and pressures.

Starting with the German firms, this chapter centers on the major players in the U.S. market in the 1920s. Because synthetic organic chemicals remained deeply embedded in the transatlantic political economy, the

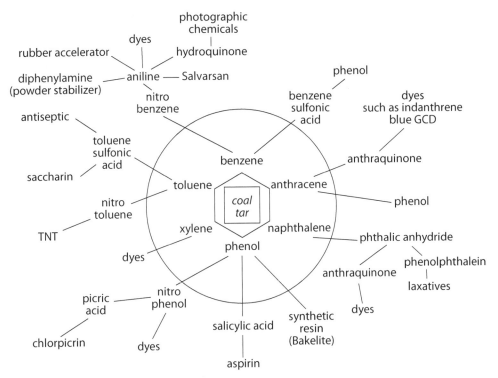

FIGURE 8.1. Selected aromatic synthetic organic chemicals (ring molecules). (Based on charts in Norton, *Artificial Dyestuffs Used in the United States* [1916]; Peter Groesbeck, graphic artist.)

strategic decisions of American chemical firms continued to carry larger significance to national identity. Although the rise of the aliphatic branch of the industry integrated advances from both sides of the Atlantic, including many "firsts" by Germans, Americans narrated the developments as an American story.

German Industry

Overcapacity in dyes and diversification in products characterized the German synthetic organic chemical industry in the 1920s. Dragged down by international political and legal impositions, as well as by domestic civil unrest and hyperinflation, the German manufacturers nonetheless fought hard to regain markets around the world. To meet overcapacity in dyes, they merged and concentrated firms and faced the international markets with a mix of ferocious competition and cooperative agreements. Increasingly

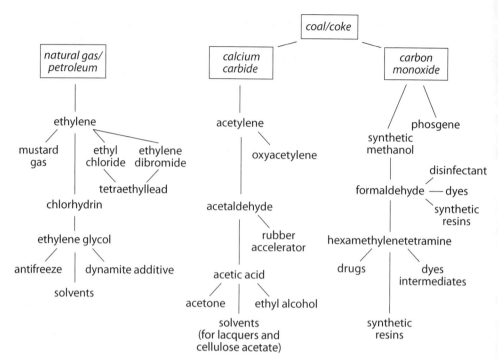

FIGURE 8.2. Selected aliphatic synthetic organic chemicals (chain molecules). By midcentury, acetylene was more commonly derived from petroleum cracking. (Based on information in U.S. Tariff Commission, *Census of Dyes and of Other Synthetic Organic Chemicals* [1930]; Peter Groesbeck, graphic artist.)

the German leadership focused their energies on new products, built on their prewar expertise with synthetic organic chemistry and with the high-pressure synthesis techniques developed for nitrogen fixation. With their chemical expertise, the Germans continued to lead the world, and they could leverage the economic power their expertise carried, despite postwar obstacles and the great expansion of industries elsewhere, particularly the United States. While the Germans continued to be the unquestioned experts in synthetic dyes through the 1920s, they acknowledged they had lost much of their former market for cheaper dyes sold in quantity and therefore concentrated on the more complicated, high-quality dyes. Confiscated patents and high tariff barriers complicated their return to the American market, but the German technical superiority was great enough to permit them to recover about a quarter of the U.S. dyes market by 1930. While Garvan and his allies tried hard to sever ties between the German firms and their American affiliates, longstanding institutional and personal ties contributed to the German return to the United States.

Looking over the world markets, the German manufacturers understood that former consumers like the United States had slammed shut their borders with restrictive protectionist measures. They accepted, if bitterly, that they would not recover their prewar position in the world's dyes and pharmaceuticals markets, and their most significant response was their increasing consolidation, punctuated by the merger into Interessengemeinschaft Farbenindustrie A.G. (I.G. Farben) in 1925. Already in a close cartel formation in 1916, the six major firms completed a full merger, eventually eliminating duplicate operations and turning the efforts of the new behemoth to newer and more profitable lines of chemicals. In particular, the I.G. Farben scientists applied high-pressure techniques to a wider range of products, including organic products like synthetic alcohols and synthetic gasoline from coal. Notably, Carl Bosch, co-developer of the Haber-Bosch nitrogen fixation process in 1913, ascended to the top position in I.G. Farben, rather than Carl Duisberg, the leader of Bayer who had expanded Bayer's share of the dyes market. While I.G. Farben stabilized their dyes industry on profitable, high-end dyes, the leadership in the 1920s thought their future lay elsewhere and invested heavily in the high-pressure technologies.[3]

The I.G. Farben merger accomplished consolidation among the German firms, but the German executives also addressed the international competition and political uncertainty with cooperation. Very quickly, German manufacturers looked to their former enemies to negotiate arrangements to curtail competition. In France, the government-supported La Compagnie Nationale des Matières Colorantes et des Produites Chimiques led the dyes industry; the country's eight firms in 1919 made less than half of French consumption and made none of the more complicated dyes. In 1921, BASF and the Compagnie Nationale executed an agreement: in exchange for BASF's technical assistance, including personnel, processes, and plant designs, the French agreed to turn over 50 percent of their profits, plus restrict their sales to France and French colonies. Such terms reflected the immense disparity between the French and German industries. In April 1922, a rumor of the agreement reached the American trade press through the British, but the French denied most of the terms, stating only that they had purchased outright the German expertise to avoid the cost of learning the processes through research. After "[a] storm of protest throughout France," reported a Department of Commerce official in 1924, the "French Government made an official denial of the existence of such an agreement." The same report noted that the number of dyes made in France went from 10 in 1920 to more than 600 in 1921 and more than 1,000 in 1922, an unacknowledged tribute to German expertise. When the French invaded the Ruhr in 1923, claiming a substantial stockpile of German dyes in the process, the agreement ended, and the Compagnie Nationale merged with a more profitable French firm, Etablissements Kuhlmann.[4]

That the French would arrive at such a decision with BASF perhaps surprised few close observers; when the news leaked that a British firm engaged in similar negotiations, many Americans in the industry felt betrayed. Already in 1921, the Germans approached the British Dyestuffs Corporation (BDC) to pursue an agreement similar to the one with the French company, an offer that the British firm's leadership initially deemed insulting. But the leaders resumed negotiations in 1923 because of profound doubts about the BDC's future. As in the United States, the domestic manufacturers could supply the majority of dyes for the home market, and one firm, Scottish Dyes, profitably made vat dyes. Including a Swiss subsidiary, nineteen firms made dyes in Great Britain. The British import licensing system effectively kept German dyes from competing with domestic products, except when a court suspended the law for 1920, during which time the British textile manufacturers stockpiled as many German dyes as they could possibly acquire. But the licensing law expired after ten years, and William Alexander and others in the leadership of the British Dyestuffs Corporation doubted a decade would be long enough to become competitive. Unlike the German firms or the big American firms, British Dyestuffs Corporation made only dyes and could not rely on other lines of business to cover losses during a learning period. Negotiating from a weak position in markets and technical expertise, Alexander attempted to conclude an agreement with the Germans, but it fell apart, not least because of significant opposition in Great Britain. Instead, with Alexander fired, the British Dyestuffs firm merged in 1926 with other British chemical firms into the Imperial Chemical Industries (ICI), a corporation that included a broader range of chemical lines, more akin to I.G. Farben, Du Pont, and Allied Chemical in scope.[5]

And the German ties to the United States? The steep U.S. tariff prevented the German manufacturers from reclaiming their prewar market share, but they were formidable competitors for the world markets, which the American manufacturers had invaded during the war. With their superior expertise still intact, the Germans also hoped to pursue agreements with American firms. Deep anger still existed on both sides—the Germans believed their property had been illegally taken, and the Americans' anti-German sentiment faded only in the early 1920s. At one extreme, Garvan and Herty kept up their drumbeat of chemical independence as necessary for national defense, and the Trading-with-the-Enemy Act required Americans who wished to do business with Germany to step carefully. Despite the political and legal landmines, the fierce competition nationally and internationally made cooperation with the German firms a temptation for several American manufacturers—and not only for those with prewar ties to the German firms. Especially as the hostility to Germany faded, Americans found they could do business with their former enemies. While antitrust law limited the extent to which the sides could engage in market-sharing agreements, at least within the United States, German expertise in both old and new product lines formed the basis for negotiations.

The American political environment in the few years immediately after the war made discussions between Americans and Germans tricky but not impossible. With TWEA, no formal peace agreement until the summer of 1921, and residual if waning anti-German sentiment, any American conducting business with Germans needed to be mindful of law and public relations. To some Americans in the chemical community, commerce with Germany appeared as a betrayal long after the government lifted most restrictions on trade in the summer of 1919. Naturally, Garvan and his allies very closely watched the importers for German firms, primarily Kuttroff, Pickhardt & Company and Herman Metz, for the extent of their contacts with their former German partners. Domestic manufacturers, too, received scrutiny, even the two dominant firms. In 1920, Berthold A. Ludwig, a German-born chemist working for National Aniline and previously employed by the Cassella agency, traveled back to Germany to visit family and former business associates. Rumors circulated that National Aniline tried to buy German intermediates through Ludwig, and Herty, as a representative of domestic manufacturers, investigated the charges until convinced by National Aniline executives that Ludwig's trip was completely personal. The rumors, however, held more than a grain of truth, as Duisberg reported to his German colleagues that Ludwig explored the possibility of market-sharing agreements between the German firms and National Aniline. Knowing that they would be excluded from much of the U.S. market, the Germans refused to negotiate about international markets, believing them necessary for the survival of the German firms, but he suggested very tentatively that the two sides might come to an accommodation about sharing the American market. Presumably National Aniline executives ignored that idea. The trade press apparently did not learn of discussions between Du Pont and German executives in 1923, as the biggest American chemical firm cautiously explored negotiating for German technical expertise in dyes. Herty and Garvan, however, caught wind of the discussions, prompting Irénée du Pont to respond to Garvan that Du Pont was not "selling out or surrendering" the American industry, but the firm would make a deal "if the Germans offered us any valuable 'know-how' or patents." They also met rebuff from the German manufacturers, who still chafed at losing their property in the United States and who had also already made significant, if somewhat quiet, headway negotiating with their former subsidiaries and importing agents.[6]

The hostility toward Germany that prevailed makes more remarkable the actions not only of the two leading American firms, but also of the executives of Sterling Products, the firm that purchased Bayer's pharmaceuticals property from the Alien Property Custodian. Although Sterling and the German Bayer company began negotiations in the fall of 1919 in Baden-Baden, Germany, even as peace negotiations occurred in France, little news of their relationship became known among domestic chemicals manufacturers. Confusion over Bayer's rights in the South American market first brought Sterling Products and the German Bayer company together. During the early part of the war, the German firm had turned the rights to the South American business over to the American Bayer corporation, which was later seized by the Alien Property Custodian and which Sterling had purchased. Through 1919 and 1920, representatives from the German Bayer and Sterling negotiated agreements that gave the American firm the exclusive right to sell aspirin in South America in exchange for 75 percent of the profits and a promise from the Germans to stay out. However, by agreement, the German firm controlled the South American market for all other pharmaceutical products.[7]

The German firm's executives expected that their agreement with Sterling over South America would lead to similar profit- and market-sharing in the United States. But Sterling's leadership had no inclination to compromise in the American market as easily as it had in South America. First, William Weiss, Sterling's leading executive, knew very well that the Alien Property Custodian would press charges if Sterling violated prohibitions against involving the German company in its old property. He thought the matter especially dangerous to make agreements before the peace treaty was signed. Weiss observed his commitment to the Alien Property Custodian not to allow the Germans to regain ownership of the Bayer Company of 1918 throughout the interwar period, much to the chagrin of the Germans. Second, Sterling executives sought a good return on its investment, particularly on Bayer aspirin, its most profitable product and one that could be made without German technical help.[8]

Early in 1919, Sterling's executives established the Winthrop Chemical Company as a subsidiary to manufacture and sell non-aspirin pharmaceuticals, and the Americans used the Winthrop subsidiary to engage in profit-sharing deals with the Germans. Weiss resisted the German company's desire to regain its prewar share of the American market for aspirin, but he found opposing them a challenge. As one Sterling representative wrote of the German Bayer executives, they "do not only consider the seizure of the U.S. business as a great wrong to them, but [they] also can not get away from the feeling... that the purchase of the seized property is morally wrong." Weiss paid a price for excluding the German Bayer company from the American aspirin market. The Germans applied pressure by competing with Sterling in the British Empire, for which Sterling owned the Bayer trademark. At one point, a German representative suggested that Sterling could find a way to make a deal on aspirin that would not "be apparent to the outside world."[9]

The German executives found negotiations with Weiss difficult, but they knew they had leverage: they could establish a competing business in the United States, if necessary. In 1922, they contracted with another American firm to produce one of the German Bayer's new products, and Weiss understood the message. The Germans also believed that Weiss's goal, "like all Americans," was to "make money," and that would lead to a deal. Aspirin continued to be the sticking point in the negotiations. In April 1923, Sterling and German Bayer executives reached their first agreement that covered Sterling's operations in the United States, and the terms resembled those offered to the French and British firms. Most importantly, Sterling shared half the profits of its Winthrop subsidiary in exchange for Bayer's promise not to sell its products in the United States through any other source. The German Bayer also promised to share technical and manufacturing expertise. Thus, in addition to agreements dividing the South American and British markets, this latest compact further joined Sterling and Bayer in profit-sharing arrangements. Notably, Sterling executives did not surrender stock ownership of the American Bayer to the German firm, and therefore stayed within the letter if not the spirit of their obligations to the Alien Property Custodian.[10]

Meanwhile, the Grasselli Chemical Company, which suffered heavily during the economic downturn of 1921, arrived at arrangements with the German Bayer similar to those between Sterling and the German firm. Grasselli had purchased the Bayer dyes property from Sterling, however, and the Alien Property Custodian never required that Grasselli adhere to the same legal obligations prohibiting sale of their stock to the former German owners. For their part, the Germans in their negotiations had the distinct impression that Grasselli executives really had no profound desire to invest the energy and capital to make dyes well and welcomed the opportunity to let the German Bayer take the lead in their dyes business. In 1924,

Grasselli and Bayer organized jointly the Grasselli Dyestuff Corporation. The German firm supplied technical assistance in the American plants and designated Grasselli as the exclusive sales agent of Bayer dyes. As a result, Grasselli lost membership in SOCMA, which had by-laws that required American ownership to be a member. But by the mid-1920s, the Grasselli and Bayer firm engaged in capital improvements to Grasselli's part of the Rensselaer plant, which made dyes, and also to Grasselli's second plant in Linden, New Jersey, which manufactured organic intermediates.[11]

Unlike the firms associated with I.G. Farben firms, the American Merck and Heyden remained independent from their former German parent companies and tended to ally with other American firms rather than German ones. The trustees appointed to Merck's board of directors by the Alien Property Custodian prevented George Merck from resuming close relations with the German Merck, but the American company began to import chemicals from E. Merck in Germany, received technical aid (including plant blueprints, apparatus, processes, and "recipes"), and generally had friendly and cooperative relations with the Darmstadt firm. In 1925, George Merck's son, George W. Merck, took the firm's helm and led the merger of Merck with one of its leading rivals, the Powers-Weightman-Rosengarten Company of Philadelphia. The new firm retained the Merck name and eventually outgrew its German parent company. By 1930, Merck was one of the most important pharmaceuticals firms in the United States and contributed a range of synthetic organic medicinals, including barbiturates, arsenic organics, and general anesthetics. Merck, with its wider experience in synthetic organic drugs and a core group of competent researchers since the war, conducted research sufficient to produce aspirin, barbital (Veronal), and other synthetic drugs. In the 1930s Merck developed a more intensive fundamental research agenda, one that explicitly targeted synthetic organic chemicals.[12]

In 1926, the Heyden Company bought the Norvell Chemical Corporation of Fords, New Jersey, a company with a similar line of products that had been founded by George Simon after he lost the Heyden Company to the Alien Property Custodian in 1919. He again became a director of the Heyden Company after the acquisition, although his ties to Germany at that point remain unclear. Heyden's production of synthetic organic chemicals in 1930 included salicylates and anesthetics.[13]

With the merger of I.G. Farben, the German manufacturers pushed their various American partners to consolidate as well. In 1925, Herman Metz helped to orchestrate the merger of the most prominent dyestuffs importers into the General Dyestuff Company. In the summer of 1925, Metz combined

Table 8.1. U.S. Synthetic Organic Pharmaceuticals Output, 1922–1929

Year	No. of Firms	Output (in Millions of Pounds)	Sales (in Millions)
1922	35	2.9	$4.2
1923	36	3.3	$4.7
1924	?	3.0	$5.2
1925	?	3.2	$6.3
1926	?	3.7	$6.7
1927	?	3.6	$6.8
1928	?	4.0	$8.6
1929	30	5.0	$8.4

Source: Figures compiled from U.S. Tariff Commission annual reports; *Census of Dyes and Coal-Tar Chemicals*, 1922–1930, rounded to the nearest 100,000; and Haynes, *American Chemical Industry*, 4:245.

his dye companies—H. A. Metz & Company and Consolidated Chemical Company—with the Grasselli Dyestuff Corporation. Metz and Hoechst understood that his dyes plants needed upgrading to manufacture competitively, and the I.G. Farben executives had no interest in duplicating their ongoing investments in Grasselli Dyestuffs. With significant overcapacity in the industry already ("sufficient quantities to supply the universe," as Metz put it), Metz willingly sold his dyes plants to the Grasselli interests. In exchange, Grasselli sold its dyes through the new General Dyestuff Company. The final addition came on January 1, 1926, when the dyestuffs section of Kuttroff, Pickhardt & Company joined General Dyestuffs, which brought in the distribution rights not only for BASF products but also several smaller domestic producers. Adolf Kuttroff became the chair of the board and Metz the president. Berthold Ludwig became the manager of General Dyestuffs. (Ludwig's career was most intriguing: a German initially employed at Schoellkopf, he worked at the Cassella agency before the war, stayed with National Aniline from 1917 to 1925, became manager of General Dyestuffs from 1925–1930, and then went back to National Aniline in 1930 to become its president.)[14]

The pharmaceutical agencies for the constituent members of I.G. Farben also merged in the late 1920s. At the time of the merger, Bayer and Hoechst both held contracts with the American manufacturing and distributing companies, Sterling's Winthrop subsidiary and H. A. Metz Laboratories, respectively. Concentrating his energies on General Dyestuff, Metz was amenable to selling his pharmaceuticals firms, Metz Laboratories, and the smaller General Drug. I.G. Farben's Bayer group pressured Weiss to relinquish control of Sterling's two subsidiaries, Bayer Company of New York

Table 8.2. U.S. Synthetic Organic Pharmaceuticals, Selected Output (in Pounds), 1922–1930

	Aspirin	Neosalvarsan	Cinchophen	Acetanilide
1922	1,482,998	2,904	—	222,517
1923	1,525,498	3,365	32,710	564,498
1924	1,366,530	3,220	56,003	425,950
1925	1,499,166	3,289	60,722	158,756
1926	1,823,748	4,113	79,632	458,927
1927	1,715,686	3,889	84,212	366,842
1928	1,816,015	4,814	94,330	480,273
1929	2,710,374	5,525	99,538	355,019
1930	—	4,561	93,765	297,778

Source: U.S. Tariff Commission, *Census of Dyes and Coal-Tar Chemicals*, 1928, p. 57, and 1930, p. 47.

and Winthrop Chemical Company. He would not compromise on the ownership of the American Bayer Company and its aspirin business, but he engaged in negotiations to expand I.G. Farben's role in Winthrop beyond profit sharing. By June 1926, Weiss and the Germans agreed to share the ownership of the Winthrop Chemical Company, and I.G. Farben put up half the money for Winthrop to purchase the Metz drug companies. Consequently, Winthrop became I.G. Farben's pharmaceuticals representative in the United States; Weiss continued in his refusal to give any ownership to the Germans of Bayer Company. Although the agreements stipulated I.G. Farben provide manufacturing expertise to Winthrop, increasingly the Germans manufactured the goods at home and exported them to the United States for Winthrop to sell. To avoid higher tariffs, products often arrived in a state of incomplete manufacture, which the American plants would finish. When Department of Justice officials in World War II reviewed the interwar relationship between Sterling and I.G. Farben, they concluded that the American end had largely become a sales agency dependent on German research and manufacturing, a situation comparable to the years before World War I.[15]

In the late 1920s, I.G. Farben moved to consolidate further its interests in the United States, and they reestablished a strong, if not dominant, position in the American dyes and pharmaceuticals market and looked ahead to new products. In 1928, Grasselli sold its heavy chemicals division to Du Pont, and I.G. Farben bought the Grasselli Dyestuff Corporation and renamed the firm the General Aniline Works. From 1926 to 1933, I.G. Farben invested $12 million in capital improvements to its American dyes plant. While the Germans typically still chose to make its high-end dyes in Germany—they estimated such dyes cost two to three times more when made in their American plants—they moved the production of some more complicated dyes to the United States and even acknowledged adopting in Germany a few technical innovations from the American plant. In 1929, the German cartel concentrated its American synthetic organic chemicals interests in a new holding company, the American I.G. The merger included the General Aniline Works, the General Dyestuffs Corporation, and a controlling interest in Afga-Ansco, the second largest manufacturer of photographic film in the United States. Although I.G. Farben's interest in Winthrop remained outside the American I.G., William Weiss of Sterling sat on the board of directors. Other directors included Metz, Adolf Kuttroff, several members of I.G. Farben, Walter Teagle of Standard Oil, and Edsel Ford of Ford Motor Company. The formation of the American I.G. marked the end of six decades of the independent existence for American importers of German organic chemicals. The Germans never regained the overwhelming market share of the prewar years, but through production at their American plants they secured a healthy quarter of the American dyes market by 1930, plus their 12–13 percent share of dyes imports into the United States. Especially after the tariff rate automatically dropped in 1924, the rate of dyes imports increased, despite growing domestic production. In 1930, the General Aniline Works produced 170 different dyes, about the same as Du Pont. By the end of the 1930s, the firm accounted for more than one quarter of synthetic dyes sales in the United States, slightly higher than Du Pont's market share. With their hold on the synthetic dyes and pharmaceuticals market stabilized and profitable, the I.G. Farben leadership sought to bring to the United States their newer high-pressure products such as synthetic methanol and gasoline.[16]

The formation of the American I.G. set off old warriors like Francis Garvan, who found the very name offensive. At one point he wrote a letter to Standard Oil's Walter Teagle and banker Paul Warburg, both on the board of the American I.G., accusing them of unwittingly serving as the Trojan horse that would let the Germans slip back into American markets and undercut the American chemical industry. In their most conspiracy-minded ruminations, Garvan and his allies imagined the Germans establishing themselves as the research laboratory of the world, maintaining a monopoly on chemical expertise and knowledge while using American capital to undertake the development work. Such an arrangement would leave Americans vulnerable if the Germans subsequently withdrew their support. That Garvan increasingly protested from the margins was evident; the American I.G. stock sold well precisely because investors trusted German expertise and experience. Garvan accused American financiers of putting profit ahead of the national interest by investing in the joint venture.[17] But Garvan increasingly belonged to a passing generation.

Beyond I.G. Farben's direct investments in the United States, the chemical industries of the two countries developed relationships around international cartels and market sharing. By the end of the 1920s, a handful of American firms had emerged as the dominant producers of synthetic organic dyes in the United States, forming an oligopoly and creating conditions for participation in international cartel agreements. In Europe, where no antitrust legislation hindered combinations, chemical firms continued their decades-old trend toward tighter cartelization. In Germany, Switzerland, France, and Britain, one firm or cartel dominated its home country's market, and I.G. Farben exercised powerful leverage with its continental neighbors, France and Switzerland. The European firms, partly as the result of overcapacity and the increasing scale of operations in the chemicals industries, needed to export their products to other parts of the world. Instead of competing in free market competition, however, the European firms "negotiated" market shares in a form of "business diplomacy" to form cartels and mitigate the risk of competing and of political uncertainties.[18]

National Aniline & Chemical Company/ Allied Dye & Chemical Corporation

At the end of the war, National Aniline & Chemical Company was the largest American manufacturer of synthetic dyes, but a merger into a larger firm reshaped the direction its executives pursued. At the opening of 1921, National Aniline joined with other firms to become the Allied Dye & Chemical Corporation, which was the second largest chemical company after Du Pont for most of the interwar years. The Allied leadership made its top priority an expansion into synthetic ammonia and nitrogen rather than into organic chemicals, particularly niche-market goods like dyes. Allied manufactured primarily large-quantity industrial chemicals rather than any products reaching a final retail consumer, and its name therefore remained relatively unknown outside of the industry. The corporation's leadership gained a reputation for decision making based on close financial analysis, and laid-off chemists and outside observers questioned the firm's commitment to research, especially when compared to Du Pont and up-and-comer Union Carbide.

National Aniline entered the peacetime market in the strongest position of any American dyes company. The firm turned profits approaching 10 percent in the early postwar years when Du Pont lost so much. Its assets at the end of 1918 totaled $45 million. In 1919, the firm manufactured 200 dyes and 84 intermediates and possibly held as much as 60 percent of the domestic market. Recalled one important board member, "The company made a great deal of money without even trying to exploit" the market "to its fullest potential." It boasted the largest dyes plant in the United States in Buffalo ("as large as a German unit of the same class," said the company president) and claimed to be the world's largest producer of aniline oil with its plant in Marcus Hook, Pennsylvania, making "five times as much aniline oil as the total consumption prior to the war." A visitor to the Buffalo plant in 1920 marveled at the plant's size, cement construction, two powerhouses, labor-saving devices, and even its cleanliness.[19]

Two key people shaped the direction of National Aniline and the larger Allied: banker Eugene Meyer and executive Orlando F. Weber, both of whom readily dismissed managers. Even before the Allied merger negotiations in 1920, the immediate postwar years provided a period of frequent reorganization and personnel changes at the top of the National Aniline hierarchy. Meyer had first backed William Beckers's dyes company and remained a National Aniline board member. Meyer brought in Weber to help run National Aniline, and he served as Meyer's representative on the board when Meyer could not attend. Growing up in Wisconsin, Weber raced bicycles and opened his own bicycle shop in Milwaukee. From bicycles, Weber moved into the automobile industry, first with the Pope Toledo Company and then with the Maxwell Automobile Company. While at Maxwell, Weber met Meyer, who sat on Maxwell's board of directors. Meyer liked Weber's hard-working, independent nature and his facility with numbers and organization.[20]

After Meyer and the other board members pushed the Schoellkopfs out of management in 1918, William J. Matheson, formerly of the Cassella importing firm, continued as the chair of the board and president of the company. According to Meyer's recollection, Matheson "made a failure of his management," too, and when the very wealthy Matheson took four months off to enjoy his real estate holdings in Florida, Weber became acting president. During that time, he "did a good deal of clean-up management work. He got rid of undesirable people," installed a better accounting system, and placed the company "on a fairly decent business basis," in Meyer's interpretation. When Matheson returned to the job, he not only was upset at Weber's changes but also "began to undo the good that Mr. Weber had done." Before Meyer joined the peace negotiations in Paris for a few weeks, he told Weber that when he returned he wanted "Mr. Matheson out and Mr. Weber in as president of the company. On my return Orlando Weber was president of National Aniline. I don't know how he handled that. I didn't ask." Sensing the growth of 1919 would not continue, Meyer resisted the plans of Weber and the other executives to expand with new factories. Instead, they followed Meyer's advice to exploit existing facilities to the maximum and thus, he believed, avoided the worst effects of the steep recession in 1920–1921.[21]

In 1920, William Nichols of General Chemical led negotiations among the four firms that held stock in National Aniline to create a merger of the five firms into a larger corporation that encompassed all necessary processes from raw materials to finished product. Meyer and Nichols engaged in some hardball negotiations, according to Meyer, and hammered out the merger agreement that organized the Allied Dye & Chemical Corporation. Comprising National Aniline, General Chemical, Solvay Process Company, Semet-Solvay, and Barrett, the Allied merger on January 1, 1921, gave the United States a second large chemical corporation with a broad range of products, organic and inorganic. With assets of $282 million, the five companies brought together processes and markets from synthetic dyes and intermediates, inorganic acids, alkalis and nitrogen products, by-product coking plants, and by-product distillation. The synthetic organic chemicals operations of National Aniline benefited from the integration of its raw materials suppliers into the new corporation. Because Allied was a holding company, its constituent parts retained their separate corporate identities, and the dyes were still known as National Aniline dyes.[22]

Orlando F. Weber gave up the presidency of National Aniline, moving up to become the president of Allied, where he quickly made his style of management felt. For the next fifteen years, he kept himself and his corporation shrouded in a layer of secrecy unparalleled among corporate contemporaries. Weber never granted interviews or gave speeches, and Allied joined neither trade organizations nor international agreements to share technical information or, apparently, to divide markets. In time, Allied gained a reputation for rapidly promoting promising young employees, although Weber was equally notorious for firing executives, which garnered him the sobriquet "Fire Chief." Eugene Meyer noted the change in Weber. While he believed Weber ran Allied "in a very clean way," Weber "underwent a transformation with success." "He got a big swelled head." He "liked dictatorial methods" and was "slightly on the Fascist side" and "slightly on the megalomania side." "He wanted to buy up all the companies in the country and run them all as subsidiaries of Allied Chemical." And Meyer liked Weber, comparatively speaking. When the recession of 1920–1921 hit Allied, Weber cleaned house, reorganizing the corporation and laying off employees throughout the company. Whether departing voluntarily or forced out,

Henry Wigglesworth and Clarence Derick were part of the exodus from the National Aniline division in 1921. Wigglesworth lost his position on National Aniline's board, and Derick, who afterward complained of "those ignorant arrogant executives," became vice president of the By-Products Steel Corporation and then a consulting chemist in 1925.[23]

Derick was not the only person to note National Aniline's abandonment of chemists in the top echelon of the corporation. Like others, Derick believed that corporations led by financial men rather than chemists undervalued research and the role it played in the chemical industry. His frustration with National Aniline led him to despair for the future of the American industry; as he privately bemoaned to Herty, the industry "has deserted true research entirely and is worshipping the god of hunches and trusting to sharp practices to hold its own." While National Aniline had the strongest position in the dyes industry shortly after the war, evidence suggests they had their technical problems. As with the rest of the American industry, the company faltered on vat dyes and appeared to struggle with its indigo production at least as much as Du Pont. In January 1919 National Aniline reportedly produced 30,000 pounds of indigo and seemed ready to capture a sizable share of the market. Although evidence comes largely from annoyed rivals, the company seems to have suffered manufacturing problems over the next two years and let any advantages from Schoellkopf and Beckers's prewar experience slip away. In 1920, another indigo manufacturer, Herbert Dow, recognized Du Pont as the producer that "has the customers" and groused when National Aniline dropped indigo prices in an attempt to lure Du Pont's and Dow's customers away.[24]

National Aniline officials made it clear already in 1921 that they had little interest in cooperating with the rest of the trade. Not only did they oppose the formation of SOCMA and refuse to join, but they also rejected any discussion that they cooperate in selling exports under the Webb Act, the law that made cartel behavior legal in selling exports. The rest of the trade gossiped pretty regularly about National Aniline. In 1925, Orlando Weber reportedly snubbed William Alexander of the British Dyestuffs Corporation when he sought pricing agreements. Calco's Crossley believed National Aniline had taken their methylene blue process; he also reported the bad quality of nitrobenzene his firm received from National Aniline in 1925. Dow and Du Pont discussed how to respond to National Aniline's frequent attempts to lower the price on indigo. Both believed they were more cost effective producers than the Buffalo firm, and they contemplated filing a complaint against National Aniline with the Federal Trade Commission, though they doubted National Aniline could sustain reduced prices.[25] And so on. National Aniline was the firm that people in the trade loved to hate, and not a few took delight in hearing of some failure, of which there appeared to be an increasing number by the mid-1920s.

In the 1920s, the National Aniline division of Allied Dye & Chemical pursued internal development of its dyes business rather than purchasing competing manufacturers. National Aniline closed down the organic chemicals plants in Marcus Hook, Pennsylvania, and elsewhere and moved all of its dyes-related processes to Buffalo to the original Schoellkopf dyes plant. Like the other leading organic chemicals manufacturers, National Aniline constructed facilities to produce synthetic anthraquinone from phthalic anhydride. This eventually led to a line of vat dyes, but National Aniline only put a few vat dyes, exclusive of indigo, on the market in 1928. Why the firm moved relatively slowly on the lucrative vat dyes remains unclear, but evidence suggests that Weber simply limited expenditures on research in the National Aniline division. The notorious secrecy of Allied Chemical, the holding company that was National's parent, spawned speculation among the other manufacturers, who were more accustomed to exchanging information at professional and trade association meetings. Herbert Dow in 1923 was convinced National Aniline had simply "cut out research work."[26] Other rumors, begun when a plant worker moved from National Aniline to Du Pont, suggested that National Aniline's dyes division consistently lost money due to insufficient technical know-how.[27] By the end of the decade, Du Pont and National Aniline had changed positions among dyes manufacturers, and National Aniline lost its technological edge among American manufacturers. The trade recognized National Aniline for satisfactory dyes of established classes but not for innovation or mastery of the more complicated dyes.

Weber, it appears, decided against expanding National Aniline's domestic advantage in synthetic dyes while directing Allied's research and other assets toward its heavy investment in nitrogen products, including its massive synthetic ammonia plant in Hopewell, Virginia ("an un-garden city around Richmond," noted *Fortune* in 1930) and otherwise broad diversification in inorganic chemicals.[28] Indeed, Weber may have quite logically concluded that the profitability of dyes, which depended heavily on the tariff, was too uncertain in the face of steep international and domestic competition. Nitrogen, on the other hand, was newer and without such entrenched rivals in the market, and seemed even to the Germans to be the product of the future, along with its high-pressure technology. And the other units in Allied had more history than National Aniline with inorganic products. The question is what was lost when Allied (apparently) neglected the National

Aniline division, particularly in research. At Du Pont, as David Hounshell and John Smith have shown, the investment in organic chemicals research that began with dyes branched out in newer and more profitable directions. Given the dyes market, Weber's decision was logical and defensible, but one that kept Allied out of the new organic products.

E. I. du Pont de Nemours & Company

Buoyed by extraordinary financial resources from World War I, Du Pont, among American chemical firms, developed an exceptionally broad base of technical and organizational expertise, and it remained the largest of the firms throughout the interwar years. Du Pont built expertise by internal development of research and managerial skills but also by engaging externally with policymakers, trade associations, universities, and other firms. After the war, Du Pont resumed and accelerated its prewar strategy to diversify beyond explosives, and in 1924, for the first time, the firm's non-explosives products generated more revenue than its explosives. Its interest in nitrocellulose products, including rayon, created another center of expertise that the firm brought to bear on synthetic organic chemicals. Du Pont continued to build up its dyes business despite heavy losses in the early postwar years, but the firm's leadership also seized the opportunities offered by the newer aliphatic chemistry. Over his career, eminent business historian Alfred Chandler highlighted Du Pont's pioneering organizational strategies, in which its executives reorganized their giant firm in the 1920s to cope more effectively with their much broader, more diversified range of products. The firm developed the ability to build and harness organizational capabilities in one part of the business and share those skills firm-wide. A key strategy of growth, Chandler argued, centered on Du Pont's ability to achieve economies of scope—that is, to use their existing knowledge and production base to diversify along related lines. Du Pont stood out among American chemical firms for the extent of its broad-based research program but also for its reach into consumer goods. Besides their skillful management inside the firm, Du Pont was also the most outward-looking of the major American chemical firms in the interwar period. Whether developing relationships with other firms (American and European), providing active leadership in trade organizations, building mutually beneficial ties to universities, or (as discussed in earlier chapters) lobbying aggressively for favorable government policies, Du Pont maintained a visible and often dominating presence in the many arenas in which the firm operated. The late 1920s marked an important transition, not only from aromatic chemicals to aliphatic chemicals, but also from catching up with past German developments to innovating in new directions. Fifteen years after World War I, the American industry had mostly fulfilled the aspirations set up during the war, and Du Pont stood at the forefront of many of the most important new developments.[29]

With synthetic dyes, Du Pont's leaders took one of their first significant steps away from the firm's traditional stronghold in gunpowder and explosives and toward a diversified chemical company, a costly step in the early postwar years. Through the war and the 1920s, Du Pont's executive leadership actively lobbied for political and legal protections for the domestic dyes industry, given the German dominance; but even under the protective umbrella, Du Pont struggled to make dyes profitably. In the early 1920s, rumors in the trade suggested that everyone in the Du Pont Company except for the du Pont family members running the firm wanted to abandon dyes production. Similarly, a Du Pont chemist later recalled that it was the du Pont brothers "who stubbornly insisted that Americans could do anything that Germans could." Du Pont and other manufacturers continuously revised procedures to improve the efficiency of their operations to reduce cost. Like other companies, Du Pont might scrap whole sets of equipment that became obsolete in revised processes, and even satisfactory processes could wear out equipment very rapidly. And, as Irénée du Pont told the Senate Finance Committee in December 1919, "the yields have been abominable. . . . The stumbling block is that in the later processes you may only get 40 per cent, or 30 per cent, or 20 per cent yield. . . . In the case of indigo there are some 15 steps. This reduction of your output is cumulative, step by step." Du Pont generated deficits in dyes manufacturing every year between 1919 and 1922.[30]

As research costs mounted, Du Pont executives again looked outside the firm for expertise, most dramatically and controversially in 1920 when Du Pont surreptitiously recruited several chemists from German manufacturers. A Du Pont representative sent to Europe, Eric Kunz, who had helped Herty on his reparations mission in the fall of 1919, was abroad principally to lure German chemists to the United States. Kunz's initial attempts failed, but in 1920 several German chemists traveled to Switzerland where other Du Pont officials interviewed them and arranged for "at least ten" to join Du Pont's dyes operations. After the interview, some chemists returned to Germany, collected various papers undoubtedly related to dyes manufacturing, and eventually made their way to Wilmington, despite four arrests in the Netherlands and in New York that officials later rescinded. The German chemists violated explicit contracts with their German employers that prohibited them from working for a competitor within three years of their departure. The chemists provided Du Pont with additional expertise and bolstered the company's leaders' confidence that they had essentially pursued a promising course of action.[31]

Du Pont's recruitment of Germans prompted a range of reactions, however, including condemnation. Already unpopular with many in the United States because of their size and power, Du Pont angered even some allies in the American chemical community. Charles Herty blasted the company: "In war," he wrote, "information is obtained as far as possible from captured opponents, but renegades are not placed in positions of high command.... It is not too late to repair the damage.... Whatever the ability of these ... chemists, however intimate their knowledge of special lines of manufacture may be—send them home and let the American industry proceed to its full development in an American way by the force of American brains." Du Pont used a frequent defense: the hiring was necessary because the German patents were "so craftily devised that only a German chemist who has had experience in the production of the articles covered by the patents can put them to practical use." The editors of the trade journal *Drug and Chemical Markets*, on the other hand, referred to the German chemists as "refugees" and taunted the Germans: "Americans 'put one over' on the great Syndikat." After angrily protesting the loss of their chemists, the German trade press (purportedly quoting Bayer representatives) suggested that Americans, who "know only concentration and mass quantity production," would need more than a few German chemists to match "the entire chemical organization of Germany."[32]

Du Pont curbed its heavy losses in 1923, and one of the firm's notable achievements during the 1920s was the production of vat dyes other than indigo, difficult products that stymied most American dyes makers. Increasingly, the textile manufacturers used vat dyes, which, although more expensive than other dyes, offered much more resistance to the fading effects of light, laundry, perspiration, and a range of other factors. The dyes also promised higher profit margins to the dyes producers in competitive markets, a great appeal for the firms with the technical ability to make them. The cheaper supply of anthraquinone, an important intermediate developed synthetically, helped Du Pont and other manufacturers to expand the range of American-produced vat dyes; the quantity of vat dyes increased from 1 million pounds in 1922 to over 9 million in 1929. Du Pont improved its position in vat dyes in 1930 when it purchased the Newport Chemical Company of Carrollville, Wisconsin. Newport had grown to become one of the leading dyes producers in the industry, largely because of two dyes experts, chemist Ivan Gubelmann, a Swiss native, and Elvin H. Kilheffer, the sales manager. By 1929, Newport's sales and profits were roughly half those of Du Pont's dyes plant. Over the 1930s, Du Pont concentrated production of vat dyes in Carrollville. In 1930, Du Pont officials estimated that their firm contributed about 25 percent of U.S. production of all dyes by value and about 20 percent in sales. Besides Gubelmann, the Newport purchase brought a bevy of talented research chemists who helped Du Pont expand their expertise in organic chemistry.[33]

With dyes, Du Pont cemented research as an important part of its strategy for growth, but research work became increasingly important for all U.S. dyes firms. In 1919, 65 of 214 dyes makers reported they maintained separate research laboratories; the figure increased to 76 in 1920. As mergers and failures trimmed the number of synthetic organic chemicals manufacturers, the number of industrial laboratories in the sector correspondingly declined during the 1920s, from 67 in 1922 to 42 in 1929. The number of chemists employed in the industry fluctuated. It peaked in 1923 at 1,882 and then gradually declined to a low of 1,358 in 1926 before rising again to 1,746 in 1930. However, the proportion of chemists to the total number of employees in the industry rose steadily from 11 percent in 1922 to 15 percent in 1930. In 1930, when Du Pont had 687 people in research and development across the corporation, 75 were in its organic chemicals unit. Although determining the appropriate organization and level of funding for research remained a perpetual challenge, through dyes, the firm's executives learned, expensively, to appreciate the benefits of integrating research into planning and operations to enhance the organization's broader capabilities. Du Pont became the leader in the scope of its research and development in the American chemical industry.[34]

Du Pont's diversification led the executives to create a new organizational structure—decentralized and multidivisional—that also shaped the conduct of research. In September 1921, Pierre du Pont led the firm's corporate decentralization, a reorganization into divisions primarily by product clusters, rather than functional departments such as purchasing or sales, to manage the operations of an increasing number and diversity of chemical products. Each division, such as dyes (later organic chemicals), explosives, and cellulose products, managed its own purchasing, production, and sales quite independently of others. Three months later in December, the executives also rearranged the firm's research organization. The executives set up many divisional laboratories focused more directly on the products and processes of their respective divisions. The heads of the divisions wanted the firm's R&D apparatus aligned to work in closer cooperation with plant production—to address the hurdles associated with scaling up laboratory

Table 8.3. U.S. Synthetic Dyes Output in Millions of Pounds and Millions of Dollars (Rounded), 1914–1930

Year	No. of Firms	Output in Pounds	Value of Output/Sales	Average Price per Pound
1914	7	6.5	$2.5	
1917	81	46	$58	
1918	78	58	$62	$1.07
1919	90	63	$67	$1.00+
1920	82	88	$95	$1.08
1921	74	39	$39	83¢
1922	87	64	$41	60¢
1923	88	94	$47	54.5¢
1924	78	69	$35	54¢
1925	75	86	$35	47¢
1926	61	88	$36	42¢
1927	55	95	$39	39¢
1928	53	96	$40	42.6¢
1929	54	111	$46	43.2¢
1930	50	86	$39	43¢

Note: In the period of 1914–1920, the census reported the value of dyes produced. The figures for the other years represent actual sales of American-made dyes, rounded to the nearest million. The number of dyes produced in a given year was recorded only for 1920, when the number was 360. Source: Thomas H. Norton, "A Census of the Artificial Dyestuffs Used in the United States," *Journal of Industrial and Engineering Chemistry* 8 (November 1916): 1045; figures compiled from U.S. Tariff Commission annual reports and *Census of Dyes and Coal-Tar Chemicals*, 1917–1930.

products to commercial production and to improve existing processes. In the stresses of war and diversification, the Chemical Department—the main research laboratory—had not been able to keep up with the growing demands of the diverse corporate units, and the executives reassigned many of its tasks to the divisional laboratories. The rump of the Chemical Department became the firm's central laboratory to conduct research that was "of an involved scientific nature and . . . on products which are not connected with existing manufacturing processes," as one internal report described. In the later 1920s, the revitalized Chemical Department undertook research crucial to Du Pont's future in synthetic organic chemistry.[35]

Dividing research activities across several divisions allowed scientists to focus research on related product sets, but it also created the challenge of sharing expertise across the divisional boundaries. By the late 1920s, the executives increasingly required standardized and regular reports on research, which at least helped to identify developments useful in another laboratory. And in the next decade, the research directors met in a steering committee to coordinate and communicate across the firm. But, before and besides reporting requirements, the firm evolved means to share expertise across organizational boundaries, especially at particular stages of development. As a promising product moved out of the laboratory, Du Pont scaled up from lab research to plant production with an intermediate step, testing out new products and processes in a "semi-works" or pilot plant. In this small-scale setting, chemists coming up from a laboratory met with chemical engineers to work out processes that would be efficient in a full-scale plant. Charles Stine, head of the Chemical Department as of 1924, particularly saw the semi-works as a node inside the firm at which multiple specialties worked together. The chemical engineers who supervised the semi-works operations, he wrote in 1932, particularly stood in a position to "accumulate a wide range of experience" from introducing new products and processes that became "quite invaluable" when the next ones came through, no matter from which division. Even within a single laboratory, "the harmonious and coordinated efforts of a corps of diversified highly-trained specialists" required the careful attention of research directors.[36]

In addition to developing an effective organization for research facilities and efforts, a central challenge lay in harnessing the scientific creativity of the individual researcher for the benefit of the larger organization. Du Pont managers generally shared a company belief that profiting from research creativity required attention to the institutional culture inside the laboratory and firm. The primary challenge for research directors, suggested Ernest Benger in 1930, lay in creating an "atmosphere" in the laboratory that allowed chemists to work their best. Particularly for young recruits, the lab director must take care that the new chemist is "inducted into the service of the organization" and "imbued with the spirit of the organization." He recommended assigning an older mentor and giving the young chemist "a number of small jobs" reflecting a "diversity of work" to build confidence and to learn the company's way of doing business. Introduce a good chemist the wrong way, however, and the firm could inadvertently bring "serious damage to the man and disappointing results" to the firm. Chemists of all levels of experience needed encouragement and other rewards to keep them engaged in their work and in the larger developments of the discipline. For example, if the firm allowed chemists to publish the results of their research when it caused no harm to company interests, the chemists' morale improved with the recognition and the firm received "legitimate advertising" of its "scientific prowess." Employees should "be permitted to visit . . . other parts of their own company" to stay intellectually engaged, he thought. The industrial chemists needed to stay abreast of "the vital, growing, and expanding sciences of chemistry and physics"; contacts with "extraneous influences and information" kept a laboratory from becoming too insular. This was one of many reasons Du Pont cultivated ties to universities.[37]

The quest for expanded technical expertise went well beyond the borders of the firms, however, and Du Pont developed relationships with several universities, not least because the company needed graduates in chemistry and chemical engineering. Young chemists emerging from American universities became a central part of building a synthetic organic chemicals industry, and manufacturers sought ways to coax more chemists out of U.S. higher education. Starting in 1918, Du Pont funded fellowships for graduate research at roughly two dozen different major universities around the country, including Harvard, MIT, and the University of Illinois, with which Du Pont particularly built closer relationships in the interwar years. With the backing of Pierre and Irénée du Pont, Charles Reese, head of the Chemical Department, led the formation of Du Pont's fellowship program in 1918; by the end of the 1920s, both Stine and Elmer Bolton of the Dyestuffs Department believed the fellowships had been enormously successful in building bridges to universities, which helped the firm to recruit top graduates. The motivation lay in having more well-trained chemists from which to recruit for Du Pont, but Reese and others made it clear that Du Pont should not choose the research problems. Robert Rose, also in the Chemical Department, echoed Reese in 1920; he warned against encouraging academic chemists to address specific problems, like developing a new dye, when "what is required is the broadening of our knowledge of the fundamental theories of organic chemistry," which is the "real research." "It is my own opinion that our chemists, on the average, show more individuality, more independence of thought, and greater adaptability and resourcefulness than those produced by any other educational system; their excellence is the real reason for the present success of the dye industry here."[38] The universities best helped industry by training students broadly and in fundamentals like logic and observation.

Du Pont also built ties to universities by engaging professors as consultants to provide specific knowledge and skills. The consultants brought the company at least three benefits. First, the academics kept abreast of the latest scientific literature and could highlight relevant developments to help the firm's chemists stay current. Second, relationships with professors gave Du Pont another means to identify and recruit the best students each year. Third, Du Pont's academic consultants came to the firm's laboratories and talked to their research chemists about their specific work, discussing problems large and small. Roger Adams and Carl Marvel, two outstanding organic chemists at the University of Illinois, served as consultants to Du Pont for decades. Du Pont hired many of their students, which placed them—particularly Adams, who became friends with Bolton at Harvard—in a position to know intimately the research and institutional culture of the firm. As of 1927, Du Pont had eight academic consultants, but the number grew and included chemists from a range of universities, including Johns Hopkins, Princeton, and Wisconsin. Some academic chemists hesitated to become involved so heavily with a firm, partly fearing restrictions on publication. Harvard's James B. Conant began consulting with Du Pont in 1929, and, like other academic consultants, he wrote reviews of recent literature in the field, visited Du Pont's laboratories, conducted research on specific problems, and simply brainstormed when a Du Pont chemist wrote to ask for ideas. But Conant remained uncomfortable with the arrangement, particularly signing over rights to new knowledge that might become proprietary, and he ended the relationship after a year.[39] Du Pont's expanding research prowess and wide network of university relationships, in addition to acquiring other firms with particular knowledge bases, placed Du Pont in a strategically strong position to lead many of the new directions in the synthetic organic chemicals industry.

In the 1920s, one of the early important synthetic organic industrial chemicals from the aliphatic (chain molecule) branch was synthetic alcohol, particularly methyl alcohol (or methanol), marking another dramatic shift from natural to synthetic. Du Pont was one of two American firms to put synthetic methanol, manufactured as a by-product to its synthetic ammonia plant and drawing heavily on newer techniques in high pressure technologies, on the market in 1927—three years after the Germans. In 1924, Germany began to export synthetic methyl alcohol into the United States and caused a panic among the traditional suppliers, the American wood distillation industry. The distillers implemented more efficient processes but also lobbied successfully in 1926 for a steeper tariff rate, which offered them little protection from the domestic synthetic methanol manufacturers. The consumption of methyl alcohol, whether synthetic or natural, expanded in the 1920s to a great extent because of its use in the manufacture of formaldehyde, which went into the synthetic resins like Bakelite. Americans celebrated the relatively quick development of an American synthetic version, but the Germans again had the lead in a new synthetic product.[40] Du Pont increasingly established itself as the frontrunner in several other new products from the aliphatic branch, however.

When the *Journal of Chemical Education* recognized some of Du Pont's accomplishments in 1925, it suggested that "perhaps, nothing . . . has contributed more greatly to the economy and pleasure of modern motoring than the practically complete freedom from tire worries." In 1905, American chemist George Oenslager added an aniline derivative to rubber vulcanization, in which the manufacturer introduces sulphur to natural rubber to bind the polymers and give it a more fixed shape. The aniline derivative operated as a catalyst to speed up the process—a "rubber accelerator." Automobile tires offered a large market for many kinds of rubber, such as more wear-resistant rubber for treads. Du Pont came out with several varieties of accelerators, each of which imparted different qualities to the rubber, and the firm set up a Rubber Chemicals Laboratory in 1922. Rubber accelerators, a Du Pont chemist later recalled, marked the point at which Du Pont moved into new synthetic organic product lines that "offered the opportunity of getting into new and relatively uncomplicated fields and 'growing up' with them, instead of starting in the middle of a well established industry, as had been the case with the dye industry." Rubber accelerators still fell on the aromatic side of organic chemicals; Du Pont would see its most spectacular developments in the newer aliphatic organic chemicals.[41]

Much of the American chemical industry benefited from the booming automobile industry in the 1920s, but none more than Du Pont. Not only did its chemicals supply finishing lacquers, artificial fabrics, and gasoline antiknock compounds, but Du Pont also owned between 23 and 38 percent of General Motors throughout the interwar period. Pierre du Pont had invested personally in GM before the war, joined the board in 1915, and then led the Du Pont executives' decision to buy GM stock in 1917. In 1919, Pierre turned over Du Pont's presidency to his brother Irénée, and at the end of 1920, he took on the presidency of GM, which he held until Alfred P. Sloan took over in 1924. Thus, as his biographers wrote, Pierre du Pont presided over the management and administrative reorganization of two of the largest and most profitable American firms in the first half of the twentieth century. He took on the GM leadership primarily to protect Du Pont's investment, but, as he noted in 1922 about the automobile business, an industry "essential to civilization warrants the attention of any man." As a firm, Du Pont reaped rewards from its large ownership in GM in at least three ways. First, although GM maintained a policy that required two suppliers for their major purchases, Du Pont supplied the lion's share of GM's purchases when it made the product, particularly lacquers (the cellulose-based Duco foremost among them) and artificial fabrics. Second, and more importantly, the GM stock was incredibly profitable, which gave Du Pont additional resources for expansion and diversification—and for issuing substantial dividends of its own. In 1927, for example, Du Pont's total income amounted to $47 million;

$16.5 million came from operations. Of the $30.5 million from investments, $29 million came from GM stock. Indeed, Du Pont's income from GM stock exceeded its income from chemical sales to GM many times over, explaining why Pierre would not insist too emphatically that GM observe agreements to buy from Du Pont if such a purchase hurt GM's bottom line. Finally, GM's research laboratory provided Du Pont with two important and lucrative synthetic organic chemicals—tetraethyllead (an antiknock compound in gasoline) and Freon (a refrigerant), the latter a big seller in the 1930s.[42]

Du Pont's first major product in the aliphatic branch of synthetic organic chemistry, tetraethyllead (TEL), started disastrously. Charles Kettering and the General Motors Research Corporation took on a technical problem—the premature combustion of gasoline in engines, or "knock." TEL significantly improved efficiency of engines and held out the promise of an enormous market, one that Du Pont captured by the late 1920s. When the company began working on TEL, however, Du Pont faced serious hurdles. First, Standard Oil leveraged its acquisition of key patents to make a three-way deal with GM and Du Pont; Standard Oil's patents made TEL with ethyl chloride, which was less expensive than GM's ethyl bromide that Du Pont began to develop. GM facilitated an accommodation for Du Pont to use the ethyl chloride process, solving the first hurdle. Second, and more seriously, making TEL by either process was dangerous, posing the hazard of lead poisoning, which could drive people to insanity and be fatal. Standard Oil launched into experimental manufacture and experienced a terrible accident in October 1924 in Bayway, New Jersey, where five people died of lead poisoning. Under pressure to produce, both Du Pont and GM suffered similar poisonings—five dead at Du Pont between late 1923 and early 1925 and two dead at GM's Research Corporation at Dayton, Ohio, in spring 1924. After Du Pont's most serious accident, the *New York Times* reported that Du Pont workers labeled the TEL facility at Deepwater "The House of Butterflies," a reference to the lead poisoning, in which "one of the early symptoms is a hallucination of winged insects." The deaths, particularly those at Standard Oil—more publicized than the others—prompted a public outcry and federal investigation on the safety of TEL. Finding that the plant dangers could be contained with better procedures and that TEL posed little threat to public health when used as an antiknock compound, the Surgeon General's report stopped a ban on TEL. Du Pont, which held the patent rights until 1948, resumed manufacturing, and it became a very profitable product.[43] But other products from the aliphatic branch—at much lower human cost—marked Du Pont's status as a worthy rival to the German industry.

As the 1920s closed out, with the help of their expansive research and manufacturing capabilities, Du Pont was at the frontier of innovation in the aliphatic branch of organic chemistry. From 1927 to 1932, Du Pont ran a fundamental research program that, despite a relatively short period of operation, led the firm to important new directions in synthetic organic chemicals. With an eye on AT&T's research success, Charles Stine, the head of the Chemical Department, convinced the executive leadership that Du Pont should create an arena for excellent chemists to conduct basic research relatively free of commercial pressures. Stine thought it possible that fundamental research might lead to profitable products, but he thought such a lab environment and its science would be more important to garner "scientific prestige," help to recruit new chemists, and improve "morale" among chemists. He wanted the Chemical Department to function like an academic laboratory, granting chemists wide latitude to pursue their interests with minimal expectation of profitable products. This vision enticed Wallace H. Carothers to Du Pont, where his research led to two major synthetic organic products.[44]

In 1928, Du Pont brought Carothers, a brilliant young Ph.D. chemist, to its new fundamental research program, luring him from his academic position at Harvard University. At Du Pont, Carothers added to ongoing exploration in the chemistry of macromolecules, as chemists called polymers in the 1920s. Scientists debated whether these particular compounds were clusters of small molecules bound by as yet unknown forces, or molecules with normal bonds but much longer than other molecules by several orders of magnitude. Carothers proved the latter hypothesis—one championed by German chemist Herman Staudinger—by building the long molecules in the laboratory. This research helped Du Pont to earn a reputation for undertaking some of the most sophisticated chemical research in the world, accomplishing one of Stine's main goals for the fundamental research program.[45] Besides the first-rate science, however, the research of Carothers's laboratory also gave Du Pont two important commercial products, neoprene and nylon.

In 1930, Carothers's polymer research had two breakthroughs that led to neoprene, a synthetic rubber, and nylon, a synthetic fiber, which the firm developed and commercialized over the following decade, securing their position in the new branch of their synthetic organic chemicals industry. The historians of Du Pont's R&D emphasize that creating profitable products from research depended on a range of capabilities at Du Pont that went well beyond Carothers's particular genius and his laboratory work. The new products also reflected Du Pont's attention to the larger world in which it operated. Neoprene occurred in the context of broader research in synthetic rubber by researchers in Germany and the United States during the 1920s. For neoprene, Du Pont used a catalyst developed by Father Julius Nieuwland of the University of Notre Dame, whose research focused on acetylene, an aliphatic raw material increasingly important in the interwar years. After its synthesis in the laboratory, neoprene went through several years of intense development to make it stable and to give it marketable characteristics, particularly durability, that differed from the much cheaper natural rubber. Production began in 1932, but Du Pont slowly expanded production throughout the decade, making improvements to processes along the way, one of which required a license from I.G. Farben. Among other things, neoprene benefited from expertise at the Rubber Chemicals Laboratory that the Dyes Department had set up to develop rubber accelerators, and nylon's development depended heavily on synthetic fibers expertise built up through rayon production. With neoprene's resistance to water, oil, and gasoline, industrial customers bought it for wire insulators and rubber hoses, and retail markets soon included shoe soles and gloves. World War II expanded production dramatically.[46]

Nylon became the most profitable product in Du Pont's history. Carothers's research led to the laboratory synthesis of a polyester synthetic fiber; subsequent revision of the formula led to nylon in the laboratory in 1934. Nylon offered superb strength and elastic characteristics, even when wet, and Du Pont quickly targeted markets traditionally filled by silk—focusing first on hosiery, which was profitable almost immediately when it went on the market in 1940. Du Pont rushed to get workable, cost-effective production processes—including making intermediates and spinning the new fiber—in place and revised them over time. Nylon production drew on technologies, domestic and international, related to high-pressure chemistry and other synthetic fibers. And Du Pont worked with its customers in the rubber and textile industries to put their combined expertise into the products' development. Besides hosiery, subsequent major products included parachutes during World War II, and lingerie, tire cords, and carpet after the war.[47]

Du Pont's outward orientation meant ongoing international relationships with European firms in addition to the networks with domestic manufacturers and academia. Du Pont had long had ties to explosives manufacturers abroad, and their ongoing international connections came in the form of patent licensing, joint ventures, and shared research, particularly with British and French firms. As the result of a purchase by I.G. Farben, Du Pont even held stock in the big German cartel for a few years. The large firms across the Atlantic recognized Du Pont's growing capabilities in synthetic organic chemicals and saw Du Pont as both a competitor and a potential ally. Imperial Chemical Industries (ICI), formed by a merger of British explosives, dyes, and inorganic chemicals firms in 1926, signed an agreement with Du Pont that lasted from 1929 to 1948 to use patents to divide world markets and share scientific and engineering information. They discussed their respective research, exchanging experts and sharing thousands of reports. The agreement gave the two firms a means to collaborate without violating the antitrust laws, to which Du Pont remained particularly sensitive. Both sides recognized the relationship as mutually beneficial, but Du Pont's executives generally felt they were the superior innovators and found ways to avoid sharing some of their most promising developments, particularly nylon (and ICI was circumspect about their developments in polyethylene). Du Pont tried to negotiate agreements with German firms at various points in the 1920s, but their efforts generally went nowhere, largely because the Germans felt they had little to gain from Du Pont and because of lingering mistrust from the wartime confiscations. A decade later, Du Pont's technical advances led to patent licensing agreements with I.G. Farben, including some patents related to synthetic rubbers and nylon, but World War II ended the agreements.[48]

Going into the 1930s, Du Pont competed fiercely with the German industry to be at the cutting edge in synthetic organic chemicals, rather than scrambling to rediscover and repeat past German developments. Du Pont stood out for the extent to which it engaged externally—whether buying out firms with expertise, creating joint ventures and sharing research with foreign firms, developing relationships with academia, participating in trade associations, or lobbying to shape government policy. Scholars have increasingly explored the "porous boundaries," or the way that values, cultures, ideas—and information—flow in and out of firms; Du Pont was exceptionally good at building and drawing in expertise from multiple sources. Du Pont's extensive networking helped it to build capabilities across research, manufacturing, and sales, and its collective expertise in synthetic organic chemicals grew to incorporate past German discoveries—dyes brought in the most sales among organic chemicals in the interwar period—and to find new product lines.[49]

Dow Chemical Company

When Herbert Dow committed to synthetic organic chemicals production during the war, his strategy centered on "mass production" of a few chemicals such as indigo and salicylic acid, and he distinguished his strategy from that associated with the German industry, that is, efficient, tightly integrated use of by-products to make small batches of interrelated dyes and pharmaceuticals. "But today," he noted to his stockholders in 1930, "the interlocking of by-products is as marked in your Company's plant as it is in any foreign plant and the mechanical equipment is undoubtedly better." Reflecting back on the expansion of his firm through the 1920s, Dow pointed to the ability of his firm to scale up plant size considerably, which particularly depended on engineering innovation, and he frequently sang the praises of his engineering and machining units. "The Dow Company is using the heavy chemical method for the manufacture of fine chemicals, and it gives them a very distinct advantage so long as they are manufacturing on a much larger scale than their competitors." And the "interlocking" bound his traditional inorganic product line to the newer synthetic organic chemicals.[50] When Herbert Dow died in October 1930, his company, like the American synthetic organic chemicals sector more generally, sat on the cusp of important breakthroughs.

In 1924, Dow knew he was a small player compared to the big firms—as he calculated it, Allied had net assets of $72 million, Du Pont $60 million, and Dow $2 million (Union Carbide, which he didn't include in his list, had net assets of about $50 million). "It is quite certain that it is now impossible to do manufacturing on a small scale, and it looks as though the chemical industry, whether we like it or not, is destined to ultimately be under the control of a very few organizations," wrote Herbert Dow in 1930, convinced his firm would be one of those few. Partly because of his firm's engineering expertise, he wrote to a colleague in 1926 that "I can see no reason why expansion will not be a normal and continuing operation with The Dow Chemical Company."[51]

The first manufacturer of indigo in the United States, the Dow Company continued in the 1920s to improve its line of synthetic organic chemicals in the shadow of its larger business in bromine, chlorine, and other inorganics. Indigo remained one of the most important synthetic dyes during the 1920s, accounting for 25 to 35 percent of all synthetic dyes produced in the United States. Faced with stiff competition from Du Pont, National Aniline, and international rivals, Dow pursued at least three strategies to retain a profitable share of the world's indigo market. First, the company expanded output. In 1923, when the price of indigo hovered about 23¢ per pound, Dow officials estimated that the firm needed to reduce costs below 18¢ per pound to meet expected price decreases. Already Dow's cost per pound had decreased from 30¢ per pound in early 1922. Only by increasing production could the firm drop its unit costs, its leaders decided, despite overcapacity in the industry. Second, Dow especially promoted its line of brominated indigos, for which it received technical assistance from the Swiss CIBA in their agreement of 1920. Although forming a relatively small share of Dow's indigo production, the CIBA dyes became the "most profitable end" of Dow's dyes production by the late 1920s. Finally, Dow began to make important intermediates for indigo, especially aniline. Through the early and mid-1920s, Dow purchased aniline from the Calco Chemical Company. By 1925, however, Dow had built an experimental aniline plant and proceeded to expand production throughout the remainder of the decade.[52]

Brought together by patent disputes, Dow developed a mutually beneficial relationship with a Swiss dye maker. The reception the Germans received in the United States might have been predicted by the experience of the Swiss after the war. Dogged by real and rumored connections to the Germans, the Swiss received a considerable amount of abuse from some American industry supporters as they looked to gain a foothold in the American market. The Swiss chemical manufacturers, relatively flush with profits from the war, purchased Ault & Wiborg in 1920, renaming it the Cincinnati Chemical Works. With protectionist sentiment running high in the United States, the foreign manufacturers no doubt viewed investment in an American plant the most viable long-term solution to maintain a share of the American market. By the armistice, too, the different firms in Switzerland—CIBA (Society of Chemical Industry), Geigy, and Sandoz—had formed a single cartel to enhance their prospects in the postwar world market, and the Swiss posed a potent commercial threat. Their purchase of the Cincinnati plant generated mixed reactions among American observers. While some thought U.S. interests were protected as long as the plant existed on American soil, others denounced foreign ownership as a dangerous trend.[53]

In 1920, Dow Chemical negotiated an agreement with CIBA of the Swiss cartel. The Swiss manufacturers charged Herbert Dow with violating their patents for brominated indigoes. Rather than enter into litigation, Dow and the Swiss agreed that Dow could manufacture under the patents for a royalty and that the Swiss would forgive any past infringements. The negotiations evolved amicably through the mediation of a mutual business partner, Alfred F. Lichtenstein, who had worked for the Geisenheimer importing firm, which had long imported Swiss chemicals into the United States. During the war, Geisenheimer retired, and Lichtenstein renamed the agency Aniline Dyes & Chemicals, Inc. It continued importing dyes and began selling dyes produced by domestic dyes manufacturers as well, primarily those made by Dow Chemical and Ault & Wiborg, which CIBA had bought in 1920. In 1930, Ault & Wiborg made 105 dyes; by 1939, the Swiss produced about 9 percent of dyes manufactured in the United States but sold, by value, 15 percent of dyes on the domestic market. In addition to patents, Dow and the Swiss exchanged technical expertise. In 1926, Dow toured the indigo plant in Basle, convincing him that he needed to make adjustments in his indigo plant, and in return offered "unlimited exchange of data in regard to aniline" from Dow's new process.[54] Dow remained suspicious of German business practices and advocated the development of a domestic, American industry, but his relationship with the Swiss CIBA worked very well for both parties, due in no small part to Lichtenstein's abilities as an intermediary.

Dow's dyes business had led to a mutually beneficial relationship with another medium-sized firm, Calco Chemical Company, but Dow's innovation undermined the basis for the relationship, a supply of aniline. Like Herbert Dow, Calco's president, Robert Jeffcott, believed that the brightest future for his firm and the American industry lay in volume production. As Jeffcott put it to Dow, "We have tried to pursue [a policy] of making a few intermediate products on a large scale and supplying these to the principal manufacturers of dyestuffs. . . . A good tonnage, of course, is essential for efficient costs." Dow manufactured aniline during the war but found it cheaper to buy it for his indigo afterward, and Dow was one of Calco's largest customers in the early 1920s. Jeffcott wanted to achieve economies of scale by selling externally, but his strategy ran into trouble as the other manufacturers, including Dow, found that the "interdependence" of chemicals—the efficiency that came with using by-products in other processes—applied to synthetic organic chemicals produced on a large scale as well as to smaller batches of dyes and pharmaceuticals. In 1925, Jeffcott learned that Dow had begun to build a new aniline plan and, "much disturbed by this advice," Jeffcott pleaded with Dow to reconsider. "It does seem to me that duplication of effort is always, in the final analysis, wasteful." Dow politely half-explained that he had "special reasons . . . to go into this venture." Dow's research team had developed a new process to make aniline, and he was exploring possible markets for brominated aniline. In the second half of the 1920s, Jeffcott shifted his strategy to developing a more vertically integrated firm with a broader range of products. In 1929, American Cyanamid, manufacturer of cyanamide fertilizers, diversified its operations by purchasing Calco, which continued to function as an independent unit and grew with purchases of about twenty small dyes and pharmaceuticals firms over the next two years. The purchases almost put

Calco in the same league with Du Pont and National Aniline, with production of 146 dyes in 1930.[55]

Dow relied heavily on the Swiss relationship to profit from his firm's dyes—the "German" chemicals, but he saw much greater growth in the products directed to the pharmaceuticals and burgeoning plastics markets, two cornerstones to the expansion of the "American" industry. Dow's second major synthetic organic product among the "German" chemicals was phenol. One use for Dow's phenol was the firm's own production of salicylic acid, the intermediate used in the manufacture of aspirin. The low cost of its phenol helped the Dow Company compete in the salicylic acid market, and Herbert Dow expected his firm to become "altogether the biggest factor in this line of manufacture." Led by chemist William J. Hale, Dow's research staff developed a new, lower-cost process for phenol they placed into production in 1927. Although initially delayed while trying to acquire licensing rights for a crucial patent, the company quickly worked the bugs out of its process, a high-pressure system. Dow Company's leaders considered the firm's engineering organization, which could build the high-pressure equipment in-house, as a major competitive advantage over its rivals, including the Germans. By the late 1920s, Dow was a major supplier of phenol to the Bakelite Company; George Baekeland believed his father's firm to be "the largest buyers of phenol in the world" in 1929, in part because Bakelite had mothballed their own phenol plant when its costs exceeded Dow's and other suppliers' prices.[56]

Dow brought similar methods to his pharmaceutical manufacturing, and he thought the other pharmaceutical manufacturers were slow to appreciate the potential of large-scale production. Traditional pharmaceutical companies made their products "in beakers and evaporating dishes, and more recently in porcelain-lined kettles, but heavy chemicals have always been made in bigger equipment." Dow knew that in products like salicylic acid, nearly all of his competition suffered through the 1920s: "The Merck plant is for sale [1926] . . . and have paid no dividends since the War"; wages in Philadelphia put pressure on Rosengarten; Semet-Solvay left the business in 1924, selling their stocks to Monsanto; and Heyden regularly skirted the edge of financial collapse. Word was out in 1924 that W. E. Kemmerick, formerly of the Bayer plant in Albany, ran Heyden's plant in Garfield, New Jersey, but generally considered it such a "hopeless" task that he was looking for another job.[57]

Herbert Dow had definite ideas on the proper way to develop the technical abilities of his organization, which he articulated with consistency over a period of years. He preferred to hire new employees straight out of college.

Over the 1920s, Dow grew more appreciative of the value of a Ph.D., particularly in organic chemistry, but he still thought graduate school could inculcate habits incompatible with the speed and efficiency required in an industrial setting: "We prefer men who have not had post-graduate work, except it be in organic chemistry. . . . We would rather they would start a life-long post-graduate course with us before they get in the class where it is difficult to teach an old dog new tricks." He also established a policy of not hiring men who had worked at other firms, not only because they also fell into the "old dog" category, but also because he found the practice of such hiring a little dubious ethically, bordering on industrial espionage. Even more than that, he thought promoting from within was vital to maintaining morale and a cooperative spirit among his employees. The company "develops the men who develop the processes," and "those men who have grown up and developed with the plant are the greatest asset that we have. But it would be a very easy matter to reduce the value of this big asset if we brought in some outside men and put them ahead of some of the men who are striving with all their might to get as good a position as possible." To the same end, Dow's profit-sharing plan depended on the performance of the entire firm to encourage cooperation rather than competition among divisions of the firm.[58]

Dow invested heavily in research and built a cooperative relationship with university researchers. As of late 1928, the entire Dow Chemical Company supported a technical staff of about 174 "graduate chemists, chemical engineers, and other engineers," about 40 of them hired just in the previous year. A year earlier, Dow had asserted that his company "has maintained a bigger research organization, proportional to its paying production, than any company with which we are acquainted." Dow thought it possible that his firm ranked fifth in the country after AT&T, General Motors, Du Pont, and General Electric, and in any case "we seem to be able to more than hold our own with every product we are producing." Dow hired from a wide range of colleges and universities, but his firm had a particularly close relationship with two. From Case School of Applied Science, Albert W. Smith, a professor of chemistry and Dow's friend since college, continued to supply Midland with a steady stream of chemistry and chemical engineering students. Increasingly, Dow Chemical also recruited heavily from the University of Michigan, where William J. Hale, Dow's son-in-law, taught chemistry until he joined the firm full-time. In addition to sending students to Dow, the two chemists consulted for the company and devised important and profitable new chemical processes. Hale played an important role in Dow's new aniline and phenol techniques.[59]

Dow Chemical's expanding capabilities in large-scale production, high-pressure technology, and organic chemistry positioned it well to take advantage of the shift from aromatic to aliphatic synthetic organic chemicals. Ethylene and the new aliphatic chemistry fit Dow well. By early 1928, Dow suggested "that there is going to be no practical limit to the amount of ethylene that we will require." Reflecting back, he also thought he had missed a key opportunity to develop an ethylene-based product line much earlier, either from some prewar research or from the ethylene plant for his wartime mustard gas operation. Two products in particular drove the expansion of Dow's ethylene plant in the 1920s. First, the tetraethyllead (TEL) developed by Du Pont and its allies created a market for ethylene dibromide, a product that drew on Dow's traditional strengths in bromine and coupled it with his newer organic capabilities. As of late 1926, the Ethyl Gasoline Company bought 62 percent of its ethylene dibromide from Dow. "They have resumed their operations in a much more modest way than prior to their trouble with the New York World [which published about the TEL deaths], but the demand is steadily increasing and they speak very optimistically of the future." The enormous demand for TEL over the decade led Dow into larger and larger production of ethylene, which he then put to other uses. A second important product in the late 1920s was acetic anhydride, which he supplied to Eastman Kodak and pharmaceuticals companies like Bayer and which had Dow thinking big about ethylene. "We are now using ethylene for the production of ethylene dibromide and to a smaller extent for some other compounds, and if the price was very much lower than it now is . . . and some of our dreams come true, it would be the biggest business we would have in Midland." In the 1930s, Dow's chemists used ethylene to make a polystyrene plastic, Styron, "the runaway No. 1 bestseller of Dow's first century."[60]

Dow believed his firm had developed its competitive advantage by playing to "American" strengths: designing equipment to make large quantities of material at low cost. "Modern chemical processes are primarily the result of recent achievements in engineering practice," explained Herbert Dow at his Perkin Medal speech at the end of 1929. Dow portrayed recent developments in the American chemical industry as part of longer American tradition in interchangeable parts, excellent machining, and "big-scale operations"—all driven by finding substitutes for expensive American labor. He took particular pride in his firm's engineering abilities. His engineering unit "has enabled us to break away from the limitations imposed by so-called standard equipment," and sometimes came up with "radically new equipment" in-house" that was better as well as bigger.[61] Dow had begun his career with inorganic chemicals, more typically produced in large quantities, and he built on that tradition with his expanding organic business.

Union Carbide & Chemical Company

In 1941, the business magazine *Fortune* wrote extensively and favorably about Union Carbide & Carbon Corporation in a series of three feature articles. "Carbide's work in natural- and petroleum-gas chemistry opened a field as tremendous as Germany's coal-tar chemistry," noted the magazine in the third article. "And, unlike other U.S. chemical firms, Carbide's name has never been linked with the great German chemical cartel, for the reason that there is no link. From first to last it has been pioneering on a solidly American frontier."[62] The article described Union Carbide as many had in the previous two decades: the firm that best represented an "American" synthetic organic chemicals industry, one intimately tied to American natural resources, American mass production and mass consumption, and American science and technology. In reality, even Union Carbide benefited from old and new German science and technology, but their product lines lacked the same cultural identification with Germany as either dyes or pharmaceuticals. Their products included many ethylene-based chemicals, from antifreeze to solvents; the firm exploited the aliphatic (chain molecule) branch of organic chemicals rather than the aromatic (ring molecule) organic chemicals upon which the Germans built their industry. By 1930, Union Carbide owned plants on dramatic American landscapes, such as Niagara Falls and the forceful Kanawha River in the West Virginia Appalachians; the firm headed the transition to petrochemicals; and, like so many other firms in the 1920s, it thrived on the enormous expansion of the American automobile industry. Union Carbide & Carbon Corporation became the second-largest American chemical firm in the interwar years, prospering with synthetic organic chemicals more intimately bound to the American economy and culture.

The Union Carbide & Carbon Corporation began in 1917 with a corporate merger of four firms that already had a working relationship, one built primarily on acetylene. The firm was a leader in the transition from the traditional aromatic feedstocks from coal tar to the aliphatic feedstocks from petroleum and natural gas, which supplied most of the industry after World War II. The simple chain compound, ethylene (C_2H_4), replaced the simple ring compound, benzene (C_6H_6), as the key building block. In the interwar period, however, before hydrocarbons could be easily captured from petroleum cracking, acetylene (C_2H_2) was a crucial starting compound for many of the new developments in the aliphatic branch. In brief, the Union Carbide Company produced calcium carbide in electric arc furnaces, bought oxygen from Linde Air Products, bought furnace electrodes from National Carbon, and sold acetylene to Prest-O-Lite. The Union Carbide Company took the lead in the merger, already owning 40 percent of Prest-O-Lite and 30 percent of Linde Air Products. The main capitalists behind Union Carbide—both the 1898 Union Carbide Company and the 1917 Union Carbide & Carbon Corporation—owed their wealth to Chicago's People's Gas, Light, and Coke Company. Their initial interest in acetylene centered on enriching the gas sold by their utility. Reportedly worth $30 million in 1913, the primary financier was C. K. G. Billings, known best in some circles for hauling horses up a midtown Manhattan freight elevator in 1903 to host a luxury dinner on horseback.[63] Billings recognized the strategic fit among the four firms, and the history and product set of the four gave Union Carbide a different profile than either Du Pont or Allied.

Originally formed in 1898, the Union Carbide Company owned patents for producing acetylene from calcium carbide. When thrown together in an electric arc furnace at 3,500 degrees Fahrenheit, limestone and coke form calcium carbide, a rock-like solid; calcium carbide then reacts with water to produce acetylene, a hydrocarbon gas. Working in North Carolina in 1892, a Canadian using a French-designed electric arc furnace accidentally discovered a commercially viable way to produce calcium carbide in quantity, and he sold his patents to a predecessor firm of Union Carbide Company. Acetylene first found a market in lighting, as the gas company backers had hoped, but Union Carbide found a more lucrative market for acetylene a few years before the war when they imported a French process for cutting and welding metals: oxyacetylene.[64]

For the oxygen in oxyacetylene, Union Carbide turned to Linde Air Products. In the 1890s, German engineer and entrepreneur Carl Linde, already the head of a successful firm that made refrigeration machinery, turned his attention to the science of refrigeration. In his Munich laboratory, Linde cooled air down to liquid and separated out the oxygen, which he sold commercially by 1901. Five years later, he journeyed to the United States to establish an American branch of his firm, which launched in 1907. Linde partnered with Charles Brush, another entrepreneurial engineer, whose fortune came from electric arc lighting and who had helped Linde secure his American patents in 1903. The American branch brought in the technology from Germany, and the firm expanded quickly, demanding more capital to keep up with the growth. Brush and other capitalists in Cleveland kept raising the capital, but they were a mixed blessing for

Linde. As the capitalization grew larger, Linde's portion of ownership grew smaller; by World War I, it was "inconsequential."[65]

By 1899, the National Carbon Company, another firm in the 1917 Union Carbide & Carbon Corporation merger, was the "carbon trust," which held a near domestic monopoly on high-grade carbons for electric arc lights, and it diversified into other carbon-based products, including dry batteries (Columbia and Eveready brands by World War I) and furnace electrodes. The carbon electrodes, powdered coke mixed with a binder and baked into rods, allowed electric arc furnaces to reach their enormously high temperatures (over 6,000 degrees Fahrenheit). Their furnace technology, reported *Fortune*, was in 1941 still "one of Carbide's most jealously guarded secrets." The magazine also reported that National Carbon's aging directors, particularly Myron T. Herrick, Andrew Squire, and James Parmelee, proposed the 1917 merger partly as their retirement plan.[66]

The fourth firm in the merger, the Prest-O-Lite Company in Indianapolis, was Union Carbide Company's largest customer of calcium carbide. Two partners, Carl G. Fisher and James A. Allison, manufactured an acetylene headlight, first for bicycles and then more significantly for automobiles. They acquired the U.S. rights to a French method of using tanks to hold acetylene—steel cylinders lined with asbestos, the acetylene dissolved in acetone—and a national distribution network of interchangeable tanks. Automobile owners bought Prest-O-Lite tanks, which sat on the cars' running boards to supply acetylene to their headlights. When a tank emptied, owners exchanged it for a full one at any distributor, which included automobile dealerships, and the company claimed to have a distributor in nearly every town with a population over 2,000 people in the United States and parts of Canada. With the growth of oxyacetylene, Presto-O-Lite also sold their acetylene to machine shops and others with a need to cut or weld metal. Their brand name—and their tanks—became widely recognized before World War I.[67]

In the middle of World War I, then, the formation of Union Carbide & Carbon Corporation brought together production expertise and experience in electrochemistry and gas manufacturing as well as extensive distribution networks for batteries and acetylene. As with so much of American industry, World War I demanded extraordinary quantities of material from Union Carbide's constituent firms, particularly oxyacetylene for war-related manufacturing. Union Carbide & Carbon Corporation came out of the war ready to lead or compete aggressively in alloy metals, in batteries and related carbon products for electrical uses, and in gases such as acetylene and oxygen. Other than acetylene, nothing in this lineup suggested a future in synthetic organic chemicals for Union Carbide; nor had the firm's leadership actively participated in any of the wartime discussions or policy initiatives centered on establishing an American synthetic organic chemicals industry. But important acetylene-based research during the war set up the firm to take the industry in an "American" direction.

Union Carbide's significant new organic research breakthroughs occurred at the Mellon Institute in Pittsburgh during the war. Funded by the Mellon banking family and opened in 1914, the institute was the brainchild of Robert Kennedy Duncan, who, impressed with the ties between German industry and science on a 1906 trip abroad, developed industrial fellowships in an attempt to promote such collaboration in the United States. Firms paid Ph.D. scientists at the Mellon Institute to tackle specific research problems. Prior to the Union Carbide & Carbon Corporation merger in 1917, Prest-O-Lite, Union Carbide Company, and Linde all independently sponsored fellowships at the Mellon Institute. Prest-O-Lite wanted their industrial fellow to develop a new way to make acetylene, one independent of Union Carbide Company's patented calcium carbide process. Union Carbide Company, meanwhile, established a fellowship to develop additional marketable uses for acetylene. And Linde, having separated oxygen from air, wanted their Mellon research to yield commercial products for the leftover nitrogen. After the merger, the new corporation's leadership continued sponsoring a combined research project at the Mellon Institute. The Pittsburgh-based chemists handed Union Carbide some remarkable developments, including the research that underpinned the firm's dramatic interwar expansion in synthetic organic chemicals.[68]

George O. Curme Jr. arrived at the Mellon Institute in November 1914, greeted by an empty room and a budget to fill it: he received the Prest-O-Lite assignment to find a new process to make acetylene. A native of Iowa, Curme received an excellent education in chemistry—a B.S. from Northwestern University in 1909, a brief stint at Harvard the following year, a Ph.D. from the University of Chicago in 1913, and then a year's study in Germany, with Fritz Haber and Walther Nernst at the Kaiser Wilhelm Institute and with Emil Fischer at the University of Berlin, all prominent German chemists. When, by happenstance, Curme left Germany just ahead of World War I, he possessed a training his American contemporaries considered among the best possible. He began at the Mellon Institute at age twenty-five; twenty years later, when the medals and awards began to arrive, his peers viewed him as pivotal in the development of the domestic synthetic organic chemicals industry.[69] From his introduction at the Chandler award ceremony in 1933:

Dr. Curme is considered one of the greatest living exponents of aliphatic chemistry. He perhaps heads the list of those who have brought the leadership in organic chemistry from Germany where they hold incontestable supremacy in the aromatic field, to the United States where the abundance of raw materials and independence of thought have permitted American chemists to strike out in entirely new directions. His genius for the application of fundamental principles to practical problems of synthetic organic chemistry has been largely responsible for the building of a distinctly American chemical industry.[70]

Using electrochemistry, Curme began his celebrated career by cracking gas oil to develop a new acetylene process for Prest-O-Lite. After Prest-O-Lite merged into Union Carbide & Carbon Corporation, his research agenda changed because the new corporation no longer needed an alternate supply of acetylene. But he had pulled chemicals other than acetylene from his process, and the quest to turn them into commercial products lured him deeply into the aliphatic branch of organic chemicals, particularly ethylene. He and his research group faced the situation of the aromatic organic chemicals of earlier generations: an enormous number of by-products that would either make a process uneconomical or, if a market could be created for them, make a profitable range of interdependent chemicals. Curme's particular genius lay not in pushing the frontiers of basic science but in generating original and cost-effective commercial processes and products, and his Union Carbide bosses took note. By the time he left Pittsburgh, he had synthesized several organic chemicals derived from ethylene.[71]

To develop salable products from Curme's ethylene research, the Union Carbide executives established a new division in the corporation in 1920, adding chemicals to its preexisting strengths in carbons, alloys, and gases. Along with creating the organizational unit, they turned to scaling up laboratory processes to commercial size. North of Charleston, West Virginia, at Clendenin in the Kanawha River valley, the firm converted a natural gas processing plant into a small-scale chemicals plant. A technical staff of about forty worked for five years testing the technical and economic feasibility of different manufacturing processes and products. While scaling up the processes, the Clendenin plant helped to cover costs through sales of gases, such as propane for home heating, aided by Linde's processes of gas separation.[72]

By 1925, officials at Union Carbide moved their fledgling synthetic organic chemicals division out of the experimental Clendenin plant into a full-scale works in South Charleston, West Virginia, and began to mass-produce synthetic organic chemicals. The plant grew quickly, spreading out to cover an eighty-eight-acre island in the Kanawha River. The plant left strong impressions on those who visited. When young Case graduate Paul S. Greer arrived at the South Charleston plant in the 1920s as a new chemical engineer, he was struck, he later recalled, by "the sense that it was a very actively developing situation." And he noticed that the new plant used its own products: piping welded by oxyacetylene was "running all over, for long distances." A concentration of chemical companies gathered in the Kanawha River valley, and Union Carbide's new neighborhood already included Du Pont's ammonia plant and coke ovens (up the river at Belle), E. C. Klipstein's dyes plant and Westvaco's chlorine plant (both South Charleston), and the giant World War I nitrocellulose explosives plant at Nitro that became home to several chemical companies. Like Union Carbide, the other chemical companies found alluring the rich coal, gas, and hydroelectric potential of the area. Natural gas made up two-thirds of the feedstock at Union Carbide's South Charleston plant, but the firm later expanded its feedstocks to include gases from Du Pont's coke ovens, an oil refinery, their own cracking plant, and stripped commercial gas.[73]

Curme and his colleagues felt like pioneers in "an entirely new industrial field," he noted in 1935, and the previous years had prepared Americans to want to see synthetic aliphatic chemistry in that light. Perhaps more than most, however, Curme acknowledged the extent to which Union Carbide mixed European, particularly German, developments with American. While creating specific and patentable processes, Curme understood that he and Union Carbide incorporated chemistry and industrial technologies developed incrementally over time and place. "It became evident," he continued, "that our achievements were but a part of the much larger program in industrial organic synthesis" in other acetylene work, plastics, high-pressure synthesis, and the earlier dyes and pharmaceuticals. Union Carbide's processes depended on the ability to reach extremely high temperatures in furnaces and extremely low temperatures for gas separation, and German inventors and firms worked out the fundamentals of both.[74]

Curme and the other Union Carbide chemists continued to pay close attention to German science, but more as rivals than as models to follow. In 1919 when William F. Barrett of the Linde subsidiary planned a trip to Europe, he received a wish list from Curme, who wanted Barrett to gather "valuable information on the most recent commercial developments" in Britain, France, Switzerland, and Germany on products related to the Mellon Institute research. In some cases, he was curious about European production processes, but he expressed more interest in markets—to what uses did Europeans put these products? Examples included ethylene glycol, "which had a particularly thorough commercial trial in Europe during the past five years owing to the great shortages and extreme demand for glycerin," and hydrogen, whether for transport or hydrogenation or beyond. Synthetic rubber "promises to be of the greatest importance" and "a study of the advance in this field promises to be valuable." Curme appeared well informed about recent developments abroad and generally expressed confidence about his own processes compared to European competitors—his ethylene process was "superior to any European." Curme and his firm expected to follow new paths rather than replicate the course taken by the German industry, the paradigm that dominated most of the discussions about a domestic industry.[75]

Despite continuities with German science, however, George Curme and Union Carbide had crossed into different territory, providing substantial justification for declarations of a new "American" industry. The petrochemical raw materials, the mass production, and the kinds of markets all marked a significant turning point in the synthetic organic chemical industry. Union Carbide's product set fit well into powerful developments in the American economy and culture of the 1920s, particularly the stunning growth of the automobile industry, the most iconic of American industries. Curme himself, looking back from 1935 during another of his award speeches, singled out the automobile industry, which had "largely influenced world history for the past generation," as a massive market for synthetic organic chemicals. He identified synthetic organic chemistry in nearly every part of the automobile from gaskets, exterior finish, and rubber tires to the hydraulic fluid and brake blocks that stopped a car, a "hard trick which most drivers modestly attribute to their own skill."[76]

Union Carbide's best-known and most profitable contribution to the automobile, however, was ethylene glycol, brought out in 1927 as Prestone antifreeze. Ethylene glycol drove growth at the South Charleston plant; by 1940, Prestone consumed about 90 percent of Union Carbide's ethylene glycol. A substitute for glycerin, ethylene glycol first found a market among dynamite manufacturers, who appreciated qualities such as a freezing point lower than glycerin, which made their dynamite safer to use in temperatures below 55 degrees Fahrenheit. Only in 1939 would the firm's patents on Prestone expire and allow rivals to challenge their position in the market.[77]

Union Carbide cracked large quantities of ethylene from natural gas and other feedstocks, and they looked to find markets for by-products and derivatives, including alcohols and other solvents. In 1930, after receiving special permission from the Commissioner on Prohibition, Union Carbide commercially produced synthetic ethyl alcohol, and Union Carbide was virtually alone in its ability to manufacture the synthetic product economically in a market still overwhelmingly dominated by natural ethyl alcohol. "The synthetic production of ethyl alcohol is considered by some as important a step in the field of aliphatic chemistry as the production of synthetic indigo was in the field of aromatic chemistry," declared the Mellon Institute's E. R. Weidlein at Curme's Perkin Award ceremony. Union Carbide also "pioneered" the synthetic production of ethyl ether and acetone. Compared to the natural products, the synthetic versions offered a steadier supply and lower price. Synthetic acetone provided their Prest-O-Lite division a reliable supply of solvent for their acetylene lamps and tanks. Although not the pioneer, Union Carbide also offered synthetic methanol among their line of aliphatic organic solvents. Union Carbide expanded its product line with other solvents, including Cellusolve (ethylene oxide) for cellulose-based chemicals, including plastics, and ethylene dichloride for tetraethyllead. Its ethylene compounds grew from a total of six in 1926 to thirty-five in 1934.[78]

Some industry observers attributed the growth of Union Carbide's chemicals division to the central role of chemists and chemical engineers throughout the Union Carbide organization, noting in that respect a closer similarity to Du Pont than to Allied among the three biggest firms. From research to sales, the management created an institutional culture that deliberately developed and shared internal expertise and promoted collaboration. Curme later recalled that he became a believer in "group research" while working at the Mellon Institute. "With the major research problems that face industry today, no investigator working alone, or even those in small groups, can hope to make progress at a rate that justifies hope of success within a period that even 'patient money' is willing to await." The firm also used "highly trained chemists" in sales who could advise customers on their products and help to identify and develop new markets. In the interwar years, Union Carbide and other top firms could design compounds with particular characteristics to solve customer problems. Management made it an institutional habit to rotate their new technical employees into different positions to allow the employee to learn the business more broadly but also to fit the employee to the task. And, like Herbert Dow, the Union Carbide management promoted from within by deliberate habit. *Fortune* in 1941 noted that Union Carbide employees stayed with their firm—in contrast to the regular supply of former Allied employees on the market.[79]

Union Carbide's exploitation of bulk aliphatic chemicals coincided with the professionalization of chemical engineering, a discipline that achieved a separate professional identity 1915 to 1925. Chemical engineers laid claim to expertise in "unit operation," such as distillation, cracking, and separating, and the coordination of the operations on an industrial scale. In 1933, the trade journal *Chemical and Metallurgical Engineering* bestowed its award for achievement in chemical engineering to Union Carbide, the first time it went to a company rather than an individual. "Chemical Engineering Builds a Synthetic Aliphatic Chemicals Industry," ran the headline on the article announcing the award. "In the building of a great synthetic aliphatic chemicals industry," noted the journal, the "purely chemical engineering nature of this enterprise" stood out. The people who "planned, operated, and managed" were chemical engineers "in fact if not in name." Union Carbide brought together existing American strengths in mass production technology and expertise with rich natural resources in petroleum and natural gas and thereby gave the domestic synthetic organic chemicals industry a distinctively "American" turn.[80]

Bakelite Corporation

In 1922, a trade journal article described Bakelite, a synthetic phenolic resin or a type of early plastic, as "one of the few chemical industrials that is wholly American in its inception and conduct, and which owes none of its success to foreign sources." But in both symbolic and literal ways, Bakelite sat at the transition between old and new in the synthetic organic chemical industry. Leo Baekeland, born and educated in Europe, found success in the new world, successfully combining phenol from the aromatic branch associated with German chemistry and formaldehyde from the aliphatic branch that increasingly spurred the American industry. By careful manipulation of heat and pressure, Baekeland could calibrate his high-pressure processes to control the reaction better than any previous attempt. Bakelite's appeal lay in the ability to mold it into stable shapes, and it thrived in the mass consumer market of the 1920s in the United States and beyond; manufacturers used it in everything from radio knobs to electrical insulators. As historian Robert Friedel has written, Bakelite had chemical characteristics that marked it as a truly synthetic plastic not dependent on the "imitation" of nature.[81] Significantly different from the traditional "German" product set, Bakelite and subsequent plastics became a driver in the American synthetic organic chemicals industry over the twentieth century.

Baekeland himself frequently reiterated the American-ness of Bakelite, but between his Belgian background and the expansion of his firm abroad, Baekeland more comfortably operated on a global stage than most other American manufacturers. Baekeland received his Ph.D. from the University of Ghent, married his Ph.D. advisor's daughter, came to the United States on a fellowship in 1889, and stayed. Four years later, he developed his Velox photographic paper, and when he sold it to Eastman Kodak in 1899 (and promised to stay out of the photographic business for twenty years), Baekeland retired to his home laboratory in Yonkers, New York. He also bought his first car in 1899, early enough to be "considered by my neighbors as an impertinent nuisance" because his car "made noise enough to awaken a whole cemetery"; making improvements on his car became a hobby, and in 1906 he and his family drove around western Europe. But he also traveled to Germany in 1900 to study electrochemistry and applied himself to additional industrial challenges, including the work on combining phenol and formaldehyde successfully. When he won the ACS's Perkin Award in 1916, he noted that in Belgium he had "excellent opportunities of education," and in the United States he received a "real intense education" when "confronted with the big problems and responsibilities of practical life." In 1910, the same year Baekeland moved his operations out of his home laboratory into his first American Bakelite plant (in Perth Amboy, New Jersey), Baekeland traveled to Germany to negotiate the production of Bakelite there. He licensed his Bakelite patents to the Rutgerswerke, a smaller coal tar firm in Berlin experienced with other resins; they initially made Bakelite at the Rutgerswerke plant, but Bakelite Gesellschaft's first separate factory remained incomplete for the duration of World War I. During and after the war, Baekeland also licensed his product and trademark to firms in Japan, Canada, Great Britain, and Italy.[82]

Already in 1909, when Baekeland introduced Bakelite to the world, he imagined a wide range of uses, a prediction borne out as Bakelite fit into every nook and cranny in the expansive consumer markets of the 1920s. By 1916, the demand had already grown so high that he noted a "paradoxical situation" in which "the greatest drawback of bakelite was its multiplicity of applications" because efforts could be diffused in so many directions. Even so, improvements in the process allowed his firm to manufacture an enormous variety of goods, helped along particularly by skyrocketing automobile and radio sales. In an automobile, Bakelite could be found in the ignition, spark plugs, paint, battery terminals, ashtrays, steering wheels, and assorted other locations. Similarly, radios employed Bakelite in the casing, knobs, dials, and electrical panel. In addition to these two expanding industries, Leo Baekeland's creation went into pipes, jewelry, enamels and lacquers, gears, grinding wheels, kitchen counters, and textile bobbins, among other things. The ubiquity of Bakelite already by 1924 put Baekeland on the front cover of *Time* magazine and sent the *Time* journalist over the edge: "Books and papers will be set up in Bakelite type. People will read Bakeliterature, Bakelitigate their cases, offer Bakeliturgies for their dead, bring young into the world in Bakelitters." The Bakelite Corporation proved adept at advertising and promoting its product to the public in a way that most of the synthetic organic chemicals industry, which primarily catered to other industrial firms, had rarely found necessary until the 1920s.[83]

Baekeland took the lead in phenolic resins, establishing his first plant in Perth Amboy, New Jersey, in 1910, but others had explored similar products, and he spent many years in patent battles, the largest one resolved by a merger. While Baekeland had opted to keep the process of his Velox photographic paper secret, he patented Bakelite in the United States and abroad. In the early 1920s, he emerged victorious from a long patent infringement suit against his two leading competitors, the Redmanol Chemical Products Company of Chicago and the Condensite Company of America of Bloomfield, New Jersey. Already before the war, almost from the beginning of his rivals' existence, Baekeland charged L. V. Redman of Redmanol and Kirk Brown of Condensite with violating his Bakelite patents. In May 1922, just after Redmanol purchased Condensite, a federal court in New York ruled for Bakelite and left Redman few choices. Redman opted to accept Baekeland's proposal to merge with General Bakelite Company into a new firm, the Bakelite Corporation, although the three firms retained their plants and trade names.[84]

From the beginning of his commercial career, Baekeland knew how to protect his intellectual property; he developed expertise and opinions, which he shared in publications and before policymakers. Even before his long fight with Redmanol, Baekeland had become a critic of the U.S. patent system. He testified before Congress in 1912 that part of the reason he left Velox unpatented was that he lacked the financial resources to defend it against infringers and therefore felt safer not disclosing the process through the patent description. The American system, he said, made "the strong stronger and the weak weaker; a game . . . played by the large corporations . . . with the money to pay the lawyers." By the time he invented Bakelite, he was among the strong and had the resources to defend his patent. The American system made it too easy to challenge and tie up patents in litigation, he believed. After the war, he echoed Garvan, in more moderate tones, expressing concern for the role patents played in international cartels; his solution was to support a working clause (the requirement to use the patent or lose it).[85]

As did the other manufacturers of synthetic organic chemicals, Baekeland embraced research to stay innovative and expand the uses for Bakelite, and the firm quickly expanded beyond Baekeland's hands-on work in his laboratory as the firm grew and built expertise. In 1926, the General Bakelite

Corporation employed "about eighty chemists, engineers, and physicists, about one-half of whom are engaged in research or development work." Leo Baekeland preferred to hire his employees on a month-by-month "trial engagement" of up to a year before deciding whether and in what department to keep his technical employees. Some development of new products occurred inside the firm, but right from 1909 Bakelite drew on the expertise of customers and subcontractors to develop processes for specific applications. Even so, some of the trial and error in pioneering workable molds for new products became part of company lore, such as the mold used "only once because the toolmaker forgot to design it so it could be opened."[86]

Baekeland's strategy was to purchase his raw materials from outside suppliers and maintain his firm's focus on producing final goods for individual and industrial consumers, rather than trying to integrate fully from raw materials to market. When a phenol shortage hit in late 1922 and early 1923, the Bakelite company deviated from practice and built its own phenol plant in Painesville, Ohio, but even during construction, the firm viewed the plant as a temporary investment until other chemical manufacturers could produce the intermediate for less, and Baekeland mothballed the plant in 1926. By the late 1920s, Bakelite purchased its phenol from Dow, Monsanto, Elko Chemical Company of West Virginia, and U.S. Steel. In addition, Bakelite maintained long-running contracts with Roessler & Hasslacher Chemical Company to purchase formaldehyde, the other necessary ingredient for the plastic; the manufacture of formaldehyde consumed large amounts of methyl alcohol. At Bakelite's twenty-fifth anniversary, they claimed more than 2,000 Bakelite products used in millions of applications. The firm held to Baekeland's strategy, remaining independent and focused on Bakelite, until it merged with Union Carbide in 1939 and integrated into its larger chemical operations.[87]

Like many of the newer aliphatic product lines, Bakelite arrived on the market in large quantities. Through the 1920s, Bakelite gained new rivals with the development of aliphatic synthetic resins, particularly synthetic vinyl and urea resins, the latter of which could offer a less brittle and more translucent product. The output of synthetic resins grew about twenty times over the course of the decade, from 1.6 million pounds in 1921 to 5.9 million pounds in 1922 to 33 million pounds in 1929.[88] Leo Baekeland, the naturalized American citizen with a lifelong Flemish accent, established one of the most distinctive synthetic organic chemical products of the 1920s, and he took the lead in proclaiming it an American achievement.

###

By 1930, the Americans had their "American" synthetic organic chemicals industry, but it wasn't the industry they set out to build in World War

Table 8.4. Synthetic Resins

	SYNTHETIC RESINS OF COAL TAR ORIGIN			OTHER SYNTHETIC RESINS
	Production (in Pounds)	Value of Sales	No. of firms	Production (in Pounds)
1927	13,452,230	$6,094,656	7	—
1928	20,411,465	$7,211,958	9	—
1929	33,036,490	$10,393,397	12	6,800,698
1930	30,867,752	$7,323,656	15	12,529,363
1939	179,000,000	—	—	33,700,000

Source: U.S. Tariff Commission, *Census of Dyes and Other Synthetic Organic Chemicals, 1930*, 49; the 1939 figures come from *Rogers' Manual of Industrial Chemistry*, 2: 1191.

I when the capacities of the great German chemical industry loomed so large and menacing. In synthetic dyes, Americans were not equal. I.G. Farben still wielded considerable advantage in the depth of its technical understanding and ability to manufacture a wide range of complicated dyes with significant cost advantages over the Americans, British, and other less experienced manufacturers. And while Carl Bosch led I.G. Farben's massive investments in the high-pressure technologies, dyes remained one of the big firm's most profitable sets of products in the 1930s. From the mid-1920s until the late 1930s, dyes made up roughly a third of I.G. Farben's total sales and anywhere from 38 to 74 percent of the firm's profits.[89] By 1930, American promotional policies had clearly allowed the American synthetic organic chemicals industry to survive and even flourish in many ways, but dyes, the products most identified with the Germans and most targeted by wartime policy, remained first and foremost a German specialty.

"Henry Ford's methods of mass production are now being applied to chemistry in this country," and Americans "would be fast mounting to the foremost position of the world," reported the *New York Times*, paraphrasing Frank Whitmore of Northwestern University in 1928.[90] Although the German chemical industry remained strong, Americans nonetheless felt that their achievements in the previous decade had sufficiently limited their vulnerability to future shortages, and that they had built a firm scientific, economic, and political infrastructure at least to compete and often to surpass the German industry. A newer generation coming of age in the 1920s entered an increasingly dynamic industry in which the activities of the Germans mattered less than American natural resources, the American consumer market, and American higher education.

Conclusion

In the early 1920s, Merck & Company corresponded regularly with E. Merck in Darmstadt, sharing technical information despite having become separate firms, but the management on each side faced a telling problem. In the spring of 1922, Rudolf Gruber at Merck & Company exchanged several letters with E. Merck in Darmstadt about which language their chemists should use to communicate across the Atlantic. The technical staff in Germany requested German, suggesting it was in the American firm's best interest to make sure the Germans could understand Americans' questions and problems. Gruber sent them an English-German technical dictionary, perhaps with an implicit message intended, but he lost that battle and agreed to send communications in German. But the Merck exchange was a harbinger of things to come: in the 1920s, for the first time, the number of chemical abstracts published in English exceeded the number published in German. German was losing out as the dominant language of chemistry, and the United States had made a patriotic mission out of matching the German chemical industry. The shift in language was part of a more dramatic transition precipitated by World War I.[1]

How much was World War I a catalyst for the growth of the U.S. synthetic organic chemicals industry? The story line begs most temptingly for that counterfactual speculation: what if there had been no World War I? In chemistry, a catalyst is a tool to change the speed of reactions, but not necessarily the outcome; I would argue that the war sped up certain trends already underway, but the fear and hostility generated in the war also altered the outcome in important ways. On the one hand, by World War I, the United States had an industrial economy of great breadth and depth, which included a strong inorganic chemicals industry, effective chemical engineering, a massive natural resource base for organic chemicals, and a university system increasingly oriented to advanced scientific research. These trends might all have led to an American synthetic organic chemicals industry, but most likely one with a different trajectory and configuration.

Perhaps the automobile and other consumer goods might still have sparked a large demand for mass-produced aliphatic organic chemicals—the solvents and antifreeze and Bakelite radio knobs. Amid the vast output of aliphatic commodity chemicals, the relatively minuscule poundage of dyes and pharmaceuticals might not have been missed in a simple measure of volume. American manufacturers of Bakelite and the aliphatics plowed ahead into relatively new territory, sometimes ahead, behind, or equal to the German industry, but nearly always in a competitive position.

On the other hand, without the war's interruption in trade, would the powerful American textile industry have supported tariffs high enough to support a domestic dyes industry? It seems unlikely. The counterfactual crystal ball is even hazier when contemplating synthetic pharmaceuticals—a smaller set of products than dyes in an economy with a strong biologically based pharmaceutical industry. The aspirin patent expired in 1917 and perhaps such a well-known and popular product would have tempted manufacturers to try, or perhaps the Germans might have expanded pharmaceutical manufacturing on U.S. soil to retain market share.

The war also brought significant attention to chemists, chemistry, and the chemical industry in a way that contemporaries believed made a difference in recruiting new chemists and expanding the amount of top-level research in organic chemistry and related fields—in academia and industry. Closer relations emerged among academic chemists, industry, and the military. The war gave chemists and related professions a crash course in synthetic organic chemistry, from the laboratory to the industrial plant, broadening, deepening, and redistributing the expertise available in the United States. The initial effort to replicate German dyes production helped to build American chemical expertise on which newer synthetic organic product lines could draw.

Without the war, the confiscation and disruption of German business in the United States is particularly difficult to imagine, despite persistent strains of American isolationism before World War I that emerged more strongly afterward. The war stopped the nineteenth-century globalization abruptly and so thoroughly that decades passed before the American economy recovered a comparable degree of global integration. Policymaking, from large laws to administrative minutiae, is shaped by the larger political, economic, and cultural context, and the war affected policies across all branches of government.

Even more speculatively, might the international political economy in the interwar years have taken a different direction, particularly in Germany, had Americans reacted differently to the war? As historians study the vexing and complicated rise of Nazism, part of the explanation usually involves the blows the German economy suffered in the 1920s, including those inflicted by the terms of peace. The American determination to build a domestic synthetic organic chemicals industry added pressure and insult to an increasingly explosive German political economy. Before the war, the German economy depended heavily on exports, and the British and American textile industries had been among the German chemical industry's largest customers. Both denied German chemicals easy access to their markets after the war, which undermined a strong pillar of the German economy. As their economy deteriorated, Germans abandoned their Weimar democratic experiment and turned to radical Nazi political leadership that embraced their own autarkic agenda through the decade leading up to World War II. After the war, I.G. Farben became one of the infamous names associated with concentration camps and the Holocaust.[2]

In *The Creative Society*, historian Louis Galambos describes the American government in the early twentieth century as a promotional state, a government that "encouraged enterprise rather than regulat[ed] it," and the U.S. synthetic organic chemicals industry was a beneficiary of expanded federal activities on behalf of business. The promotional state worked not by some grand central plan but by an accumulation of political compromises and precedents to satisfy the demands coming from different parts of American society, from different parts of the chemical industry, and even from different parts of a single corporation.[3] World War I gave a tremendous impetus to the industry, but the policymaking occurred across several sites in the legislative, executive, and judicial branches. Initiatives emerged from the private sector to lessen the risk of venturing into the "German" industry, but government officials, such as those in the Office of Alien Property, the military, or Department of Commerce, also initiated efforts to build the industry, and they shaped the collection of policies designed to promote the U.S. synthetic organic chemicals industry.

Scholars of American political history note the opportunity crises present for policy innovation, and in the synthetic organic chemicals industry, American policymakers faced the challenge of creating laws and institutional mechanisms to build technical and market expertise in one of the most complicated science-based industries of the early twentieth century. Policymakers often drew on familiar tools, particularly the tariff laws, but some policies reflected newer trends in society, including the growing faith in information and data associated with the Progressive Era. In other cases, policymakers experimented and tried alternatives to suit the immediate task at hand. Trial and error permeated the halting steps to mobilize for war. The confiscation of German chemical subsidiaries, and the establishment of the Chemical Foundation to manage former German patents, grew out of fears and hostilities cultivated by war, but it was also a creative, if not altogether just, policy means to an end. The officials lower in the hierarchy also sometimes needed to innovate in their implementation of laws and initiatives, whether in the details of customs administration or accepting reparations chemicals from Germany.

###

World War I rattled the already shaky faith Americans had in the globalization of the long nineteenth century. More than anything else, war and the threat of war undermined the confidence in the mutual benefit of international trade. Back when Congress debated the 1883 tariff that the World War I generation made into a scapegoat, the stakeholders understood the tradeoffs. Before a congressional committee in 1882, Everett Wheeler of the New York Free Trade Club argued that "the only objection" to relying on cheaper European dyes "would be . . . if we got into a war with some of the European countries. . . . I suggest that that is a very contingent and remote disadvantage; that the probabilities of such a war are insignificant."[4] And, in 1882, war indeed was quite remote, far from inevitable, and global trade flourished. In the middle of World War I, however, Americans perceived international trade not as an opportunity to obtain cheap German dyes from a lower-cost producer but as an economic, political, and military liability.

Their trust in international commerce shattered by World War I, Americans made self-sufficiency in the industry a political priority; they learned to fear both economic shortages and potential breaches in the national defense. The war destroyed much of the nineteenth-century global commercial network, including the dyes and pharmaceuticals trade out of Germany. In synthetic organic chemicals, Americans intentionally dismantled the prewar German market presence in the United States and reconstructed postwar commerce on different terms. The isolationist turn could not completely prevent resumption of global commerce in synthetic organic chemicals, partly because of basic economic factors such as costs of production, but international trade proceeded in narrower channels. Americans of the World War I generation used the government tools at their disposal to support a domestic industry through transfers of technology, market protection, and various hindrances to German competition.

###

The sense of crisis drew the nation's attention to synthetic organic chemicals, redirecting resources to the field, but survival of the industry would depend on whether the American chemical community could acquire the necessary technical expertise. Americans' acquisition of skill came both from international and domestic transfers of technical knowledge, but also increasingly from the internal development of new knowledge. University chemistry departments, which had long attended to the needs of agriculture and other practical applications of organic chemistry, responded to the new opportunities in synthetic organic chemistry after the war. Already the number of Americans receiving their chemistry Ph.D.s in Germany declined before the war, and students increasingly stayed in the United States to obtain their graduate training. Research departments in the largest firms, particularly Du Pont and increasingly Union Carbide, provided important centers for generating new innovations as well as improvements on existing ones.

The government role in the development of the industry's technical and market expertise took on many forms, from large macroeconomic policies to research on specific technical problems in government laboratories. Tariff policies and other protectionist measures created a market environment that fostered the infant industry, and U.S. government policies held the German firms at bay long enough for Americans to accumulate a pool of technical expertise through education and experience. The ability of various government agencies—the Department of War, the Department of Commerce, the U.S. Tariff Commission—to collect and redistribute information was a function that, in many cases, only the government could provide. Import statistics from customs invoices, explosives and gases manufacturing techniques, and assessments of foreign markets were all kinds of information provided to the domestic industry from sources government officials could access more easily than private entities.

The government's role in generating new science and technology mostly lay in the future, although not far. The Chemical Warfare Service foreshadowed a more extensive military role in advanced science, and with the strong urging of Charles Herty, among others, the launch of the National Institute of Health in 1930 marked a commitment to basic scientific research beyond military purposes.[5] When World War II and the Cold War arrived, the relationship among government agencies, industry, and university grew in complexity and dependence, reflected in everything from the Manhattan Project and "military-industrial complex" to Silicon Valley and biotechnology. Americans still struggle with the balance among globalizing forces, national defense imperatives, and the role of advanced science and technology in both.

###

Historians know well that chronological periodization shapes the trajectory of their stories and analyses. By stopping in 1930, we leave an American industry about to launch into enormous productivity and profitability, on the cusp of nylon, plastics, synthetic rubber, and a great array of new synthetic organic chemical lines. Americans brought mass production and their huge national market to the synthetic organic chemicals industry, and the big American chemicals firms registered substantial profits for subsequent decades. We have a picture of mostly cooperative rather than contentious relations between industry and government. For Du Pont, this twenty-year span was preceded by an antitrust suit in 1907 and followed in the mid-1930s by Senator Gerald Nye's hearings in which he labeled the company "merchants of death" for their role in the war. Du Pont's active political engagement in gaining support for the synthetic organic chemicals industry—all the tariff hearings and the patent confiscations and Chemical Foundation, in addition to explosives production—gave the firm a high profile that made them a target when the political winds shifted in the midst of the Great Depression. And Du Pont angered others when they tentatively built ties to German firms in the late 1920s and 1930s. When World War II arrived, Du Pont and the military again worked together on mobilization.[6]

The "American" synthetic organic chemicals industry gained its identity by transforming the niche markets of German dyes and pharmaceuticals into mass-produced commodity chemicals, developing the aliphatic branch of chemistry and drawing on the expertise of chemical engineering. In later decades, Americans would have reason to worry that their approach had reached a limit; the mass production that had buoyed the American industry in the interwar years became less and less profitable as the industry matured and as each firm attempted to eke out returns on the margin. Looking back on the development in synthetic organic chemicals in the decade after the war, Herbert Dow assessed the relative standing of the American and German industries. His faith in Americans' abilities to mass produce more cheaply than Germans remained unshakable. "The molecules react the same way in Germany as they do in the United States," he argued, "but machinery in the United States is distinctly different [from] the machinery in Germany." The Germans were "extremely efficient with material and extremely prodigal with labor," and the future lay with the Americans.[7]

Try as they might, Americans failed in their attempt to disengage from the international political economy after the war, and the synthetic organic chemicals industry remained deeply embedded in domestic and international politics even as most of the intense wartime animosities faded. The Smoot-Hawley tariff in 1930 renewed and reinforced most of the very high rates on synthetic organic chemicals; critics subsequently included it among aggravating factors in the Great Depression. Some of the extra provisions, such as American valuation, survived decades afterward until declared by international agreement an unfair trading practice in the 1960s. Trade policy continued to move more fully to the executive branch to provide increased flexibility to negotiate with international partners. Internationally, the instability of the Germany's political and economic systems caused a second world war. The intense competition for world markets brought the American firms, particularly Du Pont, to the negotiating table with I.G. Farben and Britain's ICI. Historians John Smith and William Reader, among others, refer to the international deal-making as a kind of diplomacy that brought stability in the arenas in which firm's executives could exert some influence in that uncertain age. At the outbreak of World War II, Americans again turned to the Trading with the Enemy Act and the Office of Alien Property to investigate German chemical holdings in the United States.[8]

Notes

Abbreviations

ADR	*American Dyestuff Reporter*
BFDC	RG 151, Bureau of Foreign and Domestic Commerce, NARA
CCOHC	Columbia Center for Oral History Collection, Columbia University
CHH	Charles H. Herty Papers, Emory University
DCM	*Drug and Chemical Markets*
DHC	Dow Historical Collection, Chemical Heritage Foundation Archives
ERS	Papers of Edward Reilly Stettinius Sr., University of Virginia
FPG	Francis P. Garvan Collection, University of Wyoming, American Heritage Center
IEC	*Industrial and Engineering Chemistry*
JIEC	*Journal of Industrial and Engineering Chemistry*
MCE	*Metallurgical and Chemical Engineering*
NARA	National Archives and Records Administration (USA)
OPDR	*Oil, Paint and Drug Reporter*
WDM	*Weekly Drug Markets*

Introduction

1. "Giant U-Boat Held to Be a Trader," *New York Times*, July 11, 1916, pp. 1–2; "Navy Experts Say Giant Submarine is Merchantman," *New York Times*, July 12, 1916, pp. 1–2; "German Undersea Merchantman Sets Long-Distance Record for Submarine Transocean Trip," *Washington Post*, July 10, 1916, p. 1. Despite its solecisms, the *Washington Post* article suggests the tone of the day: "The Germans of Baltimore are celebrating in true German style the arrival of the Deutschland instead of singing 'Deutschland Uber Alles' it is 'Deutschland Unter Alles,' since Germany has come to them under, instead of over the sea. Its no longer the 'flying Dutchman,' but 'the diving Dutchman.'" Messimer, *The Merchant U-Boat: Adventures of the Deutschland, 1916–1918*, 48–73. The *Deutschland* made a second trip in November 1916, this time to New London, Connecticut, and it carried pharmaceuticals as well as dyes on that voyage.

2. Winter, ed., *The Legacy of the Great War: Ninety Years On*, provides a helpful overview of recent scholarly literature on World War I. For an engaging one-volume history of the war and its global dimensions, see Eric Dorn Brose, *A History of the Great War: World War One and the International Crisis of the Early Twentieth Century*. Brose notes historians' skepticism that all-out unrestricted submarine warfare against merchant

ships would have defeated the British, even in the absence of American allies (207-8). Niall Ferguson's *The Pity of War* offers a controversial and interesting counterfactual speculation—that, had the British decided to stay out of the war in 1914, Europe would have followed a more peaceful twentieth-century route to integration.

3. Edgerton, *The Shock of the Old: Technology and Global History Since 1900*, 143. Fussell, *The Great War and Modern Memory*, is the classic account of the war's lasting cultural impact. The most complete account of gas warfare in World War I is Haber, *The Poisonous Cloud*. Essays covering the production of a wider range of military chemicals are found in MacLeod and Johnson, eds., *Frontline and Factory: Comparative Perspectives on the Chemical Industry at War, 1914-1924*.

4. Galambos, *The Creative Society*, especially 54-56.

5. Galambos (*The Creative Society*) is one of those scholars, as is Skowronek (*Building a New American State*).

6. *Report of the Alien Property Custodian, 1918-1919*, 9.

7. Palmer testimony in Senate Committee on Appropriations, *Urgent Deficiency Appropriation Bill* (1918), 6; Coben, *A. Mitchell Palmer: Politician*.

8. *Report of the Alien Property Custodian, 1918-1919*, 14, 25, 17.

9. "Francis P. Garvan, Lawyer, Dies Here," *New York Times*, November 8, 1937, p. 23. Garvan and his wife, Mabel Brady Garvan, became avid art collectors; Whalen discusses them in "American Decorative Art Studies at Yale and Winterthur."

10. Steen, "Patents, Patriotism, and 'Skilled in the Art.'"

11. O'Rourke and Williamson, *Globalization and History: The Evolution of a Nineteenth-Century Atlantic Economy*; James, *The End of Globalization*, 5, 12. In Obstfeld and Taylor, *Global Capital Markets: Integration, Crisis, and Growth*, 27-28, the authors note that only in the very late twentieth century did capital markets regain the level of mobility they had in 1914.

12. Beer, *The Emergence of the German Dye Industry*; Wilkins, *The History of Foreign Investment in the United States to 1914*; Steen, "Confiscated Commerce: American Importers of German Synthetic Organic Chemicals, 1914-1929."

13. Wilkins, *The History of Foreign Investment in the United States, 1914-1945*, xiii.

14. Jones, "The End of Nationality? Global Firms and 'Borderless Worlds.'"

15. U.S. Bureau of the Census. *Thirteenth Census of the United States, 1910*, vol. 1, *Population*, chap. 8, p. 875; Higham, *Strangers in the Land: Patterns of American Nativism, 1860-1925*, 208-9. David M. Kennedy discusses the United States in World War I more generally in *Over Here: The First World War and American Society*.

16. Goldstein, *Ideas, Interests, and American Trade Policy*, 122-24. Goldstein points out that some of these innovations grew out of prewar policy ideas and attempted implementation on a more limited scale. Wolman, *Most Favored Nation: The Republican Revisionists and U.S. Tariff Policy, 1897-1912*.

17. Kent, *The Spoils of War: The Politics, Economics, and Diplomacy of Reparations, 1918-1932*; Manfred F. Boemeke, Gerald D. Feldman, and Elisabeth Glaser, eds., *The Treaty of Versailles: A Reassessment after 75 Years*; Steen, "German Chemicals and American Politics, 1919-1921."

18. Krige, *American Hegemony and the Postwar Reconstruction of Science in Europe*. See also Krige and Barth, eds., "Global Power Knowledge: Science and Technology in International Affairs"; Jones, "American Chemists and the Geneva Protocol." David J. Rhees emphasizes chemists' assertiveness on several fronts in creating awareness and respect

for their profession. See Rhees, "The Chemists' Crusade: The Rise of an Industrial Science in Modern America, 1907–1922."

19. Several scholars have explored the relationship between science and the state, some including foreign relations. Crawford, Shinn, and Sörlin, eds., "The Nationalization and Denationalization of the Sciences: An Introductory Essay"; Krige, *American Hegemony and the Postwar Reconstruction of Science in Europe*; Forman, "Scientific Internationalism and the Weimar Physicists"; Mizuno, *Science for the Empire: Scientific Nationalism in Modern Japan*. Gregory Dreicer explores nationalism and the way it shaped understandings of technology in "Building Bridges and Boundaries: The Lattice and the Tube, 1820–1860."

20. The many historians exploring exchanges across borders and transnational history include Rodgers, *Atlantic Crossings*; Bender, "Historians, the Nation, and the Plenitude of Narratives."

21. Beer, *The Emergence of the German Dye Industry*; Haber, *The Chemical Industry during the Nineteenth Century*; Haber, *The Chemical Industry, 1900–1930*; Travis, *The Rainbow Makers: The Origins of the Synthetic Dyestuffs Industry in Western Europe*; Murmann, *Knowledge and Competitive Advantage*.

22. Skowronek, *Building a New American State*, quotation on 166; Richard R. John discusses the nineteenth- and twentieth-century political economy in *Spreading the News* and *Network Nation*; Carpenter, *The Forging of Bureaucratic Autonomy: Reputations, Networks, and Policy Innovation in Executive Agencies, 1862–1928*; Usselman, "Unbundling IBM: Antitrust and the Incentives to Innovation in American Computing," 250–52.

23. Hart, "Corporate Technological Capabilities and the State: A Dynamic Historical Interaction." In his scholarship, Hart has been among those who have successfully drawn on insights from both political science and the history of science and technology.

24. Galambos, "The Emerging Organizational Synthesis in Modern American History" (1970); Galambos, "Technology, Political Economy, and Professionalization: Central Themes of the Organizational Synthesis" (1983); Galambos, "Recasting the Organizational Synthesis: Structure and Process in the Twentieth and Twenty-first Centuries" (2005). A helpful overview is John, "Elaborations, Revisions, Dissents: Alfred D. Chandler, Jr.'s, *The Visible Hand* after Twenty Years."

25. Galambos, *The Creative Society*; Skowronek, *Building a New American State*; Balogh, "Reorganizing the Organizational Synthesis: Federal-Professional Relations in Modern America."

26. Norton, *Artificial Dyestuffs Used in the United States* (1916).

27. Ellis Hawley's work on Hoover is now classic: "Herbert Hoover, the Commerce Secretariat, and the Vision of an 'Associative State,' 1921–1928," quotation on 117–18; Hawley, *The Great War and the Search for a Modern Order*. David M. Hart discusses the informational failures in "Herbert Hoover's Last Laugh: The Enduring Significance of the Associative State in the United States," 420. Historian William Becker notes that Department of Commerce officials seized the opportunity to help firms build markets abroad for American goods, an effort that brought mixed results and appealed more to small firms than large firms. And while the domestic industry had few expectations about competing abroad against Germany, its need for information made it an avid consumer of the department's reports, even on foreign markets. Becker, *The Dynamics of Business-Government Relations: Industry and Exports, 1893–1921*.

28. Koistinen, *Mobilizing for Modern War, 1865–1919*; Jones, "Chemical Warfare Research during World War I"; Steen, "Technical Expertise and U.S. Mobilization, 1917–1918: High Explosives and War Gases."

29. Clarke, Lamoreaux, and Usselman, "Introduction," in *The Challenge of Remaining Innovative*, 1–38.

30. Mowery and Rosenberg, *Technology and the Pursuit of Economic Growth*, 85; Thackray, Sturchio, Carroll, and Bud, *Chemistry in America, 1876–1976*. Kleinman takes a close look at a more contemporary intersection of university science and commercial interests in *Impure Cultures: University Biology and the World of Commerce*.

31. Peter Spitz covers important parts of the aliphatic story in *Petrochemicals: The Rise of an Industry*.

32. Works that address the interwar years include Abelhauser, von Hippel, Johnson, and Stokes, *BASF: The History of a Company*; Hayes, *Industry and Ideology: I.G. Farben in the Nazi Era*.

33. U.S. Tariff Commission, *Census of Dyes and Coal-Tar Chemicals, 1927*, 129.

Chapter One

1. Reinhardt, *Forschung in der chemischen Industrie*, 220–24; Reinhardt and Travis, *Heinrich Caro and the Creation of Modern Chemical Industry*, ix. A scholarly history of BASF is Abelhauser, von Hippel, Johnson, and Stokes, *BASF: The History of a Company*. On indanthrene blue GCD, see Norton, *Artificial Dyestuffs Used in the United States* (1916), 189.

2. Beer, *The Emergence of the German Dye Industry*, 134–35; Cain and Thorpe, *The Synthetic Dyestuffs and the Intermediate Products from Which They Are Derived* (1913), 128–31, 252, 284–85. Anthony Travis, historian and chemist, says that the process originally used 2-amidoanthraquinine rather than the 2-aminoanthraquinone Cain and Thorpe described in 1913. Furnas, *Rogers' Industrial Chemistry* (1942), 2:1155–56. There are pictures of BASF's plant in 1893 and 1895 in Reinhardt and Travis, *Heinrich Caro*, 255–56. The letters in a dye's name could give an indication regarding shade: B for blue, R for red, and G for yellow (*gelb*), but naming practices varied widely. Norton, *Artificial Dyestuffs Used in the United States*, 32–33.

3. U.S. Tariff Commission, *Census of Dyes and Other Synthetic Organic Chemicals, 1922*, 156; Norton, "A Census of the Artificial Dyestuffs Used in the United States," *JIEC* 8 (November 1916): 1042–44; Norton, *Artificial Dyestuffs Used in the United States*, 186–91.

4. Beer, *Emergence of the German Dye Industry*, 97–98; see also a company history of Hoechst in Bäumler, *A Century of Chemistry*, 33–38; Marsch, "Strategies for Success: Research Organization in German Chemical Companies and I.G. Farben until 1936," 26.

5. Parascandola, "The Theoretical Basis of Paul Ehrlich's Chemotherapy"; Travis, "Science as Receptor of Technology: Paul Ehrlich and the Synthetic Dyestuffs Industry"; "Doctors Have Faith in Ehrlich's '606,'" *New York Times*, February 9, 1912, p. 18; Herman A. Metz, letter to editor, "Dr. Ehrlich's Remedy," *New York Times*, November 8, 1910, p. 8; Metz, "Solving Medical Mysteries by Help of Animals," *New York Times*, January 28, 1912, p. SM6; Bäumler, *A Century of Chemistry*, 33–38; "Neo-Salvarsan is No. '914,'" *New York Times*, April 28, 1912, p. C3.

6. "An Act to Reduce Tariff Duties and to Provide Revenue for the Government, and for Other Purposes" (1913), 115, 153, 181–92.

7. For more on the importers' relationship to the German firms before World War I, see Steen, "Confiscated Commerce: American Importers of German Synthetic Organic Chemicals, 1914-1929."

8. Cain and Thorpe, *Synthetic Dyestuffs*, 37-38; Furnas, *Rogers' Industrial Chemistry* (1942), 2:1155-56. "The name 'vat-dyestuffs' has been given to those dyestuffs which are insoluble in water, but give soluble products on reduction, which may be coloured or otherwise, and which have a distinct affinity for textile fibres. For dyeing it is sufficient to allow the material to remain in the vat, remove it, and expose to air; the original insoluble substance is again formed inside the cells of the fibre, on which it is thus fixed." Wahl, *The Manufacture of Organic Dyestuffs*, 220.

9. Bäumler, *A Century of Chemistry*, 33-38; "Fault Found in Technique," *Los Angeles Times*, April 3, 1914, 1:11; "Doctors Have Faith in Ehrlich's '606,'" *New York Times*, February 9, 1912, p. 18; "Obviate '606' Disadvantages," *New York Times*, March 19, 1912, p. 3.

10. Murmann uses the example of the dyes industry in Germany, Great Britain, and the United States to develop an evolutionary model of economic growth, and his work is a very useful synthesis of much of the historical literature on the industry. The number of German firms (by 1914) comes from the database he and Ernst Homberg compiled on firms in the synthetic dyes industry. Murmann, *Knowledge and Competitive Advantage*, 41-45. In addition to the Big Six firms, two smaller but notable German companies produced dyes. In 1896, J. W. Weiler & Company merged with the Kommanditgesellschaft E. ter Meer to form Weiler-ter Meer, a small manufacturer of dyes located near Cologne. The Chemische Fabrik Griesheim, a soda manufacturer that made its mark in electrochemistry in the 1890s, purchased the dyes firm owned by K. G. R. Oehler of Offenbach in 1905 and became the eighth sizable German dyes manufacturer. Another twelve firms, each much smaller, also made synthetic dyes on the eve of the war. Haber, *The Chemical Industry during the Nineteenth Century*, 128, 131-32; Haber, *The Chemical Industry, 1900-1930*, 20, 120-21, 162; Bäumler, *A Century of Chemistry*, 43-50.

11. Contemporary experts used a range of classification systems for dyes; they depended either on chemical structure or method of application. Common usage drew from both systems, a practice historians subsequently borrowed to discuss the major dyes innovations. Quoting Bernard Hesse, Thomas Norton listed seventeen classes in *Dyestuffs for American Textile and Other Industries*, 40. Cain and Thorpe listed seven, with subcategories, in *The Synthetic Dyestuffs* (1920), 40-41. Wahl's textbook, *The Manufacture of Organic Dyestuffs* (1914), listed fifteen, p. 84. The six categories by application method sometimes varied depending on the source, but one such grouping listed acid, basic, sulphur, direct, mordant, and vat dyes. U.S. Tariff Commission, *Census of Dyes and Other Synthetic Organic Chemicals, 1921*, 35-44; Cain and Thorpe, *The Synthetic Dyestuffs*, 36-38, listed seven: acid, basic, direct, mordant, vat, ingrain, and developed. Haber, *The Chemical Industry during the Nineteenth Century*, 83-86.

12. Craig, *Germany, 1866-1945*, 1-178.

13. The cartels formed part of a bureaucratized economic system in Germany that historian Jürgen Kocka and others have called "organized capitalism." Kocka, "Entrepreneurs and Managers in German Industrialization," 1:562-70. Chandler also discusses the impact of cartels in German industry, *Scale and Scope*, 587-92. The two cartel groups would establish closer ties through an agreement in 1916 and then merge in 1925 to form I.G. Farben. The firms that manufactured only pharmaceuticals, such as E. Merck and Heyden, stayed outside the cartels. Travis points out that predecessors to the 1881 alizarin convention

included an agreement made in 1869 between Perkin and BASF that concluded almost immediately upon obtaining their respective alizarin patents. Beer, *The Emergence of the German Dye Industry*, 117–31; Travis, *The Rainbow Makers*, 240.

14. Beer, *The Emergence of the German Dye Industry*, 105–8; Wertheimer, "The German Patent System," *JIEC* 4 (June 1912): 464–65; Gispen, *Poems in Steel: National Socialism and the Politics of Invention from Weimar to Bonn*, 25–34. Gispen discusses the role of Werner Siemens, the industrialist in electricity, in the formation of the 1877 law. Among other things, the law gave employers great power over the inventions of their employees. On U.S. patents held by German companies, see Liebenau, "Patents and the Chemical Industry: Tools of Business Strategy"; B. Herstein, "Patents and Chemical Industry in the United States," *JIEC* 4 (1912): 328–33; Hesse, "Coal-Tar Dyes and the Paige Bill," *JIEC* 7 (1915): 967–69. The reference work was Schultz, *Farbstofftabellen* (1914). Hesse's primary purpose in this article was to denounce proposed legislation to introduce a compulsory or working clause into the U.S. patent system. Some reformers hoped such a clause would address the power of corporations, especially foreign corporations, in the use of patents (963–64).

15. Haber, *The Chemical Industry, 1900–1930*, 38–50; Johnson, *The Kaiser's Chemists: Science and Modernization in Imperial Germany*, 107–79; Haber, *The Chemical Industry during the Nineteenth Century*, 63–68; Travis, *The Rainbow Makers*, 163–68; Ihde, *The Development of Modern Chemistry*, 310–19; Rocke, *The Quiet Revolution: Hermann Kolbe and the Science of Organic Chemistry*, 156–264; Wotiz and Rudofsky, "The Unknown Kekulé," in *Essays on the History of Organic Chemistry*.

16. Excellent scholarship on the research laboratories exists, and collectively it describes the laboratories with detail and nuance. The historians disagree somewhat on the timing of the creation of dyes firms' research laboratories, however. Recent historians have suggested John Beer's starting date of the 1860s was too early and that Meyer-Thurow, by focusing on Bayer, placed it too late. Homburg places the founding of most labs between 1877 and 1883 and suggests the patent law in 1877 spurred many firms to establish research laboratories, partly to use the patent system more effectively. Reinhardt believes that BASF, at least, founded its laboratory in 1874 in response to discoveries in organic chemistry and to market competition, reasons that Beer also gave. Beer, *The Emergence of the German Dye Industry*, 70–93; Meyer-Thurow, "The Industrialization of Invention: A Case Study from the German Chemical Industry"; Homburg, "The Emergence of Research Laboratories in the Dyestuffs Industry, 1870–1900"; Reinhardt, "An Instrument of Corporate Strategy: The Central Research Laboratory at BASF, 1868–1890." For an overview of the research laboratories, see Reinhardt, *Forschung in der chemischen Industrie*.

17. Haber, *The Chemical Industry during the Nineteenth Century*, 84–86; Travis, *The Rainbow Makers*, 220–27.

18. Reinhardt, "An Instrument of Corporate Strategy," 256.

19. Rocke, *The Quiet Revolution*, 304–9. Veronal was a barbiturate, one of the more frequently used types of sedatives. Merck and Bayer independently developed Veronal and subsequently shared production information and the trade name, which Merck created. Haber, *The Chemical Industry during the Nineteenth Century*, 134–35; Haber, *The Chemical Industry, 1900–1930*, 121–22, 131–32; Bäumler, *A Century of Chemistry*, 32–40; Furnas, *Rogers' Industrial Chemistry* (1942), 2:1323. Like aspirin, which was originally a trade name, phenacetin eventually became a generic name.

20. Beer, *The Emergence of the German Dye Industry*, 94–96. See also Schröter, "Die Auslandsinvestitionen der deutschen chemischen Industrie, 1870 bis 1930."

21. Metz testimony in Senate Committee on the Judiciary, *Hearings, Alleged Dye Monopoly* (1922), 749–50. See chapter 5 below for discussion of the charges brought against the importers. The story of the Bayer subsidiary is told best by Reimer, "Bayer & Company in the United States." Because of the rich archival sources Reimer unearthed, his dissertation provides an important account not only of Bayer but of the dyes industry in the United States more generally. His discussion of the prewar years is a particularly welcome addition because of the relative paucity of sources on the early U.S. industry. Ellwood Hendrick, "Record of the Coal-Tar Color Industry at Albany," *IEC* 16 (April 1924): 411–13; Norton, "A Census of the Artificial Dyestuffs Used in the United States," *JIEC* 8 (November 1916): 1046. On Merck, see Galambos and Sturchio, "Transnational Investment: The Merck Experience, 1891–1925." Also see Steen, "Confiscated Commerce."

22. Ellwood Hendrick, "Record of the Coal-Tar Color Industry at Albany," *IEC* 16 (April 1924): 411–13. Bayer had a track record of engaging in bribery; Hoechst and BASF generally avoided it. Some importers had sought cooperation following the German cartel agreement in 1904 but failed because of market disputes and personality differences. Metz testimony in Senate Committee on the Judiciary, *Hearings, Alleged Dye Monopoly* (1922), 753–54; S. H. and Lee J. Wolfe (CPA), "Report on National Aniline and Chemical Company," October 31, 1918, NARA, RG 131, entry 190, box 1; Testimony in *Rambo & Hegar v. Cassella Color Company*, District Court in the Eastern District of Pennsylvania, December 1, 1913, in RG 131, entry 190, box 1; "Report on Kuttroff, Pickhardt & Company, Inc.," January 11, 1919, Defendant's Exhibit No. 54 in *USA v. The Chemical Foundation, Inc.*, 11 (exhibits vol. 4): 2389; Reimer, "Bayer & Company in the United States," 116–24, 137–54.

23. Baekeland testimony in House Committee on Patents, *Amendment of the Patent Laws, Dyestuffs: Hearings* (1915), 41.

24. Beer, *The Emergence of the German Dye Industry*, 122–23, 127–28, quoting from Fritz Lauterbach, *Geschichte der in Deutschland bei der Färberei Angewandten Farbstoffe mit besonderer Berücksichtigung des Mittelalterlichen Waidbaues* (Leipzig, 1905), 99. The dollar figure is a conservative approximation gathered from Haber's estimate that BASF spent about £1 million in indigo development and Bäumler's estimate that Hoechst spent at least 14.5 million marks on the same. Using the gold standard exchange rates, the addition of the two amounts exceeds $8 million. Haber, *The Chemical Industry during the Nineteenth Century*, 84; Bäumler, *A Century of Chemistry*, 28. Hayes, *Industry and Ideology: I.G. Farben in the Nazi Era*, 3; Reinhardt, *Forschung in der chemischen Industrie*, 142.

25. Johnson, "Hierarchy and Creativity in Chemistry, 1871–1914."

26. Classic economic development literature includes Rostow, *The Stages of Economic Growth*, and Gerschenkron, *Economic Backwardness in Historical Perspective*. In *Scale and Scope*, Alfred D. Chandler Jr. compares the rise of big business in the United States, Great Britain, and Germany. For examples of different approaches to technology transfer literature, see Jeremy, *International Technology Transfer: Europe, Japan, and the USA, 1700–1914*, and Rasmussen, "What Moves When Technologies Migrate?"

27. Norton, "A Census of the Artificial Dyestuffs Used in the United States," 1046. The numbers on the German firms are Haber's estimates. Haber, *The Chemical Industry, 1900–1930*, 120–21. Haber's figure for value is £13.5 million; at the gold standard exchange rate, the number is $65.6 million. U.S. Tariff Commission, *Census of Dyes and Coal-Tar Chemicals, 1917*, 42; George A. Prochazka, "American Dyestuffs: Reminiscently, Autobiographically, and Otherwise," *IEC* 16 (April 1924): 413–16; Beckers's testimony in House Committee on Ways and Means, *Hearings, To Establish the Manufacture of Dyestuffs* (1916), 198–99.

28. Haynes, *American Chemical Industry*, 1:309-10; Schoellkopf testimony in House Committee on Ways and Means, *Hearings, To Establish the Manufacture of Dyestuffs* (1916), 188. Given that Schoellkopf discussed profits in the context of lobbying for higher tariffs, he might have been tempted to underestimate them. Based on comments by contemporary August Merz, Feldman suggests that the Schoellkopf patriarch kept the dyes firm financially afloat. Feldman, "Asleep at the Swatch: The Non-Emergence of the American Synthetic Dye Industry, 1856-1916," 32-33. Derick, "A Foundation Stone of Our Chemical Industry: Schoellkopf Research Laboratory" (1961); Cain and Thorpe, *The Synthetic Dyestuffs* (1920), 167. About 1909, the firm organized a separate corporation, National Aniline Company, to market its dyes. Between 25 and 33 percent of National Aniline's products came from the Schoellkopf factory in Buffalo; the rest came from other manufacturers, including from firms abroad. Over the next several years, the selling agency turned a higher rate of profit than the dyes manufacturing firm. This National Aniline should not be confused with the National Aniline & Chemical Company formed in 1917 as the result of a merger with Schoellkopfs and several other firms. Schoellkopf testimony in House Committee on Ways and Means, *Hearings, To Establish the Manufacture of Dyestuffs* (1916), 178, 188; Metz testimony in Senate Committee on the Judiciary, *Hearings, Alleged Dye Monopoly* (1922), 751-52.

29. Tichenor report in Hesse, "Lest We Forget!," 702; Copeland discusses the cotton manufacturers' associations. Copeland, *The Cotton Manufacturing Industry of the United States* (1917), 155-56; Galambos, *Competition and Cooperation: The Emergence of a National Trade Association*, 55-85.

30. B. Herstein, "Patents and Chemical Industry in the United States," *JIEC* 4 (May 1912): 328-33; Hesse, "Compulsory Working of Patents in the United States, Germany, and Great Britain," *JIEC* 7 (April 1915): 304-7; Hesse, "Coal-Tar Dyes and the Paige Bill," *JIEC* 7 (November 1915): 963-74; Leo Baekeland, "Protection of Intellectual Property in Relation to Chemical Industry," *JIEC* 5 (January 1913): 51-57; Louis C. Raegener, "Some Defects in the Practice of Our Patent System and Suggested Remedies," *JIEC* 1 (March 1909): 201-3; McTavish, "What's in a Name? Aspirin and the American Medical Association." Reimer, "Bayer & Company in the United States," discusses Bayer's role in this struggle over ethical patented medicines with doctors and pharmacists.

31. Swann, *Academic Scientists and the Pharmaceutical Industry*, 32; Forrestal, *Faith, Hope and $5000: The Story of Monsanto*, 16-23; Haynes, *American Chemical Industry*, 1:326-30.

32. Haynes, *American Chemical Industry*, 1:243-58, 264-68, 276-80, 6:210-11, 391-95; Warren, "Technology Transfer in the Origins of the Heavy Chemicals Industry in the United States and the Russian Empire." Warren notes the importance of U.S. tariff policy in fostering domestic alkali production. West, *1700 Milton Avenue: The Solvay Story, 1881-1981*, 1-11; Reader, *Imperial Chemical Industries*, 1:44-46, 95; Carl Meissner, "Some Recent Developments in By-Product Coke Ovens," 15; Lamoreaux, *The Great Merger Movement in American Business, 1895-1904*, 1-5; Dewing, *Corporate Promotion and Reorganizations*; Reynolds, "Defining Professional Boundaries: Chemical Engineering in the Early Twentieth Century."

33. Bernard C. Hesse, "The Industry of the Coal-Tar Dyes: An Outline Sketch," *JIEC* 6 (December 1914): 1013-14; Foy, *Ovens, Chemicals, and Men! Koppers Company, Inc.*, 12; *By-Product Coke and Gas Oven Plants*, 9-14; T. C. Clarke, "The Present Status of the American By-Product Coke Oven Industry," *MCE* 14 (May 1916): 503. The total value of

coal tar products manufactured in the United States by all industries was $1,421,720 (1899), $3,984,821 (1904), and $4,286,119 (1909). U.S. Bureau of the Census, *Thirteenth Census of the United States, 1910,* vol. 10 *Manufactures, Report for Principal Industries,* 543; Warren, *The American Steel Industry,* 112-13; Meissner, "Some Recent Developments in By-Product Coke Ovens," 13-15. As Fredric L. Quivik has written, another advantage of by-product ovens was the elimination of the heavy smoke that hung over southwestern Pennsylvania as a result of the beehive ovens, although that factor played virtually no role in the steel manufacturers' considerations. The by-product ovens, on the other hand, offered a different and perhaps more lasting pollution in the form of chemical waste sites. Quivik, "Industrial Pollution on the Southwest Pennsylvania Countryside"; "By-Product Coke Ovens in America," *MCE* 10 (September 1912): 522.

34. Haynes, *American Chemical Industry,* 1:248, 303-14; 6:45-49, 53-54; H. W. Jordan, "The Development in the United States of the Manufacture of Products Derived from Coal," *Chemical and Metallurgical Engineering* 14 (February 1, 1916): 146. On rubber manufacture, see Rae, *The American Automobile,* 50-51; French, *The U.S. Tire Industry: A History,* 13-35. As Haynes notes, Benzol Products Company was the name for two unrelated firms. In addition to the one mentioned here, another, formed in 1924 and based in Newark, New Jersey, manufactured fine organic chemicals, particularly pharmaceuticals. *The General Chemical Company after Twenty Years, 1899-1919,* 50; Henry Wigglesworth, General Chemical Company, to Joseph Davies, Commissioner of Corporations, January 2, 1915, recorded in Herty testimony in House Committee on Ways and Means, *Hearings, To Establish the Manufacture of Dyestuffs* (1916), 112-14. Because most of the information about Benzol Products exists from the war years, the prewar history of the firm is filtered through the lens of a domestic industry trying to create support for higher protection by vilifying all German trade activities. Many accounts excluded the role of the British representative altogether, implicating only Germany in international cartel behavior that outraged an America with antitrust legislation.

35. Van Gelder and Schlatter, *History of the Explosives Industry in America,* 932-44; Reader, *Imperial Chemical Industries,* 1:137-53, discusses the role of the military in high explosives in the context of smokeless powder. Dynamite, containing nitroglycerin, is a high explosive, but more important for construction than for the military; it is also an aliphatic rather than aromatic organic and therefore less chemically similar to dyes and pharmaceuticals. Ammonium nitrate, another high explosive, is an inorganic chemical. Du Pont began to consider cellulose-based fibers just before World War I. In partnership with a French company after the war, the firm expanded heavily in rayon and related materials; they were joined by Hercules and others in the field. Hounshell and Smith, *Science and Corporate Strategy: Du Pont R&D, 1902-1980,* chap. 8, and Dyer and Sicilia, *Labors of a Modern Hercules,* 204-16.

36. Du Pont and the other makers of gunpowder and explosives formed a trust in the early 1870s that lasted until 1904. Hounshell and Smith, *Science and Corporate Strategy: Du Pont R&D, 1902-1980,* 11-13. The new generation of du Ponts is discussed in Chandler and Salsbury, *Pierre S. du Pont and the Making of the Modern Corporation;* Chandler, *Strategy and Structure,* 52-67; Dyer and Sicilia, *Labors of a Modern Hercules: Evolution of a Chemical Company,* 33-41; Taylor and Sudnik, *Du Pont and the International Chemical Industry,* 22-25, 30-34.

37. In the 1860s, Alfred Nobel's multinational dynamite firm established a subsidiary in the United States, but it failed, partly because managing from Europe posed strategic

difficulties. Reader, *Imperial Chemical Industries*, 1:20-21, 125-37, 156-61, 194-204, 212-15; Chandler and Salsbury, *Pierre S. du Pont and the Making of the Modern Corporation*, 197-200; Charles L. Reese, "Twenty-Five Years' Progress in Explosives," *Journal of the Franklin Institute* (December 1924): 761; Van Gelder and Schlatter, *History of the Explosives Industry in America*, 944-45; Hounshell and Smith, *Science and Corporate Strategy*, 51, 76-77. Reese's trip to Europe in 1908 yielded another significant innovation in diphenylamine, a synthetic compound used to stabilize nitrocellulose smokeless powders. Like many dyes and pharmaceuticals, diphenylamine was a derivative of aniline. Reese had little difficulty persuading the military of diphenylamine's benefits to smokeless powder, and Du Pont included the stabilizer in its smokeless powder after 1908. Du Pont also purchased its diphenylamine from Germany before the war.

38. Charles F. Chandler, "Presentation Address," *JIEC* 8 (February 1916): 178-82; Baekeland, "The Synthesis, Constitution, and Uses of Bakelite," *JIEC* 1 (March 1909): 149-61; Kettering, "Biographical Memoir of Leo Hendrik Baekeland, 1863-1944"; Mercelis, "Leo Baekeland's Transatlantic Struggle for Bakelite: Patenting Inside and Outside of America"; Kaufmann, "Grand Duke, Wizard, and Bohemian: A Biographical Profile of Leo Hendrik Baekeland, 1863-1944." As Robert Friedel has written, Bakelite became a substitute in a wide range of products "pioneered" by celluloid plastics and yet had chemical characteristics that marked it as a truly synthetic plastic not dependent on the "imitation" of nature. Friedel, *Pioneer Plastic: The Making and Selling of Celluloid*, 103-9. Also see Bijker, "The Social Construction of Bakelite: Toward a Theory of Invention." On acetylene: "Union Carbide I: The Corporation," "Union Carbide II: Alloys, Gases, and Carbons," and "Carbide and Carbon Chemicals. Union Carbide III," *Fortune* 23, no. 6 (June 1941): 61ff.; 24, no. 1 (July 1941): 48ff.; 24, no. 3 (September 1941): 56ff.; Spitz, *Petrochemicals: The Rise of an Industry*, 69-82.

39. Haber, *The Chemical Industry, 1900-1930*, 62, 68; Servos, *Physical Chemistry from Ostwald to Pauling: The Making of a Science in America*; Geiger, *To Advance Knowledge: The Growth of American Research Universities, 1900-1940*, 1-93. Reynolds discusses the profession of analytical chemists in "Defining Professional Boundaries: Chemical Engineering in the Early Twentieth Century"; Thackray, Sturchio, Carroll, and Bud, *Chemistry in America, 1876-1976*, 247, 265, 379; Haynes, *American Chemical Industry*, 1:396; Kahn, *The Problem Solvers: A History of A. D. Little, Inc.*; Stevenson, "'Scatter Acorns That Oaks May Grow': Arthur D. Little, Inc., 1886-1953."

40. Thackray, Sturchio, Carroll, and Bud, *Chemistry in America, 1876-1976*, 249-50. Founded in 1909, the *Journal of Industrial and Engineering Chemistry* (*JIEC*) provided a forum for chemists to publish articles focused on applications of chemistry to industry. Early issues contained articles on the chemistry of food, mineralogy, sewage, and fertilizers, among others. Haynes, *American Chemical Industry*, 1:397-98; *JIEC* 1 (1909). The establishment of *JIEC* and the Division of Industrial Chemistry and Chemical Engineering was the American Chemical Society's response to the creation of the American Institute of Chemical Engineers. The ACS feared losing industry members to the new association. Reynolds, *75 Years of Progress: A History of the American Institute of Chemical Engineers, 1908-1983*, 5-7; Skolnik and Reese, *A Century of Chemistry*, 14-15, 35-36, 236-37, 309, 456. The move to Washington was precipitated not by any grand political plan but because ACS's secretary, Charles L. Parsons, left his post at New Hampshire College (later the University of New Hampshire) to accept a job at the Bureau of Mines. While Parsons never became particularly associated with the campaign for the synthetic organic chemicals

industry, his thirty-eight years as the permanent ACS secretary (1907–1945) meant that he played a leading role in the professional institution building in chemistry during the crucial early years of the industry, bringing an administrative continuity in the midst of rapid change—despite the cantankerous side of his personality. Marston Taylor Bogert, "Charles Lathrop Parsons," *IEC* 24 (March 1932): 362–63.

Chapter Two

1. "Drug and Chemical Markets Hardest Hit by War," *Pharmaceutical Era* 47 (September 1914): 407.

2. Delivery by transatlantic parcel post worked until October 1915. "Medicinal Supplies by Parcel Post from Germany," *WDM* 1 (July 14, 1915): 3; "Medicines from Europe Come by Parcel Post," *WDM* 1 (August 25, 1915): 4; "Germany Discontinues Parcel Post to U.S.," *WDM* 2 (October 27, 1915): 8. The details of these arrangements can be found in the records of the Department of Commerce and in a Senate hearing. NARA, RG 40, General Correspondence, 72155, boxes 355 and 357; Senate Committee on the Judiciary, *Hearings, Alleged Dye Monopoly* (1922), primarily in Metz's testimony, 755–810. Also, "*Matanzas* Brought No Drugs," *WDM* 1 (November 18, 1914): 9.

3. Redfield to Senator John W. Kern, April 8, 1915, NARA, RG 40, General Correspondence, 72155/42, box 357. The British official's quotation comes from F. Leverton Harris to [unknown first name] Drummond, March 24, 1915, The National Archives of the United Kingdom, PRO, FO 368.1294. Metz had purchased the shipment of drugs prior to March 1, 1915. "Shipments of Salvarsan and Neo-Salvarsan Are Coming," *WDM* 2 (December 15, 1915): 4; "Big Shipment of Salvarsan Reaches U.S.," *WDM* 2 (May 31, 1916): 14; "More Salvarsan Received in U.S.," *DCM* 3 (October 4, 1916): 36.

4. "Germany Will Release Dyes If She Gets Our Cotton, Says Statement," *WDM* 2 (December 15, 1915): 18; "Trying to Hold Customers," *WDM* 1 (September 30, 1914): 6. Metz reported confidentially to the U.S. ambassador on his observations and talks with high-level officials, including Chancellor Theobald von Bethmann-Hollweg. No reader of his report should have been surprised to learn that "Germany is much worked up over America selling munitions of war to France and England" and that Germany very definitely wanted Americans to send cotton and wool. Knowing the U.S. dependence on dyes, the Germans hoped "to use it as a club" to get the cotton and wool in exchange. Herman A. Metz to Ambassador James W. Gerard, excerpted, n.d., c. December 1914, in Link, *The Papers of Woodrow Wilson*, 31:536–37. The warning about Metz came from Heinrich F. Albert to "Benefactors and Friends," November 16, 1914, included as Defendant's Exhibit No. 61, *USA v. The Chemical Foundation, Inc.*, 11 (exhibits vol. 4): 2426–27.

5. On reserve inventories, see, for example, W. D. Livermore, American Woolen Company, testimony, House Committee on Ways and Means, *Hearings, To Establish the Manufacture of Dyestuffs* (1916), 149–53; Metz's quotation comes from Senate Committee on the Judiciary, *Hearings, Alleged Dye Monopoly* (1922), 809; J. Merritt Matthews testimony, House Committee on Ways and Means, *Hearings, To Establish the Manufacture of Dyestuffs* (1916), 59; H. B. Thompson testimony, House Committee on Ways and Means, *Hearings, To Establish the Manufacture of Dyestuffs* (1916), 121; Frank Cheney testimony, House Committee on Ways and Means, *Hearings, Dyestuffs* (1919), 50; Grinnell Jones, U.S. Tariff Commission, testimony, House Committee on Ways and Means, *Hearings, Dyestuffs* (1919), 9.

6. B. A. Ludwig to Molnar, July 25, 1916, exhibit 46 in S. H. and Lee J. Wolfe, "Preliminary Report on National Aniline and Chemical Company," n.d. (but 1918), NARA, RG 131, entry 190, box 1.

7. Alien Property Custodian, "Report on Kuttroff, Pickhardt, & Co. (Inc.)," January 11, 1919, in *USA v. The Chemical Foundation, Inc.*, Defendant's Exhibit No. 54, 11 (exhibits vol. 4): 2391–2403; "Proceedings of a Hearing Accorded by the War Trade Board Section of the Department of State to the Firm of Kuttroff, Pickhardt & Company, Inc., at Washington," January 26–28, 1920, Garvan Papers, box 44, p. 165; *Badische Company v. Thomas W. Miller and Frank White*, District Court of the United States for the Southern District of New York, "Summary of Interrogatories and Comments" and "Chronology" and "Protokoll," in Klage Kuttroff v. APC, Freigabe des Eigentum in Amerika, B4, f. 493, BASF archives. The $1.5 million included profits, but also repayment of prewar debt and patent royalties, according to both the Alien Property Custodian and BASF executives.

8. Shaw to Matheson, May 9, 1917, Exhibit 28, and Shaw to Matheson, May 24, 1917, Exhibit 35 in S. H. and Lee J. Wolfe, "Report on National Aniline and Chemical Company," October 31, 1918, NARA, RG 131, entry 190, box 1.

9. Cassella Color Company to Cassella in Germany, November 12, 1915, and December 21, 1915, telegrams confiscated by the British and quoted in War Trade Intelligence Department [British], "The Cassella Colour Co., New York, May 8, 1918," NARA, RG 131, entry 190, box 1; S. H. and Lee J. Wolfe, "Preliminary Report on National Aniline and Chemical Company," n.d. (but 1918), NARA, RG 131, entry 190, box 1.

10. Lybrand, Ross Brothers, & Montgomery, CPA, "The Bayer Co., Inc., Synthetic Patents Co., and Hudson River Aniline Color Works. Report on Examination of Accounts for the Period June 3, 1913 to June 30, 1918," November 1, 1918, p. 13, NARA, RG 131, entry 155, box 34; "Memorandum of Statement Made by Dr. Arthur Franz Felix Mothwurf," August 28, 1918, NARA, RG 131, entry 157, box 1; F. D. Lockwood to Frederick Rosengarten, September 18, 1918, Merck & Co. archives, Row 6, 1.7.2.

11. "In the Matter of the Bayer Company. Report on Williams & Crowell Color Company, Inc.," July 15, 1918; "standardizing" quotation from "Examination of Mr. Seebohm," July 12, 1918; "Statement from Dr. Christian Stamm," July 17, 1918, NARA, RG 131, entry 7, box 7; "Statement of Mr. H. C. A. Seebohm," July 30, 1918, NARA, RG 131, entry 157, box 1.

12. Lybrand, Ross Bros. & Montgomery (CPA), "The Bayer Co., Inc., Synthetic Patents Co., and Hudson River Aniline Color Works. Report on Examination of Accounts for the Period June 3, 1913 to June 30, 1918," November 1, 1918; Emery, Booth, Janney & Varney (lawyers), "Synthetic Patents Co., Inc. and The Bayer Company, Inc., Patent and Trade Mark Report," November 14, 1918, NARA, RG 131, entry 155, box 34; McTavish, "What's in a Name? Aspirin and the American Medical Association," 356–61; Mann and Plummer, *The Aspirin Wars*, 38.

13. Greeley and Giles, "Merck & Co." report for the Alien Property Custodian, February 21, 1919, NARA, RG 131, entry 134, box 27.

14. Cassella Color Company to Cassella in Germany, October 20, 1915, letter confiscated by the British and quoted in War Trade Intelligence Department [British], "The Cassella Colour Co., New York, May 8, 1918," NARA, RG 131, entry 190, box 1. Metz's correspondence with Hoechst clearly describes the German firms' unhappiness with their American partners' manufacturing ventures. They tried to discourage Metz, but they also discussed the German Bayer executives' dismay and inability to prevent the Bayer Company's expansion. See three letters from C. Blank of Hoechst to Metz, August 5, September

28, and November 12, 1915, reproduced in Lee J. Wolfe, "Preliminary Report on Farbwerke-Hoechst Co. of New York," in *USA v. The Chemical Foundation, Inc.*, n.d. (but late 1918 or early 1919), Defendant's Exhibit No. 52, 11 (exhibits vol. 4): 2350-53; Metz testimony in Senate Committee on the Judiciary, *Hearings, Alleged Dye Monopoly* (1922), 811.

15. Lee J. Wolfe, "Preliminary Report on Farbwerke-Hoechst Co. of New York," in *USA v. The Chemical Foundation, Inc.*, n.d. (but late 1918 or early 1919), Defendant's Exhibit No. 52, 11 (exhibits vol. 4): 2356. See chapter 5 for more on Gustave Metz's trip.

16. "Sterling Products, Inc.: Resumption of German Relations," NARA, RG 60, Central Files, Case 60-21-56, Enclosures, box 1329.

17. E. Merck to Merck & Co., April 1, 1915, Merck & Co. archives, Row 6, 2.8.2.

18. Hugo Schweitzer to Ambassador Johann H. G. von Bernstorff, c. September 1916, included as Defendant's Exhibit No. 77, *USA v. The Chemical Foundation, Inc.*, 11 (exhibits vol. 4): 2535; "U.S. Dye Industry Hasn't Gotten Far, Says H. A. Metz," *WDM* 2 (January 12, 1916): 10; B. A. Ludwig to Molnar, July 25, 1916, Exhibit 46 in S. H. and Lee J. Wolfe, "Preliminary Report on National Aniline and Chemical Company," n.d. (but 1918), NARA, RG 131, entry 190, box 1.

19. Sabel and Zeitlin make the point that experimenting with a range of strategies is a typical pattern in business history. "Stories, Strategies, Structures: Rethinking Historical Alternatives to Mass Production," 3-5. Scranton writes about the significance of specialty manufacturing in American industrial history; not a few Americans in the nascent synthetic organic chemicals industry believed their success would depend precisely on finding niche markets. Scranton, *Endless Novelty: Specialty Production and American Industrialization, 1865-1925*.

20. Schoellkopf testimony, House Committee on Ways and Means, *Hearings, To Establish the Manufacture of Dyestuffs* (1916), 178-80; "American Dyestuff Manufacture," *JIEC* 8 (October 1916): 951-52; W. H. Watkins (Schoellkopf Aniline & Chemical Works) to Herty, March 22, 1916, CHH, box 81; "'Give Us Protective Tariff' Say Dye Manufacturers," *WDM* 1 (July 21, 1915): 4.

21. Derick, "A Foundation Stone of Our Chemical Industry: Schoellkopf Research Laboratory," August 15, 1961, unpublished manuscript owned by the Derick family.

22. Haynes, *American Chemical Industry*, 3:234-35 and 6:292-93; William G. Beckers to Charles H. Herty, March 18, 1916, CHH, box 81; William J. Matheson to Robert Shaw, April 16, 1917, NARA, RG 131, entry 190, box 1; Beckers testimony in House Committee on Ways and Means, *Hearings, To Establish the Manufacture of Dyestuffs* (1916), 199; "Eugene Meyer Dies," *New York Times*, July 18, 1959, pp. 1, 15; "Reminiscences of Eugene Meyer" (February 11, 1952), on pages 190-91 in the Columbia Center for Oral History Collection (hereafter CCOHC). I read this interview in the Eugene Meyer Papers at the Library of Congress; Meyer marked suggested changes in the wording on his copy, but I use the text here and throughout as it appears in the CCOHC; Pusey, *Eugene Meyer*, 117-25. Later in his life, Meyer bought the *Washington Post*, which his daughter, Katharine Graham, subsequently resuscitated. He also served as the first president of the World Bank.

23. Liebenau, *Medical Science and Medical Industry*.

24. Forrestal, *Faith, Hope and $5000: The Story of Monsanto*, 23, 25-30; quotation of Jules H. Kernen, a long-term employee recalling Queeny, on 35; U.S. Tariff Commission, *Census of Dyes and Coal-Tar Chemicals, 1918*, 43-44.

25. Ernest H. Volwiler, "Abbott Science at Seventy-Five," April 24, 1963, Adams Papers, box 27, folder: Abbott Laboratories; Roger Adams to Julius Stieglitz, June 22, 1918,

Adams Papers, box 4, folder: 1918–1919; U.S. Tariff Commission, *Census of Dyes and Coal-Tar Chemicals, 1918,* 43–44; Marx Hirsch to Francis M. Phelps, June 2, 1919, FPG, box 64; "American 'Procaine' Replaces Novocain," *DCM* 4 (March 6, 1918): 10.

26. U.S. Tariff Commission, *Census of Dyes and Coal-Tar Chemicals, 1918,* 43–44; Haynes, *American Chemical Industry,* 6:219–20, 3:216–17, 237–38; Neidich also made salicylic acid. On Sherwin-Williams and Ault & Wiborg, see Haynes, *American Chemical Industry,* 6:386–87, 224–25. The chemistry is described in Furnas, *Rogers' Industrial Chemistry* (1942), 2:968–69.

27. Vanderbilt, *Thomas Edison, Chemist,* 246–55. Vanderbilt notes that chemists and the chemical industry remained unimpressed with Edison's production. His suppliers were Bakelite and Condensite.

28. This story is told well in Hershberg, *James B. Conant,* 38–39; "Agreement," April 8, 1916, Papers of James B. Conant, UAI 15.898, box 137. The firm also used the name LPC Laboratories, after the names of the three partners. *Merck Index,* 12th ed., s.v. "Benzyl Chloride." Conant's papers (box 137) include drafts of the patent text describing the details of the process. Its patentable feature centered on using bleaching powder, rather than chlorine gas, as the source of chlorine. Stanley B. Pennock earned his fame as an outstanding guard on Harvard's football team, even subsequently making Knute Rockne's "All Time" listing of 1906–1926. Pennock's father worked for the Solvay Process Company, and he put up the primary capital for the new venture. Like Pennock, the second partner, Chauncey C. Loomis, also graduated from Harvard in 1915 with a major in chemistry. "Star Killed. Stanley Pennock in Explosion," *Boston Globe,* November 28, 1916, clipping in Pennock's student record file at Harvard University, UA III 15.88.10, Harvard University Archives; Daniel Frio, "Pennock, Stanley Bagg," in *Biographical Dictionary of American Sports,* ed. David L. Porter, 2:465. Conant would rise through the ranks, becoming one of the best young organic chemists in the 1920s before moving to administrative posts. Conant became Harvard's president in 1933, loaned his expertise to the Manhattan Project, and advised on post–World War II science policy. In 1916, however, Conant stood at the beginning of his career and chose the professoriate over the potentially more lucrative business opportunity. Robert H. Hold, of Gaston, Snow, and Saltonsall, to Richardson Morris, April 5, 1917, Conant papers, box 137. The Papers of James B. Conant are used courtesy of the Harvard University Archives and Theodore Conant.

29. Travis, *Dyes Made in America, 1915–1980,* 37–64; Haynes, *American Chemical Industry,* 3:213–18, 254; U.S. Tariff Commission, *Census of Dyes and Coal-Tar Chemicals, 1918,* 39–43.

30. Dow to A. E. Convers (president of Dow Chemical Company), October 9, 1915, DHC, folder 15027. The Dow Historical Collection (DHC) is used courtesy of the Chemical Heritage Foundation Archives. Haynes, *American Chemical Industry,* 6:114–15. This was the second of two firms called the Midland Chemical Company and operated as a separate entity from Dow Chemical until 1914. The first Midland Chemical was Dow's original bromine company in Michigan. Campbell and Hatton, *Herbert H. Dow,* 19, 69.

31. Dow to J. H. Osborn, October 1, 1915, DHC, folder 150083; Dow to Albert W. Smith, May 20, 1915, DHC, folder 150017; Campbell and Hatton, *Herbert H. Dow,* 103; "Statement of Mr. Herbert H. Dow, of the Dow Chemical Company, Midland, Michigan. Before Federal Trade Commission," Detroit, Michigan, July 21, 1915, p. 1, DHC, folder 150003.

32. Dow to Cone, October 1, 1915, DHC, folder 150053; "Dr. Smith Dies of Heart Trouble," *Case Alumnus* (March 1927): 5, DHC, folder 270040. Another Case graduate, James T.

Pardee, served as an executive officer in Dow Chemical for years. Campbell and Hatton, *Herbert H. Dow*, 10; Smith to Dow, November 2, 1915, DHC, folder 150021; Dow to Smith, October 12, 1915, DHC, folder 150027.

33. Dow to Smith, October 28, 1915, DHC, folder 150020; Dow to J. H. Osborn, October 1, 1915, DHC, folder 150083; Dow to Grellet Collins, February 23, 1916, DHC, folder 160131; Dow to W. A. Erwin, September 21, 1916, DHC, folder 160128; Dow to Harold Hibbert, October 14, 1916, DHC, folder 160059; Dow to John Oberlin (Dow's patent attorney), March 10, 1916, DHC, folder 160070; Dow to Smith, March 20, 1916, DHC, folder 160003; Dow to Hiram H. Walker (unsent letter), May 24, 1917, DHC, folder 170114; Dow to A. W. Smith, April 12, 1917, DHC, folder 170013; Cone to Dow, July 10, 1917, DHC, folder 170115; John F. Oberlin to Cone, November 7, 1918, DHC, folder 180142; Dow to A. F. Lichtenstein, August 15, 1917, DHC, folder 170117.

34. "Dow Indigo Not Yet on the Market," *DCM* 3 (April 25, 1917): 8; Dow to A. E. Convers, January 18, 1916, DHC, folder 160009.

35. U.S. Tariff Commission, *Census of Dyes and Coal-Tar Chemicals, 1918*, 43–44; Dow to Convers, August 7, 1917, DHC, folder 170002; Dow to Collings, July 2, 1918, DHC, folder 180012; Dow to Austin Company, October 22, 1918, DHC, folder 180150; Dow to Frederic Rosengarten, December 28, 1918, DHC, folder 180076.

36. A. F. Lichtenstein to Herbert Dow, August 2, 1917, DHC, folder 170117; Dow to G. E. Collings, November 27, 1917, DHC, folder 170005; Van Winckel to Dow, May 30, 1918, DHC, folder 180115; Van Winckel to Dow, July 25, 1918, DHC, folder 180115. The Aniline Dyes & Chemicals, Inc., had been named Geisenheimer & Company until mid-1918 when a change in partnership occurred.

37. "Dyestuffs Manufacturers Now Number Thirty-Four," *WDM* 2 (January 26, 1916): 14.

38. Edward C. Worden, "Certificate of Analysis: Report on the Plant of the Federal Dyestuff and Chemical Company, Kingsport, Tenn.," June 8, 1916, Hagley Museum and Library, Papers of Lammot du Pont Jr., Accession 676, box 27.

39. Ibid. A Brown University graduate of 1885, Hebden's work experience included textile firms and more than twenty years with the Cassella and A. Klipstein dyes importing agencies. "J. C. Hebden Dead; Brooklyn Inventor," *New York Times*, June 5, 1929, p. 29; "Alizarin Dyes Soon to Be Turned Out by Federal Plant in Kingsport, Tenn.," *DCM* 3 (November 22, 1916): 32. Herbert Dow remembered Hebden differently; a purchaser of Dow's chlorbenzol early in the war, Hebden had "built a plant at Kingsport, Tenn. embodying the ideas he got here," believed Dow. Dow, "An Analysis of the Effect of the Public Knowing about our Processes and Methods," April 21, 1928, DHC, folder 280021. Charles R. DeLong, later chief of the USTC's chemical division, worked at Federal Dyestuff for a year. Senate Committee on the Judiciary, *Hearings, Alleged Dye Monopoly* (1922), 611.

40. "Big Shipment of American Dyes Made," *WDM* 2 (June 21, 1916): 9; "Alizarin Dyes Soon to Be Turned Out by Federal Plant in Kingsport, Tenn.," *DCM* 3 (November 22, 1916): 32; "Federal Dyestuff Stock Offered," *DCM* 3 (December 13, 1916): 31.

41. "Federal Dyestuff Plan Outlined," *DCM* 4 (September 26, 1917): 9; "Federal Dyestuff Plan Delayed," *DCM* 4 (December 12, 1917): 6; "Terms of Federal Dyestuff Plan," *DCM* 4 (December 19, 1917): 9; "Federal Dyestuff Notes Deposited," *DCM* 4 (January 2, 1918): 21; "Another Federal Dyestuff Plan," *DCM* 4 (January 30, 1918): 23; untitled note, *DCM* 4 (May 22, 1918): 19; "Federal Dyestuff Reorganization," *DCM* 5 (September 11, 1918): 7.

42. "Du Ponts Have Eye on Federal Plant," *DCM* 3 (April 25, 1917): 9; undated (c. July 1916), unsigned eight-page report beginning, "As the result of four days at the plant,"

Hagley Museum and Library, Papers of Lammot du Pont Jr., Acc. 676, box 27. The author of the report also thought Federal Dyestuffs lacked focus, diverting their research personnel to projects too far removed from dyes and related products, including Rittman's petroleum cracking process. Rumors circulated that Federal Dyestuffs offered W. F. Rittman $25,000 to establish and run a research laboratory. Charles L. Parsons to Charles H. Herty, March 14, 1916, CHH, box 81.

43. "Organization Agreement," April 5, 1917; S. H. and Lee J. Wolfe, "Report on National Aniline & Chemical Company," October 31, 1918, NARA, RG 131, entry 190, box 1.

44. Shaw to Matheson, April 27, 1917; Matheson to Shaw, May 10, 1917, NARA, RG 131, entry 190, box 1.

45. Shaw to Matheson, May 12, 1917; "Agreement between William J. Matheson and Robert A. Shaw and National Aniline & Chemical Company, Inc.," September 11, 1917, NARA, RG 131, entry 190, box 1.

46. "National Aniline & Chemical Company, Inc.," announcement of formation listing officers and directors, n.d. but c. 1917, NARA, RG 131, entry 190, box 1. Derick recalls much later his distaste for the new management in Derick, "A Foundation Stone of Our Chemical Industry: Schoellkopf Research Laboratory"; Hampton Hanna, a nephew to a business partner of Jacob Schoellkopf, hinted obliquely at causes but clearly states employees' dissatisfaction in an interview granted in 1918. "Memorandum of Interview with Hampton Hanna, Esq., September 17, 1918," NARA, RG 131, entry 190, box 1; "Reminiscences of Eugene Meyer" (February 11, 1952), on page 193 in the CCOHC; Haynes, *American Chemical Industry*, 3:243.

47. National Aniline & Chemical Co., Inc., "News Bulletin," March 13, 1918; "Memorandum of Interview with Hampton Hanna, Esq., September 17, 1918," NARA, RG 131, entry 190, box 1. A description of National Aniline's headquarters at 21 Burling Slip is in "National Aniline Company Opens New Building," *ADR* 2 (June 24, 1918): 14.

48. U.S. Tariff Commission, *Census of Dyes and Coal-Tar Chemicals, 1918*, 39-43.

49. Ibid.; Hounshell and Smith, *Science and Corporate Strategy*, 76-77, 80. The following section on Du Pont's dyes manufacturing draws extensively on chapter 3 of their book.

50. Hounshell and Smith, *Science and Corporate Strategy*, 78-81; A. D. Chambers to R. R. M. Carpenter, July 26, 1915, recorded in *USA v. Du Pont, General Motors, et al.*, Civil Action No. 49 C 1071, 1948, Defendants' Trial Exhibits, vol. 1, DP 78, Hagley Museum and Library, Longwood Manuscripts, Group 10, Series A, File 418-26, box 417.

51. Hounshell and Smith, *Science and Corporate Strategy*, 80-84; du Pont testimony in Senate Committee on Finance, *Hearings, Dyestuffs* (1919), 157; du Pont testimony in Senate Committee on the Judiciary, *Hearings, Alleged Dye Monopoly* (1922), 593. On Levinstein, see Reader, *Imperial Chemical Industries: A History*, especially vol. 1, chap. 12.

52. Hounshell and Smith, *Science and Corporate Strategy*, 81, 85-88, 91. The authors note about forty organic chemists apiece in Du Pont's Experimental Station and the Eastern Laboratory, 86.

53. Hounshell and Smith, *Science and Corporate Strategy*, 82-83, 88. Poucher initially approached Du Pont with a proposal that he, Poucher, form a company that would provide Du Pont intermediates to British dyes makers. W. S. Carpenter to Du Pont's Executive Committee, recorded in *USA v. Du Pont, General Motors, et al.*, Civil Action No. 49 C 1071, 1948, Defendants' Trial Exhibits, vol. 1, DP 85, Hagley Museum and Library, Longwood Manuscripts, Group 10, Series A, File 418-26, box 417.

54. "American Dyestuffs for Khaki," *DCM* 3 (April 25, 1917): 8; "Some Important German Colors Still in Experimental Stage Here," *DCM* 4 (October 10, 1917): 17. The second article

notes that the failure of the dyes in flags in 1917 stemmed from "German dyes ... adulterated with salt," rather than faulty American dyes.

55. "Textile Consumption of Dyestuffs," *DCM* 4 (February 6, 1918): 5-6; "Colors Used by Dyers and Finishers," *DCM* 4 (February 27, 1918): 11-12.

56. R. R. M. Carpenter, Development Department, Du Pont, "Bayer Company," November 25, 1918, Hagley Museum and Library, Acc. 2091, box 14, p. 12; Aronson quoted in Ruth B. Mork, Appointment Office, Harvard Alumni Association, to G. P. Baxter, February 16, 1918, Harvard University Archives UAV 275.42, Department of Chemistry.

57. "American Chemists Producing Dyes as Strong and Fast as German Colors," *DCM* 3 (January 3, 1917): 5-6; "American Coal-Tar Dyes Equal to Best German Product, Says I. F. Stone," *DCM* 3 (November 29, 1916): 8.

58. "Fatal Experimental Work," *WDM* 1 (November 18, 1914): 11; Haynes, *American Chemical Industry*, 3:234-35; "Plant of Midvale Chemical Company Making Aniline Dyes, Is Destroyed," *WDM* 2 (February 16, 1916): 26; Travis, *Dyes Made in America, 1915 to 1980*, 43; Herbert H. Dow to A. W. Smith, October 31, 1916, DHC, H. H. Dow Series folder 160006; A. W. Smith to Dow, November 7, 1916, DHC, folder 160007.

59. John F. Queeny, "The Coal Tar Industry," *Pharmaceutical Era* 50 (January 1917): 6.

60. Hawley, *The Great War and the Search for a Modern Order*, 7-8; Fearon, *War, Prosperity, and Depression*, 12.

61. "Address of Mr. H. Gardner McKerrow," 4-5 in "Official Report of Convention of American Dyestuff Manufacturers," *ADR* 2 (February 4, 1918): 4-8; also, "A Proposed National Association of Chemical Industries," *ADR* 1 (October 29, 1917): 8-9; Haynes, *American Chemical Industry*, 3:253; "Dyestuff Industry Organizes," *DCM* 4 (January 23, 1918): 5-6. On Hemingway, see Haynes, *American Chemical Industry*, 3:118, 253. Sherwin-Williams purchased Frank Hemingway, Inc., in 1920. Blaszczyk has noted that the end users of dyes—those in the fashion industry—also pursued standardization of colors through the Textile Color Card Association to aid buyers. Blaszczyk, *The Color Revolution*, 77-93.

62. "Harmony Should Prevail in Dyestuff Convention," *ADR* 2 (March 4, 1918): 1, 3. The dealers formed their own trade association, the U.S. Dyestuff and Chemical Importers' Association, which included importers for Britain, France, and Switzerland but excluded the German affiliates. Walter F. Sykes testimony in House Committee on Ways and Means, *Hearings, Dyestuffs* (1919), 406-7; "Color Manufacturers Invite Competition," *DCM* 4 (March 6, 1918): 9.

63. "Dye Makers and Dye Dealers May Form Two Separate Organizations," *OPDR* 93 (March 4, 1918): 28; "Dye Manufacturers to Organize as Producers, Traders Eliminated," *OPDR* 93 (March 11, 1918): 27; U.S. Tariff Commission, *Census of Dyes and Coal-Tar Chemicals, 1917*, 31-35, notes just one Du Pont dye, sulphur black; "American Dye Manufacturers Bar Concerns with German Affiliations," *DCM* 4 (April 10, 1918): 13; "Dye Manufacturers Plan Active Control," *DCM* 4 (March 20, 1918): 5-6; Robert C. Jeffcott testimony in Senate Committee on the Judiciary, *Hearings, Alleged Dye Monopoly* (1922), 452; "Putting the Dye Industry on a Basis of Permanence," *OPDR* 93 (January 28, 1918): 25; "American Dyestuff Manufacturers Name Permanent Officers for Ensuing Year," *OPDR* 93 (April 8, 1918): 28; "Color Importers Organize," *ADR* 2 (June 24, 1918): 6; "The Proposed Dyestuff Importers' Association," *ADR* 3 (July 1, 1918): 10-11.

64. Other open-price societies existed, including one among hardwood lumber companies, which in 1921 the Supreme Court declared in violation of the Sherman antitrust legislation. Arthur Eddy, a lawyer and leading proponent of such societies, advised the

ADI and the lumber companies. Robert C. Jeffcott testimony in Senate Committee on the Judiciary, *Hearings, Alleged Dye Monopoly* (1922), 451–52. "American Dyes Institute, An Open Price Society, Now in Progress of Formation," *OPDR* 93 (March 25, 1918): 28. On open-price societies, see Berk, "Communities of Competitors: Open Price Associations and the American State, 1911–1929." For an account of an open-price society among textile manufacturers, see Galambos, *Competition and Cooperation: The Emergence of a National Trade Association*, 78–83. The ADI's open-price procedure was fully publicized: "Dye Manufacturers Plan Active Control," *DCM* 4 (March 20, 1918): 5–6; Robert C. Jeffcott testimony in Senate Committee on the Judiciary, *Hearings, Alleged Dye Monopoly* (1922), 452, 491, 496.

65. "Open Price Plan of Dye Manufacturers Arouses Opposition in the Trade," *DCM* 4 (March 27, 1918): 10.

66. Although not a participant in the open-price society, Herman Metz attended a meeting of an open-price session about beta naphthol. After seeing the output figures submitted by other manufacturers, he informed them that his firm alone produced four times the amount to which they had admitted producing. Metz, for one, was convinced not that he was the leading producer of beta naphthol but that those in the open-price society participated half-heartedly. Metz testimony in Senate Committee on the Judiciary, *Hearings, Alleged Dye Monopoly* (1922), 852–53.

67. The number of exchanges peaked with five in June 1918, Minutes, American Dyestuff Institute, July 10, 1918, recorded as Plaintiff's Exhibit No. 521 in *USA v. The Chemical Foundation, Inc.*, 9 (exhibits vol. 2): 1090; Robert C. Jeffcott and Harry E. Danner testimonies, Senate Committee on the Judiciary, *Hearings, Alleged Dye Monopoly* (1922), 453, 497; Minutes, American Dyestuff Institute, July 10, 1918, recorded as Plaintiff's Exhibit No. 521 in *USA v. The Chemical Foundation, Inc.*, 9 (exhibits vol. 2): 1093; Minutes, American Dyestuff Institute, June 5, 1918, recorded as Plaintiff's Exhibit No. 520 in *USA v. The Chemical Foundation, Inc.*, 9 (exhibits vol. 2): 1080; Minutes, American Dyestuff Institute, September 4, 1918, recorded as Plaintiff's Exhibit No. 523 in *USA v. The Chemical Foundation, Inc.*, 9 (exhibits vol. 2): 1106; Minutes, American Dyes Institute, Executive Committee, December 9, 1918, recorded as Plaintiff's Exhibit No. 517 in *USA v. The Chemical Foundation, Inc.*, 9 (exhibits vol. 2): 1061–63; Senate Committee on the Judiciary, *Hearings, Alleged Dye Monopoly* (1922), 451–58; "Manufacturers' Association and Dyes Institute Unite," *ADR* 3 (December 16, 1918): 12. The investigation of monopoly, the Shortridge Hearings, occurred in 1922. Senate Committee on the Judiciary, *Hearings, Alleged Dye Monopoly* (1922), 452.

68. U.S. Tariff Commission, *Second Annual Report, 1918*, 8–9; U.S. Tariff Commission, *Third Annual Report, 1919*, 20.

Chapter Three

1. The Ordnance Department was in the Department of War, which was the official name and cabinet-level department of the U.S. Army. The separate Department of the Navy had its own ordnance unit, but it was smaller and less influential in shaping the production of synthetic organic explosives and, especially, gases in World War I. Over the years, historians have developed a sizable literature on explosives and U.S. mobilization in World War I. Particularly central to understanding mobilization are Koistinen, *Mobilizing for Modern War*, from which I have drawn extensively; Cuff, *The War Industries Board*; Rockoff, *America's Economic Way of War: War and the U.S. Economy from the*

Spanish-American War to the Persian Gulf War. An old but useful reference on explosives is Van Gelder and Schlatter, *History of the Explosives Industry in America* (1927). Also see firm histories, particularly Hounshell and Smith, *Science and Corporate Strategy: Du Pont R&D, 1902-1980*, and Dyer and Sicilia, *Labors of a Modern Hercules*. Benedict Crowell, an Assistant Secretary of War and Director of Munitions, assembled an official military history of World War I munitions, which is a very good starting point, although it glosses over problems. Crowell, *America's Munitions, 1917-1918* (1919). For the list of goods: "Memorandum" from J. P. Morgan & Company to Wilson administration (probably Colonel E. House), May 19, 1917, enclosed in Stettinius to Wainwright, Assistant Secretary of War, April 17, 1922, NARA, RG 107, Office of the Secretary of War, entry 191, General File, File 12.

2. The explosion of a shell occurred in a series of blasts. Typically made of mercury fulminate, a detonator triggered on contact with the target, which set off a "booster" charge of ammonium picrate that, in turn, set off the main charge of TNT or other "shell filler." Crowell, *America's Munitions, 1917-1918*, 103, 116-19. Van Gelder and Schlatter, 930, 940-48. R. S. McBride, C. E. Reinicker, and W. A. Dunkley, "Toluol Recovery," *Technologic Papers of the Bureau of Standards*, No. 117, Department of Commerce, December 19, 1918 (Washington: GPO, 1918), 5.

3. Williams, *History of the Manufacture of Explosives for the World War, 1917-1918* (1920), 7-10; Van Gelder and Schlatter, *History of the Explosives Industry in America*, 945-46, 952-53. Amatol was a mixture of TNT and ammonium nitrate, as discussed below. Tetryl was trinitrophenylmethylnitramine, a derivative of aniline.

4. Du Pont took up nitroglycerine and dynamite in 1880. Hounshell and Smith, *Science and Corporate Strategy*; Chandler and Salsbury, *Pierre S. du Pont and the Making of the Modern Corporation*, 291-300; and Van Gelder and Schlatter, *History of the Explosives Industry*, 541-44, 563, 944-48; Dyer and Sicilia, *Labors of a Modern Hercules*, 88-96, 101-2, 145-50. Atlas submitted a proposal to make TNT in July 1917, but there were no available supplies of toluene then, and the Ordnance officers postponed a contract until supplies were available. J. H. Burns to Atlas Powder Company, August 7, 1917, NARA, RG 156, entry 764, 471.868/46, box 141. Had the war gone into 1919, Atlas would have presided over a large government-owned TNT plant in Perryville, Maryland. See A. W. Hixson and Major C. F. Backus, "History of Explosives Section; Production Division; Explosives, Chemicals, and Loading Division," part 2, vol. 1, December 1, 1918, appendix, NARA, RG 156, entry 528, box 2, ex. 18.

5. Van Gelder and Schlatter, *History of the Explosives Industry*, 940-44; "List of Powder, Explosives, & Loading Plants, Government Owned and Privately Owned, Operated on War Department Contracts," November 10, 1918, 7-9, NARA, RG 156, entry 524, box 13, Pl. 50; Williams, *History of the Manufacture of Explosives for the World War, 1917-1918*, 66, 69 Haynes, *American Chemical Industry*, 6:367-68; Van Gelder and Schlatter, *History of the Explosives Industry in America*, 948. At both Aetna and Semet-Solvay, obliterating explosions sharply reduced TNT output. A. W. Hixson and Major C. F. Backus, "History of Explosives Section; Production Division; Explosives, Chemicals, and Loading Division," part 2, vol. 1, December 1, 1918, appendix, NARA, RG 156, entry 528, box 2, ex. 18, 15.

6. By June 1915, Albert W. Smith of the Case School, Herbert Dow's primary chemical advisor, believed the war would not end soon and expected an investment in a phenol plant would be well worth the effort. Dow agreed, but also, interestingly, expressed a concern that his board of directors might object to the production of phenol because it would be used in picric acid. Smith to Dow, June 10, 1915; Dow to Smith, June 11, 1915, DHC, folder

150018. The French government purchased the largest share of Dow's phenol, followed by Butterworth-Judson, manufacturers of picric acid, and then Japan, where firms used it to produce picric acid subsequently sold to Russia. Dow to J. H. Osborn, June 24, 1916, DHC, folder 160140; unsigned memo, May 19, 1917, DHC, folder 170088. Dow to A. E. Convers, October 9, 1915, DHC, folder 150027; Dow to A. W. Smith, November 22, 1915, DHC, folder 150021.

7. Reader notes that the British government, with Nobel's cooperation, negotiated with Du Pont to build smokeless powder plants in Britain during the war. He does not mention similar arrangements for high explosives, but the incident suggests the degree of communication across the Atlantic. Reader, *Imperial Chemical Industries: A History*, 1:301–2. Memorandum on Aetna Explosives Company contracts with France, January 1918, ERS, Acc. 2723, box 44, folder 773; F. Francois Lacombs, Haut Commissariat de la République Française aux Etats-Unis, to Col. Hoffer, November 15, 1917, with multiple enclosures, NARA, RG 156, entry 764, 471.862 Aetna, box 138.

8. "Million Dollar Deal in Picric Acid," *WDM* 1 (July 21, 1915): 10; Weems, *America and Munitions: The Work of Messrs. J. P. Morgan & Co. in the World War*, 23–26, 99–103.

9. Horn, *Britain, France and the Financing of the First World War*; Koistinen, *Mobilizing for Modern War*, 121–26; Burke, *Britain, America, and the Sinews of War, 1914–1918*, 23; Weems, *America and Munitions*.

10. "Memorandum Regarding Activities of J. P. Morgan & Co. as Purchasing Agents for the British and French Governments," enclosed in Stettinius to Wainwright, April 17, 1922, NARA, RG 107, entry 191, General File, File 12. Koistinen makes the point about purchasing policies encouraging economic concentration in *Mobilizing for Modern War*, 121–26; Burke, *Britain, America, and the Sinews of War*, 23; Williams, *History of the Manufacture of Explosives for the World War, 1917–1918*, 39–59.

11. Koistinen, *Mobilizing for Modern War*, 139–65.

12. Rodgers, *Atlantic Crossings: Social Politics in a Progressive Age*, 285–86; Koistinen, *Mobilizing for Modern War*. The following paragraphs draw on chapters 8–10; Stettinius to Hardy, Chicago, October 27, 1917, ERS, box 21, folder 376.

13. Koistinen, *Mobilizing for Modern War*, 152, 160–63, 177, 181.

14. Ibid., 139–40, 166–82, 198–216. The Council of National Defense was initially championed by Hollis Godfrey, an engineer (rather than businessman) who was familiar with prewar attempts in both Great Britain and the United States to establish such a mechanism for efficiently managing the economy in wartime. Ibid., 156. Besides Koistinen, the mobilization boards are also discussed in Cuff, *The War Industries Board*. See also Baruch, *American Industry in War, a Report of the War Industries Board*.

15. Koistinen, *Mobilizing for Modern War*, 230–44; Haynes, *American Chemical Industry*, 2:45–52, 355; Baruch, *American Industry in War*, 165–207; and Charles H. McDowell, "The Work of the Chemical Section of the War Industries Board," *JIEC* 10 (October 1918): 780–83.

16. Kevles, *The Physicists: The History of a Scientific Community in Modern America*, 102–18, 137–38. "War Activities of the Division of Chemistry and Chemical Technology," National Academy of Sciences, NAS-NRC Central File, 1919–1923, Division of NRC, Chemistry and Chemical Technology, Activities; Charles E. Munroe to George Ellery Hale, March 21, 1919, NAS-NRC Central File, 1914–1918, Division of NRC, Chemistry and Chemical Technology, Committee on Explosives Investigations, 1919.

17. Koistinen, *Mobilizing for Modern War*, 182–87.

18. The Department of the Navy also had an Ordnance Department, but it played a less significant role in the management of high explosives production than the Department of War's agency. "History of Explosives Section: Explosives, Chemicals, and Loading Division," part 1, vol. 1, "Introduction," January 28, 1919, p. 4, NARA, RG 156, entry 528, box 2; "History of Explosives Section: Explosives, Chemicals, and Loading Division," part 2, vol. 6, "Manufacture of TNT," December 16, 1918, p. 2, RG 156, entry 528, box 3.

19. Stettinius to Dunn, March 30, 1917; Stettinius to Burns, October 26, 1917; Crozier to Stettinius, February 4, 1917, ERS, box 82. Leland L. Summers was a former Export Department employee who became one of Bernard Baruch's chief assistants in the War Industries Board, where he dealt particularly with chemicals. Koistinen, *Mobilizing for Modern War*, 137, 163, 175-76. Stettinius to Major J. H. Burns, Ordnance Department, October 26, 1917, ERS, box 82; Hixson and Backus, "History of Explosives Section; Production Division; Explosives, Chemicals, and Loading Division," part 2, vol. 1, December 1, 1918, appendix, NARA, RG 156, entry 528, box 2, ex. 18, p. 19.

20. Crowell, *America's Munitions, 1917-1918*, 13-18.

21. One example of the Ordnance officers' concern to avoid interrupting existing flows of explosives is contained in Hoffer to Dunn, March 26, 1917, enclosed in Dunn to Stettinius, March 29, 1917, ERS, box 82. Also, when the Ordnance officers met with firms, they explicitly asked about current and future obligations under Allied contracts. In August and September 1917, the officers made several inquiries with firms about available capacity. NARA, RG 156, entry 764, 471.862, box 137. When J. P. Morgan & Company and the British ended their purchasing agreement, Stettinius pointed out that the U.S. government's ability to set prices was a primary reason the British should work through the U.S. government rather than through J. P. Morgan. J. P. Morgan & Company to Cecil Spring-Rice, British Ambassador to the United States, May 25, 1917, ERS, box 84. On at least one occasion, Ordnance officers threatened to commandeer a plant when its managers balked at the official price for picric acid. See Major J. H. Burns, memorandum, "Picric Acid: New England Chemical Company," October 6, 1917, New England Chemical Company, NARA, RG 156, entry 764, 471.862/2, box 139. "Contracts for $7,830,000 in Picric Acid," *DCM* 4 (October 10, 1917): 9; Lacombs to Hoffer, November 15, 1917, with multiple enclosures, Aetna, NARA, RG 156, entry 764, 471.862/40, box 138.

22. Military Attaché, London, to Chief of Ordnance, December 10, 1917, 471.868/144; Backus to Production Division, March 7, 1918, 471.868/152, NARA, RG 156, entry 764, box 141. A British plan for the NRC to provide thirty-five American scientists to help and learn from the British never received funding. MacLeod, "Secrets among Friends: The Research Information Service and the 'Special Relationship' in Allied Scientific Information and Intelligence, 1916-18"; Gilby (?) to Welles, January 17, 1918, NAS-NRC Central File, 1914-1918, Policy Files, Administration, Divisions of the NRC, General Relations, Research Information Service, Appointments—Scientific Attachés—State Department; George Ellery Hale, "War Activities of the National Research Council," address given in the Engineering Societies Building, New York, May 28, 1918, pp. 919-23, NAS-NRC Central File, 1914-1918, Administration, Public Relations, Inauguration (Speeches); Graham Edgar, NRC, to H. A. Bumstead, American Embassy, Office of the Scientific Attaché, April 25, 1918; Van Manning, Bureau of Mines, to Hale, May 13, 1918; Hale to Bumstead, June 13, 1918, NAS-NRC Central File, Policy Files, Administration, Executive Board, Projects, Scientists for Service in Allied Laboratories (Proposed).

23. J. H. Burns to Picatinny Arsenal, June 11, 1917, 471.86/1901 Miscellaneous; Burns to E. I. du Pont de Nemours and Company, June 25, 1917, 471.86/1924 Miscellaneous;

William Crozier to Charles B. Gordon, British Mission in the United States, August 1, 1917, 471.86/1994 Miscellaneous, NARA, RG 156, entry 36, box 2682; Hixson and Backus, "History of Explosives Section; Production Division; Explosives, Chemicals, and Loading Division," part 2, vol. 1, December 1, 1918, appendix, RG 156, entry 528, box 2, ex. 18, pp. 10, 15; Lieutenant Egbert Moxham, "Propellants and Explosives in the War," December 17, 1918, 471.86/880, RG 156, entry 764, box 135; Van Gelder and Schlatter, *History of the Explosives Industry in America*, 954–56. Not to leave the Russians out: Ordnance officers consulted with the Russians about their purchases of TNA (trinitroaniline) in the United States. Shinkle to Tschekeloff, April 25, 1917, 471.86/1042 Miscellaneous, NARA, RG 156, entry 36, box 2682.

24. For example, W. W. Edwards left the Aetna research labs and helped build the General Explosives Company in Missouri. Edwards to Peters, June 6, 1918, 471.86/2525, NARA, RG 156, entry 36, box 2682; Moxham, "Propellants and Explosives in the War," December 17, 1918, 471.86/880, NARA, RG 156, entry 764, box 135; Backus to New York District Office, May 7, 1918, 471.862/4 Everly M. Davis; Goodrich Lockhart Company to Peters, August 17, 1918, 471.862/34 Goodrich Lockhart, NARA, RG 156, entry 764, box 138; Stephen Morey to Backus, June 20, 1920, 471.862/266, NARA, RG 156, entry 764, box 137, p. 3; W. C. Cope, Eastern Laboratory, Du Pont, "Investigation of TNA," May 1, 1917, 471.86/1123 Miscellaneous, entry 36, box 2682.

25. Edward Mallinckrodt to G. P. Baxter, November 30, 1917; J. Preston Wills to Professor Grinnell Jones, November 11, 1917; Arthur W. Phillips to G. P. Baxter, December 9, 1917; Edward O. Holmes Jr. to Gregory P. Baxter, August 1 [1917?], UAV 275.45, Department of Chemistry, Harvard University Archives.

26. Crowell, *America's Munitions, 1917–1918*, 105, 107.

27. In addition to Calco's in-house research, the firm relied on a frequent consultant, Treat B. Johnson of Yale, to conduct investigations of TNA. The plant was incomplete at the armistice. P. H. Ashmead et al., "Report on Calco Chemical Company's Plant for the Manufacture of TNA," c. March 1919, pp. 1–58, NARA, RG 156, entry 524, box 7. Quotations come from pp. 48 and 33. See also Travis, *Dyes Made in America, 1915–1980*, 47–50.

28. Burns to Semet-Solvay Company, October 18, 1917, RG 156, entry 764, box 141, 471.868/64; Morey to Backus, June 20, 1918, RG 156, entry 764, box 137, 471.862/266; Backus to New York District Ordnance Office, August 2, 1918, RG 156, entry 764, box 137, 471.862/309; Hixson and Backus, "History of Explosives Section; Production Division; Explosives, Chemicals, and Loading Division," part 2, vol. 1, December 1, 1918, appendix, NARA, RG 156, entry 528, box 2, ex. 18, pp. 18–20; Backus to French, Federal Dyestuffs & Chemical Corporation, June 19, 1918, RG 156, entry 764, box 143, 471.868/7 Federal Dyestuff.

29. J. H. Burns [?], memorandum, "Method of Purchasing Refined TNT," July 2, 1917, 471.868/25; J. H. Burns, memorandum, "TNT Federal Dyestuff & Chemical Corporation," July 20, 1917, 471.868/36, NARA, RG 156, entry 764, box 141. TNT prices would sink to about 30¢ per pound at the end of the war. Hixson and Backus, "History of Explosives Section; Production Division; Explosives, Chemicals, and Loading Division," part 2, vol. 1, December 1, 1918, appendix, RG 156, entry 528, box 2, ex. 18, p. 20.

30. C. C. Williams to E. I. du Pont de Nemours Powder Company, September 28, 1918, NARA, RG 156, entry 764, box 143, 471.868/243 Du Pont; E. G. Buckner to C. F. Backus, August 7, 1918, RG 156, entry 764, box 143, 471.868/180 Du Pont; Williams to Du Pont, October 1, 1918, RG 156, entry 764, box 143, 471.868/249 Du Pont. Once Du Pont finally sent the

Barksdale plans, Ordnance immediately sent a copy to Atlas. Backus to R. M. Cook, Atlas Powder Company, September 27, 1918, RG 156, entry 764, box 143, 471.868/242 Du Pont.

31. References to these reports occur in many of the Ordnance records cited in this section. For example, see J. H. Hunter, "High Explosives and Research, EE #11," January 1, 1919, NARA, RG 156, entry 528, box 2; "History of Construction and Operation of Ordnance Department Light Oil Recovery and By-Product Coke Oven Plants," May 17, 1919, part 2, p. 15, NARA, RG 156, entry 524, box 4; "History of Explosives Section: Explosives, Chemicals, and Loading Division," part 1, vol. 1, "Introduction," January 28, 1919, p. 9, NARA, RG 156, entry 528, box 2.

32. As discussed in chapter 1, the American chemical industry had been producing sulphuric and nitric acid well before the war. War shortages of nitric acid, however, prompted the government to build nitrogen plants, including the attempt to establish the Haber-Bosch process in Sheffield, Alabama. Haynes, *American Chemical Industry*, 2:101–11; Noyes, "The Supply of Nitrogen Products for the Manufacture of Explosives"; "History of Explosives Section: Explosives, Chemicals, and Loading Division," part 4, vol. 2, "History of Toluol," February 20, 1919, pp. 2, 8–22, 42–49 (on the petroleum cracking), 50, 63–78, NARA, RG 156, entry 528, box 4; "History of Construction and Operation of Ordnance Department Light Oil Recovery and By-Product Coke Oven Plants," May 17, 1919, part 1, pp. 1–5, 14–22, NARA, RG 156, entry 524, box 3. The Bureau of Standards was also interested in the quality of gas standards, which led to chemical studies of the toluene recovery plants, including R. S. McBride, C. E. Reinicker, and W. A. Dunkley, "Toluol Recovery," *Technologic Papers of the Bureau of Standards*, No. 117, Department of Commerce, December 19, 1918 (Washington: Government Printing Office, 1918). The cracking process noted was developed by the Texas Company in Bayonne, New Jersey. A process by the General Petroleum Corporation of Los Angeles also provided toluene on a commercial scale by July 1918. Horace C. Porter to Myron S. Falk, January 16, 1919, "History of Service in Ordnance Department," pp. 1–6, RG 156, entry 528, box 9.

33. Backus to New York District Ordnance Office, August 2, 1918, NARA, RG 156, entry 764, box 137, 471.862/309; Hixson and Backus, "History of Explosives Section; Production Division; Explosives, Chemicals, and Loading Division," part 2, vol. 1, December 1, 1918, appendix, RG 156, entry 528, box 2, ex. 18, pp. 15, 18–20. The famous medical investigator of public health hazards, Alice Hamilton, found plenty to investigate in the high explosives industry. In addition to the risks of explosions, workers suffered from poisoning from the toluene and phenol, among other things. Hamilton, *Exploring the Dangerous Trades: The Autobiography of Alice Hamilton, MD*.

34. Stephen R. Morey to C. F. Backus, "Picric Acid, Goodrich-Lockhart Plant, Labor Shortage" report, August 2, 1918, NARA, RG 156, entry 764, 471.862/309, box 137; O'Brien, O'Brien Synthetic Dye Company, November 6, 1918, 471.862 O'Brien Synthetic Dye Company (50–100), box 139; Badgley to Backus, November 12, 1918, 471.862 O'Brien Synthetic Dye Company (50–100), box 139; Holloway to Burns, July 23, 1917, 471.868/39, box 141; Burns to Holloway, August 1, 1917, 481.868/40, box 141, RG 156, entry 764; T. E. Pickett, manager, Goodrich-Lockhart Company, to Major C. B. Peters, September 16, 1918, RG 156, entry 764, 471.862/366, box 137; C. C. Williams to Senator Joseph S. Frelinghuysen, August 2, 1918, NARA, RG 156, entry 36, 471.86/2574 Miscellaneous, box 2682; Wollenberg to Backus, March 21, 1918, NARA, RG 156, entry 764, box 142, 471.868 Aetna (1–50); Wollenberg to Palmer, Department of Labor and Industry of Pennsylvania, March 21, 1918, 471.868 Aetna (1–50); Palmer to War Industries Board, May 23, 1918, 471.868/45.

35. In relation to TNT, officers believed their reports could be "extremely valuable to any manufacturer just beginning," but they suspected that picric acid processes still needed improvements, which they expected to come in 1919. See Hixson and Backus, "History of Explosives Section; Production Division; Explosives, Chemicals, and Loading Division," part 2, vol. 1, December 1, 1918, appendix, RG 156, entry 528, box 2, ex. 18, pp. 16–18.

36. Chief of Gas Service, AEF, to Chief of Staff, AEF, "Gases, Kinds, Protection, How, When and Where to Use," April 23, 1918, NARA, RG 175, entry 8, box 6; extract from letter, Col. Amos A. Fries to Col. Charles L. Potter, November 11, 1917, NARA, RG 175, entry 8, box 1. This section relies on the scholarship of Ludwig F. Haber and Daniel P. Jones. In the most complete historical investigation of gas in World War I, Haber takes up the challenge of sifting through poison gas's fearsome legacy to try to determine the impact of gas on casualties, strategic military efficacy, and industrial mobilization, among other things. In the process, Haber reflects on the significance of the psychological dimensions and also on the role of Fritz Haber, his father and leading chemical advisor to the German military. In the end, he concludes that poison gas was a "failure" with respect to military goals. Haber, *The Poisonous Cloud*. Haber's book primarily compares the experiences of Britain, France, and Germany, but Jones focuses on chemical warfare in the United States, and readers of his work will recognize his influence here. Jones, "The Role of Chemists in Research on War Gases in the United States during World War I" and "Chemical Warfare Research during World War I: A Model of Cooperative Research." Benedict Crowell, an assistant secretary of war and director of munitions, assembled an official military history of World War I munitions and covers war gases as well as explosives. Crowell, *America's Munitions, 1917–1918*. The military's archival records are more forthcoming on problems and failures than Crowell's account. Another look focusing on the chemicals and materials is Freemantle, *Gas! Gas! Quick, Boys! How Chemistry Changed the First World War*. Despite its obvious importance, I do not discuss the casualties or related aspects of the war itself. A good overview on gases' harm can be found in Gilchrist, *A Comparative Study of World War Casualties from Gas and Other Weapons*.

37. For example, George Burrell, "The Research Division, Chemical Warfare Service, USA," *JIEC* 11 (February 1919): 93. Historians of gas warfare have questioned the efficacy of gas as a weapon, and Jones, for one, suggests officers were justifiably skeptical of gas until mustard gas's arrival. Jones, "The Role of Chemists," 91.

38. Jones, "The Role of Chemists," chaps. 1–3, and "Chemical Warfare Research," 169–74; National Research Council, Committee on Noxious Gases, "Meeting of April 21, 1917," NARA, RG 70, entry 80, box 2; Manning, "Historical Report to the Secretary of the Interior on the Origin and Development of the Research Work of the Bureau of Mines on Gases Used in Warfare, February 1, 1917 to March 1, 1918," NARA, RG 70, entry 46, box 110; West, "The Chemical Warfare Service," 149–50.

39. National Research Council, Committee on Noxious Gases, meeting transcript, April 7, 1917, p. 3, NARA, RG 70, entry 46, box 110; Burrell, "The Research Division, Chemical Warfare Service, USA," *JIEC* 11 (February 1919): 94, 101; Haber, *The Poisonous Cloud*, 131; Manning, "Report to the Secretary of the Interior on the Research Work of the Bureau of Mines on War Gas Investigations, July 1, 1917 to May 15, 1918," pp. 3–4, NARA, RG 70, entry 80, box 3.

40. Jones, "Chemical Warfare Research," 170; NRC, Committee on Noxious Gases, April 7, 1917; "Meeting of April 21, 1917," NARA, RG 70, entry 80, box 2.

41. Working quickly and patterning gas masks after the British design, the bureau provided 25,000 masks by July. Although effective against chlorine, the American mask provided no protection from chlorpicrin, and the British outfitted the new American troops with the superior British mask instead. Jones, "The Role of Chemists," 92–97; Haber, *The Poisonous Cloud*, 128–29; Burrell, "The Research Division, Chemical Warfare Service, USA," *JIEC* 11 (February 1919): 93–104. The military funded the Washington laboratory. Manning, "Historical Report . . . February 1, 1917, to March 1, 1918"; Jones, "The Role of Chemists," 115–18, 122; Jones, "Chemical Warfare Research," 173–77.

42. Victor Lefebure of Britain helped to organize the Inter-Allied Gas Conference in September 1917 in which Britain, France, and the United States shared information on gas research. The contingent from the United States was mostly a recipient of information rather than a contributor. Haber, *The Poisonous Cloud*, 131–32. The American Expeditionary Force also established a technical laboratory in Paris in January 1918, although its equipment arrived only in May. Sixty-five chemists, officers and enlisted, worked in the laboratory to inspect and evaluate German shells, test water supplies, and conduct various other analytical and practical research. Raymond F. Bacon, "The Work of the Technical Division, Chemical Warfare Service, AEF," *JIEC* 11 (January 1919): 13–15; Jones, "The Role of Chemists," 133–40. Britain's William Pope and the U.S.A.'s James Conant identified the use of sulphur monochloride nearly simultaneously; Conant had the results of Pope's earlier work on sulphur dichloride. Jones, "The Role of Chemists," 138–39; also Conant, "Progress Report, Organic Section," February 18, 1918, NARA, RG 175, entry 8, box 1. Gomberg worked briefly at the gas program's laboratories in Washington in late 1917, but returned to Ann Arbor and stayed in his laboratory, continuing to teach his classes while doing mustard gas research in his laboratory. His diplomatic letters to the gas program leadership showed a reluctance to move to Washington; in part, he felt the distance from the central lab gave him a buffer from the frequent shifts in emphasis and direction in the research program. George A. Burrell to Gomberg, December 31, 1917; James F. Norris to Gomberg, January 26, 1918; Burrell to Gomberg, January 31, 1918; Norris to Gomberg, February 4, 1918; Gomberg to Norris, February 7, 1918; Norris to Gomberg, February 12, 1918; Burrell to Gomberg, February 13, 1918; Gomberg to Norris, February 18, 1918; Gomberg to Norris, February 20, 1918; Norris to Gomberg, February 18, 1918; Gomberg to Norris, April 2, 1918; Norris to Gomberg, May 10, 1918; Norris to Gomberg, May 22, 1918; Gomberg to Norris, May 25, 1918; Norris to Gomberg, May 31, 1918, University of Michigan, Moses Gomberg Papers, Correspondence (85503 Aal), File: Correspondence, 1918–1919. *Dictionary of Scientific Biography* 5, ed. Charles Coulston Gillispie (New York: Charles Scribner's Sons, 1981), s.v. "Moses Gomberg." Gomberg's report on his mustard gas experiments is "Report on the Preparation of Dichlorethyl Sulphide," n.d. (c. late February/early March 1918), NARA, RG 175, entry 8, box 1.

43. Crowell, *America's Munitions, 1917–1918*, 396–99; Charles H. Herty, "Gas Offense in the United States, A Record Achievement," *JIEC* 11 (January 1919): 5–12.

44. Earl J. W. Ragsdale testimony in Senate Committee on Finance, *Hearings, Tariff Act of 1921*, 1:753–54.

45. Crowell and Wilson, *The Armies of Industry*, 2:491–502. Baruch, *American Industry in the War*, 178–80. The plant was designed to hold up to 8,400 people, but the number peaked at 7,400 at the armistice. On the arsenal, also see *The Edgewood Arsenal*, published as an issue in *The Chemical Warfare* 1, no. 5 (March 1919).

46. Other gas-related programs included gas masks under the direction of the Surgeon General (head of the Medical Department) and gas shell under the Artillery Ammunition

Section in the Ordnance Department. *The Edgewood Arsenal*, 7–9; "Chemical Warfare Service," *JIEC* 10 (September 1918): 675–84. The Navy supervised its own research done with the Chemical Warfare Service. O. M. Hustvedt testimony in House Committee on Ways and Means, *Hearings, Dyestuffs* (1919), 24. Crowell, *America's Munitions, 1917–1918*, 396–99; Herty, "Gas Offense in the United States," *JIEC* 11 (January 1919): 5–12; *The Edgewood Arsenal*, 7–9; "Chemical Warfare Service," *JIEC* 10 (September 1918): 675–84; "Memorandum Regarding Conference Held in the Office of the Secretary of War, from 3:00 pm to 4:45 pm, May 25, 1918, Regarding the Proposed Transfer of the War Gas Investigations of the Bureau of Mines to the War Department, under Major General Sibert, Chief of the Gas Service," quote on p. 9, NARA, RG 70, entry 80, box 4. George Burrell, Bureau of Mines, to William Walker, February 28, 1918; and Walker to Burrell, March 7, 1918, NARA, RG 175, entry 8, box 1, discuss explicitly the "duplicate research laboratory" at Edgewood. Walker told Burrell the scope of the Edgewood laboratory "depends entirely on the promptness and efficiency with which our problems are handled by your staff." William H. Nichols et al., to Franklin K. Lane, Secretary of the Interior, May 16, 1918, NARA, RG 70, entry 46, box 110; Woodrow Wilson, "Executive Order," June 26, 1918, NARA, RG 70, entry 80, box 4. In *Poisonous Cloud*, 109–10, Haber wrote that, in all countries, "scientists appeared to be working aimlessly and writing copiously" and generally working without "military guidance on end uses," a condition that would have supported Sibert's consolidation.

47. Burrell, "The Research Division, Chemical Warfare Service, USA," *JIEC* 11 (February 1919): 99; Jones, "The Role of Chemists," 141–42.

48. Ragsdale testimony in Senate Committee on Finance, *Hearings, Tariff Act of 1921* (1922), 752, contains the first quote; Amos Fries and William Sibert testimony in House Committee on Ways and Means, *Hearings, Dyestuffs* (1919), 33–37.

49. William McPherson, "Report of the Director of Outside Plants," Enclosure 9 (Bound Brook), p. 1, NARA, RG 175, entry 8, box 14; Sibert testimony in House Committee on Ways and Means, *Hearings, Dyestuffs* (1919), 33. On Du Pont, see Haber, *Poisonous Cloud*, 167; on Semet-Solvay, see Ragsdale testimony in Senate Committee on Finance, *Hearings, Tariff Act of 1921* (1922), 753. See also the references in the following note.

50. Crowell, *America's Munitions*, 397; Herty, "Gas Offense in the United States, A Record Achievement," *JIEC*, 11 (January 1919): 7, 10. The hazards also made the railroads (and eventually their government operators) reluctant to carry gases, and they required special arrangements. Jones, "The Role of Chemists," 130–31.

51. Filling shells, however, seems only to have been done at Edgewood. Shells were frequently in short supply, and the military shipped some gas to Europe in bulk containers, although the Allies' shell was also often inadequate. Crowell and Wilson, *The Armies of Industry*, 2:493–95, 504–8; Herty, "Gas Offense in the United States, A Record Achievement," *JIEC* 11 (January 1919): 8.

52. William McPherson, "Report of the Director of Outside Plants," Enclosure 7 (Niagara Falls), pp. 1–7, NARA, RG 175, entry 8, box 14. Amos Fries later testified that Oldbury received initial technical help from the British; Fries notes that Lidbury, the general manager was British. Fries testimony in Senate Committee on Finance, *Hearings, Dyestuffs* (1920), 30.

53. McPherson, "Report of the Director of Outside Plants," Enclosure 7 (Niagara Falls), pp. 1–7, and attached letters, E. J. W. Ragsdale to F. A. Lidbury, Oldbury Electro-Chemical Company, August 4, 1917, and Ragsdale to Captain Hamilton, September 22, 1917, NARA, RG 175, entry 8, box 14. For the quote, see Ragsdale testimony in Senate Committee on Finance, *Hearings, Tariff Act of 1921* (1922), 752.

54. McPherson, "Report of the Director of Outside Plants," Enclosure 8 (Bound Brook), pp. 1–8, and attached enclosures, "Notes on Process," (undated); William Walker to Hemingway, October 24, 1918; Hemingway to C. F. Long, May 17, 1918; Walker to Hemingway, May 23, 1918; W. R. Chappell, "Bound Brook Plant," a review of technical difficulties over 1918, c. November 1918; "Allotments and Disbursements," NARA, RG 175, entry 8, box 14.

55. McPherson, "Report of the Director of Outside Plants," Enclosure 8 (Bound Brook), pp. 1–8, and attached enclosures, "Notes on Process," (undated); William Walker to Hemingway, October 24, 1918; Hemingway to C. F. Long, May 17, 1918; Walker to Hemingway, May 23, 1918; W. R. Chappell, "Bound Brook Plant," a review of technical difficulties over 1918, c. November 1918; "Allotments and Disbursements," NARA, RG 175, entry 8, box 14.

56. William McPherson, "Report of the Director of Outside Plants," Enclosure 3 (Stamford), pp. 1–5, NARA, RG 175, entry 8, box 14; on chlorpicrin, see Crowell, *America's Munitions, 1917–1918*, 400. Crowell points out that later plants used calcium picrate rather than picric acid.

57. Free was thirty-five years old in the spring of 1918, equipped with a Ph.D. in biology from Johns Hopkins University and with several years' experience in the Department of Agriculture. *Who's Who in the Chemical and Drug Industries*, ed. Williams Haynes (New York: Haynes Publications, 1928), s.v. "Free, Edward Elway."

58. E. E. Free to William Walker, March 18, 1918, NARA, RG 175, entry 8, box 1, File: Batchite. The humidity comment comes from Free to Commanding Officer, Gunpowder Arsenal, April 4, 1918, p. 5, RG 175, entry 8, box 9. The discussion of security occurs in Free to Major McPherson, March 16, 1918, "Protection of the Plant of the American Synthetic Color Company, Stamford, Conn.," enclosed in the April 4, 1918, report.

59. Free to Commanding Officer, Gunpowder Arsenal, April 4, 1918, p. 5, NARA, RG 175, entry 8, box 9; Free to McPherson, May 4, 1918, RG 175, entry 8, box 8. The waste disposal is discussed in the latter report, pp. 15–16.

60. Crowell and Wilson, *The Armies of Industry*, 2:497–98.

61. Much of this account about Dow's experience comes from a summary written by Dow officials. "Statement of Situation at Midland, Michigan, Effecting [sic] the Production of Mustard Gas," c. fall 1918, DHC, folder 180013. Mustard gas is dichlorodiethyl sulfide or $(ClCH_2CH_2)_2S$. Furnas, *Rogers' Industrial Chemistry* (1942), 1:473. Despite the capacity, the plant's output was considerably less in June. William McPherson, Edgewood Arsenal, to Smith, May 31, 1918; Walker to Dow Chemical, July 25, 1918; Smith to William S. Rowland, American University Experiment Station, February 21, 1918; Rowland to Smith, February 25, 1918, DHC, folder 180002A. Also see Chemical Warfare Service, *Report of Midland Section, Development Division*, 1918.

62. "Statement of Situation at Midland, Michigan, Effecting [sic] the Production of Mustard Gas," c. fall 1918, p. 3, DHC, folder 180002A.

63. On his reaction to the explosion at the chlorbenzene plant in 1916, see Dow to Smith, October 31, 1916, DHC, folder 160006; Smith to Dow, November 7, 1916; Dow to Smith, November 8, 1916, DHC, folder 160007. The two men discuss the disincentive the explosion would provide to workers trying to sneak a smoke in the plant. Dow to G. E. Collings, July 8, 1918, DHC, folder 180012, notes the "two light casualties a day for several months." Roos made his first appearance about May 24 but left at some point and returned July 4. "Statement of Situation at Midland, Michigan, Effecting [sic] the Production of Mustard Gas," c. fall 1918, pp. 4–5, DHC, folder 180002A.

64. William Walker to Dow Chemical Company, May 23, 1918; William McPherson to Smith, June 7, 1918, DHC, folder 180002A. The letters note the increased pressure to produce mustard gas as soon as possible. A full account of the accident is in an untitled report beginning "On Wednesday, June 26th, 1918," NARA, RG 175, entry 5, box 2; the other military records of subsequent events is in the same place. These include powerfully gruesome photographs of the gas burns. "Statement of Situation at Midland, Michigan, Effecting [sic] the Production of Mustard Gas," c. fall 1918, p. 11, DHC, folder 180002A.

65. "Statement of Situation at Midland, Michigan, Effecting [sic] the Production of Mustard Gas," c. fall 1918, p. 11; unsigned [William Hale?] to Rep. G. A. Currie, September 6, 1918; unsigned to Currie, September 8, 1918, DHC, folder 180002A.

66. "Midland, Michigan. Branch of the Edgewood Arsenal," c. fall 1918, p. 2, DHC, folder 180002B.

67. "Statement of Situation at Midland, Michigan, Effecting [sic] the Production of Mustard Gas," c. fall 1918, pp. 4–12, DHC, folder 180002A; "Affidavit" of Eva C. Venner, August 26, 1918; "Affidavit of Miss Rose Morrow," August 17, 1918, DHC, folder 180002B.

68. Walker to Dow, August 13, 1918, reprinted in "Statement of Situation at Midland, Michigan, Effecting [sic] the Production of Mustard Gas," c. fall 1918, Dow papers, folder 180013; also, William McPherson to Dow, August 14, 1918, DHC, folder 180027; "Statement of Situation at Midland, Michigan, Effecting [sic] the Production of Mustard Gas," c. fall 1918, p. 1; unsigned [William Hale?] to Rep. G. A. Currie, September 6, 1918; unsigned to Currie, September 8, 1918, DHC, folder 180002A.

69. Jones, "Chemical Warfare Research during World War I," 176–79. *The Story of the Development Division, Chemical Warfare Service*, 180–82, gives a brief overview of the Development Division, which used General Electric facilities. (The report notes incorrectly that Lewisite was dubbed "G-34," which was actually the code name for mustard gas. To fool the potential spies, Lewisite and related chlorarsines were "new G-34," P1, and P2. Chlorpicrin was S1; phosgene L3.) The CWS also began construction on another plant for Lewisite (diphenylchloroarsine) in Croyland, Pennsylvania. Crowell, *America's Munitions*, 399; McPherson, "Report of the Director of Outside Plants," pp. 3–4, NARA, RG 175, entry 8, box 14. Harry M. St. John, "Standard Methods for the Manufacture of New G-34, Development Division, Chemical Warfare Service, United States Army, Nela Park, Cleveland, Ohio," approved by F. M. Dorsey, March 29, 1919, a copy of which is held at the U.S. Army Military History Institute, Carlisle, Pennsylvania; "Most Powerful War Gas Made," *DCM* 5 (December 11, 1918): 19. After the war, Amos Fries noted that "three full drops [of Lewisite] absorbed into the skin . . . will probably cause death," but it was not difficult to rub the gas off before it could penetrate. And, it "is not much more poisonous than phosgene." Fries testimony in Senate Committee on Finance, *Hearings, Tariff Act of 1921* (1922), 388.

70. Walker continued, "The planes were made and successfully demonstrated, the containers were made and we were turning out mustard gas in the requisite quantities in September." "Poison Gas Worth $60,000,000," *DCM* 5 (January 1, 1919): 8.

71. Edward O. Holmes Jr. to Gregory P. Baxter, December 12, 1918, UAV 275.42, Department of Chemistry, Harvard University Archives.

72. Bankruptcy prevented Aetna from following through on any plans to make dyes. "Munitions Plants to Make Dyestuffs," *DCM* 3 (October 25, 1916): 11; Davis and Dyer, *Labors of a Modern Hercules*, 103; *Encyclopedia of American History*, 6th ed., s.v. "War Revenue Act."

Chapter Four

1. Horne, "German Atrocities, 1914: Fact, Fantasy, or Fabrication?"; Eksteins, *Rites of Spring: The Great War and the Birth of the Modern Age*, 89, 158.

2. E. F. Roeber, "The By-Product of a Chemical Reaction," *MCE* 12 (December 1914): 738; Ferguson, *The Pity of War*, 223.

3. Leo H. Baekeland, "Applied Chemistry," *MCE* 13 (October 1, 1915): 677; Leo H. Baekeland diary, October 24, 1914, Diary No. 15, box 3, Series 8, Baekeland papers, Archives Center, National Museum of American History, Smithsonian Institution, Washington, D.C. The final quotation comes from the article.

4. Forman, "Scientific Internationalism and the Weimar Physicists."

5. Wolff, "Physicists in the 'Krieg der Geister,'" 340–44; Ferguson, *The Pity of War*, 226–35.

6. Quoted in Wolff, "Physicists in the 'Krieg der Geister,'" 339, 353, and in Badash, "British and American Views of the German Menace in World War I," 99.

7. Townes R. Leigh, "Germany's Stolen Chemistry," *DCM* 4, nos. 33–36 (April 24, pp. 5–6; May 1, pp. 5–6; May 8, pp. 7–8; May 15, pp. 13–14). The article also appeared in *ADR* 2 (June 3, 1918): 1–5. Wöhler's education was in Germany; Liebig spent two important years in Paris (1822–24) after receiving his Ph.D. from the University of Erlangen. *Dictionary of Scientific Biography* (1981), s.v. "Wöhler, Friedrich" and "Liebig, Justus."

8. William H. Nichols, "Address," *JIEC* 9 (October 1917): 924; Bernhard C. Hesse, "Contributions of the Chemists to the Industrial Development of the United States—A Record of Achievement," *JIEC* 7 (April 1915): 297–98.

9. The envy was mutual. The Germans had established the Kaiser Wilhelm Institute in part because they worried over the resources of new American research institutes, the Rockefeller Institute for Medical Research and the Carnegie Institution. Johnson, *The Kaiser's Chemists*, 18. Marston T. Bogert, in "The Dyestuff Situation—Papers and Discussion," *MCE* 13 (November 1, 1915): 782; C. Alfred Jacobson, "The Need of a Large Government Institution for Chemical Research," *JIEC* 8 (January 1916): 70–72.

10. Skolnik and Reese, *A Century of Chemistry*, 456; Thackray, Sturchio, Carroll, and Bud, *Chemistry in America, 1876–1976*, 250.

11. Milton C. Whitaker, "On Our Opportunities," *JIEC* 6 (October 1914): 794. At the turn of the century, Whitaker (1870–1963) received his M.S. in chemical engineering from the University of Colorado and then joined the faculty at Columbia University. When he left the journal, he also left academia, first taking a position at U.S. Industrial Alcohol Company and eventually becoming an executive at American Cyanamid. He won both the Perkin Medal (1923) and Chandler Medal (1950) from the ACS; he also worked on the Manhattan Project. "M. C. Whitaker, Retired Chemist," [obituary] *New York Times*, April 4, 1963, p. 47. Until 1919, the industrial division's name was the Division of Industrial Chemists and Chemical Engineers.

12. Whitaker, "On Our Opportunities," *JIEC* 6 (October 1914): 794.

13. Readers will recognize my debt to David J. Rhees throughout this section, but particularly in the ACS's publicity efforts. Rhees, "The Chemists' Crusade: The Rise of an Industrial Science in Modern America, 1907–1922"; "First National Exposition of Chemical Industries," *MCE* 13 (October 1, 1915): 690–700; "Second National Exposition of Chemical Industries," *MCE* 15 (October 1, 1916): 418–30; "Exposition of Chemical Industries," *MCE* 13 (June 1915): 351; E. F. Roeber, *MCE* 13 (August 1915): 465; "Chemical Exhibit Opens," *New York Times* (September 21, 1915), business section, p. 14.

14. Whitaker editorial, "Analysis of the Coal-Tar Dye Industry" and the committee's report, "Recommendations of the New York Section of the American Chemical Society on the Enlargement of the Coal Tar Chemical Industry in the United States," *JIEC* 6 (December 1914): 972-75.

15. "Report of the Chemical and Dyestuff Committee," *MCE* 12 (December 1914): 754-55; Chemicals and Dyestuffs Committee, New York Section, American Chemical Society to Representative E. J. Hill, January 7, 1916, recorded in House Committee on Ways and Means, *Hearings, To Establish the Manufacture of Dyestuffs* (1916), 6-11.

16. *Metallurgical and Chemical Engineering, Journal of Industrial and Engineering Chemistry.*

17. "Bernard Conrad Hesse, 1869-1934," *The Percolator* (Chemists' Club), April 21, 1934; "Interview with Dr. Hesse," 1899; "Bernard Conrad Hesse, '89p, '93," August 1934, all clippings in the necrology file of Bernhard C. Hesse, University of Michigan alumni collection, Bentley Historical Library, University of Michigan. Although Bernhard in most of his publications, he seems to have become Bernard by the time of his death.

18. Some of his articles include "Relieving the Dyestuff Crisis," *JIEC* 6 (November 1914): 953; "The Industry of the Coal Tar Dyes: An Outline Sketch," *JIEC* 6 (December 1914): 1013-27; "Contributions of the Chemist to the Industrial Development of the United States—A Record of Achievement," *JIEC* 7 (April 1915): 293-302; "Research, Scientific and Industrial in the Coal Tar Dye Industry," *JIEC* 8 (September 1916): 845-48. One of Hesse's most cited, if least typical, articles is "Lest We Forget! Who Killed Cock Robin? The U.S. Tariff-History of Coal-Tar Dyes," *JIEC* 7 (August 1915): 694-709. The quotation is from "The European War and the Chemical Industries of This Country," *MCE* 12 (September 1914): 553.

19. "Report of the Chemical and Dyestuff Committee," *MCE* 12 (December 1914): 754-55; Hesse, "Research, Scientific and Industrial in the Coal-Tar Dye Industry," *JIEC* 8 (September 1916): 845.

20. The clause, also part of British and French patent systems, mandated that a patent filed by a non-citizen must be in use within a specified time—usually one or two years—or the patent becomes void, leaving the product or process open to exploitation by others. See comments by Arthur Prill and by I. F. Stone, in "Symposium on American Dye Industry," *JIEC* 6 (November 1914): 945, 949; Edward Gudeman, "Aspects of Some Chemical Industries, in the United States, Today," *JIEC* 7 (February 1915): 152. Opponents suggested the clause worked in neither theory nor practice with regard to the American dyes industry, citing the lack of strong dyestuffs industries in France and England, which possessed working clauses. Hesse, "Coal-tar Dyes and the Paige Bill," *JIEC* 7 (November 1915): 974-77, quotation on 968. For a discussion of German chemical companies' ownership of American patents, see Liebenau, "Patents and the Chemical Industry"; I. F. Stone, "Coal Tar Colors of America," *JIEC* 6 (November 1914): 950.

21. Julius Koebig, "American Chemical Industries Must Be Built Up by American Methods," *JIEC* 8 (April 1916): 384-85; William Walker, "Education for Research," *JIEC* 7 (January 1915): 2.

22. Reed, *Crusading for Chemistry: The Professional Career of Charles Holmes Herty.* As the title suggests, Reed provides a full biography of Herty's life in chemistry.

23. Ibid., 72-73. Also, Herty to Thomas H. Norton, August 7, 1915, CHH, box 81; Herty, "Cooperation in Matters Chemical," *Journal of the American Chemical Society* 37 (October 1915): 2231-46; Rhees, "The Chemists' Crusade."

24. Herty speech, "The Dyestuff Situation," April 4, 1916, before the American Cotton Manufacturers Association; Herty to Charles R. Miller, *New York Times*, April 10, 1916; Herty to W. L. Saunders, Ingersoll-Rand Company and Naval Consulting Board, May 3, 1916, CHH, box 81.

25. Herty, "By Way of Introduction," *JIEC* 9 (January 1917): 3; Herty, "The Dyestuff Situation," April 4, 1916, before the American Cotton Manufacturers Association, Atlanta, CHH, box 81.

26. Gruber, *Mars and Minerva: World War I and the Uses of the Higher Learning in America*, 1–5, 108–16, 123, quotations on 82 and 95.

27. Chambers, *To Raise an Army*, 1–12, 266–70.

28. Ibid., 266–70; "Copy of General Draft for Mr. Green," c. 1917, Row 6, 1.7.2, Merck & Company Archives; Baker testimony, House Committee on Military Affairs, *Hearings, Selective Service Act* (1918), 5.

29. Charles R. Mann testimony, Senate Committee on Military Affairs, *Hearings, Amending the Draft Law* (1918), 91–92, 101; Gruber, *Mars and Minerva*, 213–37; McMaster, *The United States in the World War, 1918–1920*, 41–43; Geiger, *To Advance Knowledge: The Growth of American Research Universities, 1900–1940*, 102–7; Baker testimony in House Committee on Military Affairs, *Hearings, Selective Service Act* (1918), 8–9, 12, 16; "Special Circular of the Department of Chemistry, 1916–1927," *University of Illinois Bulletin* 24, no. 52 (August 30, 1927), Urbana, University of Illinois, 1927, Papers of the Liberal Arts and Sciences, Chemistry, Department Publications, 1878, 1899–, Series 15/5/805, box 2; "History of the Decade, 1916–1927, Publications of the Department, Courses of Study, Faculty, Advanced Students, and Alumni," Papers of the College of Liberal Arts and Sciences, Chemistry & Chemical Engineering, Departmental History, 1868–1961, Series 15/5/801, Box 1. University Archives, University of Illinois, Urbana, Illinois.

30. Nicholas Murray Butler to Marston Taylor Bogert, August 15, 1917; Bogert to Butler, August 13 and 18, 1917; Bogert to Butler, January 28, February 12, March 3, 1918; Bogert to George B. Pegram, Columbia, March 9, 1918; Bogert papers, folder 317/16, Columbia University.

31. Bogert to Butler, November 23, 1915, March 28, May 3, 1916, April 3 and 7, 1917, April 6, 1918, Bogert to Pegram, March 12, 1918, Bogert papers, folder 317/16; "Dr. M. T. Bogert, 85, Eminent Chemist," *New York Times*, March 22, 1954, p. 27.

32. James F. Norris (?) to George A. Burrell, September 28, 1917; James F. Norris (?) to A. Lawrence Lowell, president of Harvard University, October 3, 1917, HUG 1495.1, War Department Correspondence, Elmer Peter Kohler papers, Harvard University Archives. Lauson Stone, government form re: salary increase, May 20, 1918, HUG 1495.1, War Department Correspondence, Elmer P. Kohler papers, Harvard University Archives; G. P. Baxter to William A. Shanklin, Wesleyan University president, April 10, 1919, UAV 275.42, Department of Chemistry, Harvard University Archives. The instructor was G. Albert Hill. Around Harvard, Kohler had earned a reputation as an exceptional teacher and lecturer, but throughout his life he refused to present a lecture or paper outside the classroom. His fame in the larger world rested on his research work and producing graduate students who themselves earned exceptional distinction. Louis F. Fieser, "Elmer Peter Kohler (1865–1938)," *Proceedings of the American Academy of Arts and Sciences* 74, no. 6 (1940), 130–41, reprint in HUG 1495.5, biographical items, Elmer P. Kohler papers, Harvard University Archives.

33. "Library for Edgewood Arsenal Laboratory," *JIEC* 10 (October 1918): 868; Herty, "Turn About Is Fair Play," *JIEC* 10 (September 1918): 672.

34. House Committee of Ways and Means, *Hearings, Chemical and Optical Glassware and Scientific Apparatus* (1919), 6–11; Senate Committee on Finance, Subcommittee, *Laboratory Glassware and Scientific and Surgical Instruments* (1919), 13, 54–55; U.S. Tariff Commission, *Information Concerning Optical Glass and Chemical Glassware*, 17, 23; Arthur H. Thomas, "The Manufacture of Chemical Apparatus in the United States," *JIEC* 8 (May 1918): 437–41. On the ceramics and glass industry more generally, see Blaszczyk, *Imagining Consumers: Design and Innovation from Wedgwood to Corning*.

35. As manufacturers began to make many of the chemicals during and after the war, the university continued its program but focused its efforts on chemicals used extensively in teaching organic chemistry. The university made the reagents project permanent and integrated it into their curriculum; a 1941 department brochure touted its "business operation . . . [which] gives extremely beneficial training in larger scale production of chemicals" and noted that thirty-eight men worked on reagents in the summer of 1940. The department's "Organic Syntheses" also grew from pamphlets to an annual commercial publication. Adams, "The Manufacture of Organic Chemicals at the University of Illinois," *Science* 47 (March 8, 1918): 225–28; "History of the Decade, 1916–1927," *Special Circular of the Department of Chemistry* 24, no. 52 (August 30, 1927): 11–13; Adams, Oliver Kamm, and Carl S. Marvel, "Organic Chemical Reagents," 16, no. 43 (June 23, 1919): 3–4, "Organic Chemical Reagents II," 18, no. 6 (October 11, 1920): 3–4, Adams and Marvel, "Organic Chemical Reagents III," 19, no. 6 (October 9, 1921): 3, all in *University of Illinois Bulletin*; Virginia Bartow, "Developments in the Chemistry Department, University of Illinois, 1926–1941," Department of Chemistry pamphlet (Urbana: University of Illinois, 1941), 35; *Bulletin* and pamphlet in Papers of the Liberal Arts and Sciences, Chemistry, Department Publications, 1878, 1899–, Series 15/5/805, box 2, University Archives, University of Illinois, Urbana, Illinois.

36. Jenkins, *Images and Enterprise: Technology and the American Photographic Industry, 1839 to 1925*, 244, 313; Sturchio, "Experimenting with Research: Kenneth Mees, Eastman Kodak, and the Challenges of Diversification"; C. E. K. Mees, "Report on the Production of Synthetic Organic Chemicals in the Research Laboratory of the Eastman Kodak Company for the year 1918–1919," *JIEC* 11 (December 1919): 1141–42; H. T. Clarke and C. E. K. Mees, "The Production and Supply of Synthetic Organic Chemicals," unpublished copy of paper delivered at the Society of Chemical Industry in England, March 19, 1920, Chemical File, Eastman Kodak Company, Rochester, New York.

37. Forrestal, *Faith, Hope, and $5000*, 36; Dow to James N. McBride, July 15, 1918, DHC, folder 180203; H. G. Chickering to G. P. Baxter, April 22, 1918, UAV 275.42, Department of Chemistry, Harvard University Archives; Hamilton, *Industrial Poisoning in Making Coal-Tar Dyes and Dye Intermediates*, 6; Baekeland alerted his foreman to prepare to hire women but appears not to have hired any. Leo H. Baekeland diary, May 7, 1918, Diary No. 24, box 4, Archives Center, National Museum of American History, Smithsonian Institution, Washington, D.C.

38. For one view of the relationship, see E. Emmet Reid, "Relation of the University to the Dye Industry," *IEC* 18 (December 26, 1926): 1325–27.

39. Employee agreements, c. 1914–1924, in a box of miscellaneous items, Row 6, 3.4.1, Merck Archives, Merck & Company, Whitehouse Station, N.J. The names of the twenty-nine employees hired in 1914–1918 whose records exist: Willis W. Angus, Frank C. Chase, John Nolan Dalton, T. H. Davis (Rahway Coal Tar Products Company), Joseph W. Dunning, Lynn Edgecomb Ellis, Frank L. Falvey, A. W. Frame, Boyd C. Getty, Edmund McKendree

Hayden, J. Allen Horton, Alfred John Kinsman, Henry Koster Jr., E. E. Lauer, Harold F. Millman, Fred W. Misch, Leon Mason Monell, Robert E. Morse, Henry N. Oellrich, Frank E. Prentice, Frank H. Reichel (fired), Edward Schmidt, Ralph B. Shivel, Theodore F. Spear, Gerald Thorp, Earl Albert Tyler, Charles Wadsworth III (Harvard Ph.D.), Roger Churchyard Williams, and Louis Harrington Zepfler.

40. W. A. Noyes to G. P. Baxter, July 16, 1916, UAV 275.45, Department of Chemistry, Harvard University Archives; W. A. Noyes, "Report of the Department of Chemistry for the Academic Year 1925-1926," May 6, 1926, box 1, Series 2/6/4, Annual Reports, Papers of President David Kinley, University Archives, University of Illinois.

41. Swann, *Academic Scientists and the Pharmaceutical Industry*, 57-65; Hounshell and Smith, *Science and Corporate Strategy*, 290, 296-300; Herty, "An Army without Reserves," *JIEC* 10 (July 1918), 510.

42. *Report of the President, 1914-1915*, Bulletin of Yale University, 11th series, no. 11, August 1915 (New Haven, Conn.: Yale University, 1915), 210-11; *History of the Class of 1897, Sheffield Scientific School, Yale University*, Decennial Record, 1897-1907, ed. G. Barrett Rich Jr. (New Haven, Conn.: Tuttle, Morehouse, & Taylor Co., 1907), 41; *Quarter Century Record of the Class of 1897*, Sheffield Scientific School of Yale University, ed. Samuel E. Hoyt (New Haven, Conn. [?], 1923), 87; *Report of the President, 1915-1916*, Bulletin of Yale University, 12th series, no. 10 (New Haven, Conn.: Yale University, July 1916), 226-27, 229; "Joseph Sumner Bates," in *History of the Class of 1912, Sheffield Scientific School, Yale University*, ed. Richard E. Bishop et al. (New Haven, Conn.: Published under the Direction of the Class Secretaries Bureau, 1926), 193-94. Travis's overview of Calco employees suggests Bates's generalization would not have included some key people. Travis, *Dyes Made in America, 1915-1980*, 37-53.

43. James F. Norris, "Second Census of Graduate Research Students in Chemistry," *IEC* 17 (July 1925), 755-56. Presumably many of these organic chemists specialized in agricultural or other fields not focused on synthetic organic chemicals. Jeffrey Johnson has looked at publications and the proportion of Ph.D.s in organic chemistry among German women chemists in "Hierarchy and Creativity in Chemistry, 1871-1914," especially table 1 on p. 216; Johnson, "German Women in Chemistry, 1895-1945" (in two parts), tables on pp. 8, 11, and 68; personal correspondence from Johnson to author, March 2, 2001.

Chapter Five

1. For example, "Aniline Dyes," *WDM* 1 (September 16, 1914): 6-7.

2. Goldstein, *Ideas, Interests, and American Trade Policy*, 91-122, especially 116.

3. "W. C. Redfield Dead; In Wilson Cabinet," *New York Times*, June 14, 1932, p. 21; "Manufacturers See Low Tariff Ghosts," *New York Times*, April 7, 1912, p. 11; "World Trade Vital, Declares Redfield," *New York Times*, June 24, 1921, p. 26; Redfield, *The New Industrial Day*, quotations on 42 and 79; Becker, *The Dynamics of Business-Government Relations*, 113-56, especially 123.

4. Rockefeller declined, but Meyer invested in Beckers's plant. Redfield to Ivy Lee, Rockefeller aide, May 20, 1915, and Lee to Redfield, June 21, 1915; Eugene Meyer Jr. to Redfield, April 1, 1915, and Redfield to Meyer, April 6, 1915; Redfield to Bryan, March 29, 1915; Redfield to Woodrow Wilson, May 4, 1915, NARA, RG 40, General Correspondence, 72155/42, box 357. Redfield explained his commitment to the industry and to holding the Democratic line on tariffs on several occasions. "To Prevent 'Dumping' of Foreign Goods

after War," *WDM* 2 (October 6, 1915): 12. The same sentiment is reported in "U.S. Pans Protection for American Industries," *WDM* 2 (November 3, 1915): 12. One manufacturer's skepticism: Charles Robinson Smith to Henry Wigglesworth, both of General Chemical Company, October 26, 1915, RG 40, General Correspondence, 72155, box 358.

5. Redfield [?], twelve-page typed draft, fall 1915, probably of *New York Times* article; Redfield to Fuller E. Callaway, La Grange, Georgia, March 7, 1916 (Redfield wrote a very similar to letter to Charles H. Herty, March 7, 1916); Redfield to Woodrow Wilson, January 4, 191; Redfield to Wilson, May 10, 1915, NARA, RG 40, General Correspondence, 72155, box 358. One of Redfield's subordinates, E. E. Pratt, also stated Redfield and the department's policy. "U.S. Chemical Industry in Danger From Europe," *WDM* 2 (October 27, 1915): 10; "Conference Is Held to Consider Dye Problem," *WDM* 2 (March 29, 1916): 15.

6. Norton, "Some Aspects of the Establishment of the American Dyestuffs Industry," *Chemical Age* 1 (December 10, 1919): 236. By World War I, most of Norton's forty years of educational, industrial, and governmental employment lay behind him. As with most American chemists of the period, Norton pursued his graduate degree in Europe, receiving his doctorate summa cum laude from Heidelberg in 1875. For the next few years, a Paris chemical firm employed him as a manager, and then he became chair of the chemistry department at the University of Cincinnati, where he established the Cincinnati section of the ACS. In 1900, Norton joined the Department of State as a consul in Turkey and Germany, where he produced reports on scientific subjects, including studies of the chemical industries in smaller European countries. Norton brought a wealth of experience and chemical knowledge to his role as Commercial Agent for the Department of Commerce. *National Cyclopedia of American Biography*, 13:478, s.v. "Norton, Thomas Herbert."

7. Norton, *Dyestuffs for American Textile and Other Industries*. On the ridicule, Hesse to Charles Herty, October 23, 1915; William M. Grosvenor to Herty, October 29, 1915, CHH, box 81; Herbert H. Dow to G. E. Collings, May 15, 1916, DHC, folder 160020. On the Swiss cooperation, Norton, *Dyestuff Situation in the United States November, 1915*, 15–17; Herty to Norton, August 7, 1915; Norton to Herty, August 11, 1915; clipping, "Helping the Swiss Dyes Works," *OPDR* (August 16, 1915), CHH, box 81. William J. Matheson, the importing agent for the German Cassella firm, wrote to his German colleagues that he believed Norton's prognosis was altogether too optimistic, but he believed Norton's reports helped to attract investors. Cassella Color Company to Leopold Cassella & Company, October 20,1915, letter seized by French intelligence and recorded in the War Trade Intelligence Department's report of the Cassella Color Company, NARA, RG 131, entry 190, box 1.

8. The census listed all the imported dyes by a classification (Schultz) number, chemical name and formula, year of original synthesis, manufacturer, quantity, and wholesale invoice value, as well as commercial names. The Schultz number was named after Gustav Schultz, a German organic chemist who devoted most of his career to organizing and classifying synthetic dyes. Norton, "A Census of the Artificial Dyestuffs Used in the United States," *JIEC* 8 (November 1916): 1041, 1044; Norton, "A Census of Colors," *Scientific American* 114 (June 3, 1916): 578. The strongest protests to the release of the detailed customs information came from the importers. Herman A. Metz to William C. Redfield, September 26, 1916; A. J. Peters, Assistant Secretary, Department of Treasury to Redfield, September 28, 1916; Leo H. Baekeland complained to Redfield that "foreign" interests "suppressed" the report, September 30, 1916, NARA, RG 151, 232 Dyestuffs, box 1033. On behalf of a coalition of interests, J. Merritt Matthews, a well-respected textile chemist, penned another report based on information he gathered in a survey from manufacturers. That fewer than

half of manufacturers responded to the survey dulled the impact of Matthews's rebuttal. David Kirschbaum (chairman of the Joint Conference Committee) to Redfield, enclosing Matthews's report, February 4, 1916, CHH, box 81; "New Angle to Dye Situation," *New York Times* (February 13, 1916), section 7, p. 9; "The Dyemakers Say Things," *New York Times* (February 20, 1916), section 7, p. 8. Redfield might have been responsible for the *New York Times* editorials. On the inaccuracies of the free list, Albert W. Smith to Herbert H. Dow, November 17, 1916, DHC, folder 160007; Norton, *Artificial Dyestuffs Used in the United States*. In 1922, the department issued a supplement to the census containing the invoice information removed in the revision. R. Norris Shreve to O. P. Hopkins, Bureau of Foreign and Domestic Commerce, March 23, 1922, NARA, RG 151, 232 Dyestuffs, box 1033; Norton, *Artificial Dyestuffs Imported into the United States, 1913-1914: Supplemental Statistics to Accompany Special Agents Series No. 121*, Department of Commerce, Bureau of Foreign and Domestic Commerce, 1922. Also, "Good Prospects in the Dye Manufacturing Field," *DCM* 3 (November 1, 1916): 3-4.

9. By early 1916, the Wilson administration was clearly thinking about the implications for the election in 1916. Redfield to Wilson, January 4, 1916; Henry S. Quick, Davis & Quick, Brooklyn, New York, to Redfield, March 1, 1916, NARA, RG 40, 72155, box 358; Redfield to Wilson, April 3, 1916; Wilson to Redfield, April 4, 1916, in Link, *The Papers of Woodrow Wilson* 36:406, 410; "Low Tariff Spells Business Disaster, Says Dr. A. R. L. Dohme," *WDM* 2 (February 23, 1916): 10.

10. "Bill to Encourage American Dye-Stuffs Industry Introduced," *WDM* 2 (December 15, 1915): 4. Hill based his numbers of the ACS report in November 1914; E. J. Hill, U.S. Representative from Connecticut, to Redfield, October 29, 1915, NARA, RG 40, General Correspondence, 72155, box 355; "Protection for American Dye Industry to be Urged," *WDM* 2 (January 12, 1916): 9; J. F. Schoellkopf, Herbert Dow, and William Beckers testimony, House Committee on Ways and Means, *Hearings, To Establish the Manufacture of Dyestuffs: Hearings* (1916), 169, 174, 101, 199-200; "Chemical Manufacturers Urge Tariff Protection," *WDM* 2 (January 19, 1916): 8.

11. Charles H. Herty, George W. Wilkie, and Joseph E. Ralph testimonies, House Committee on Ways and Means, *Hearings, To Establish the Manufacture of Dyestuffs* (1916), 106, 119, 213-14.

12. A paraphrase of Stone in "Need Tariff Protection to Build Dye Industry," *WDM* 2 (September 22, 1915): 3. I. F. Stone worked for the National Aniline and Chemical Company, the long-time sales distributor for Schoellkopf. This National Aniline is not to be confused with the firm of the same name created in 1917. On the role of the textile industry, for example, William Beckers to Charles H. Herty, March 18, 1916, CHH, box 81.

13. J. Merritt Matthews, Charles L. Anger, Horace B. Cheney, and Fuller E. Callaway testimony in House Committee on Ways and Means, *Hearings, To Establish the Manufacture of Dyestuffs* (1916), 57-59, 156-60, 80, 78, 82, 81. The chemical community especially worried over the influence of Caesar Cone, who manufactured denim in North Carolina, consumed huge quantities of indigo, and had made it known he opposed the legislation. Members of the chemical community knew that Cone carried an extraordinary amount of influence with Kitchin and the other southern Democrats, and they attempted to demonstrate that Cone was out of step with his fellow manufacturers. "Dye Makers Ask Tariff on Indigo and Alizarin," *WDM* 2 (July 26, 1916): 5. Also on Cone's opposition: "Protection for Coal Tar Medicinals Is Necessary Says John F. Queeny," *WDM* 2 (July 26, 1916): 7-8.

14. Beckers, Schoellkopf, and Matthews testimonies, House Committee on Ways and Means, *Hearings, To Establish the Manufacture of Dyestuffs* (1916), 197, 171, 192, 63-64. Charles Parsons argued the case more publicly in "Dye Industry Could Make Ammunition in Event of War," *WDM* 2 (February 16, 1916): 4-5.

15. W. L. Saunders, Ingersoll-Rand Company and Naval Consulting Board, to Charles H. Herty, March 24, 1916, abstracting large portions of a recent letter from Metz to Saunders; also Herty's reply, Herty to Saunders, March 28, 1916; Bernhard C. Hesse to Herty, February 7, 1916, CHH, box 81. "Chemical Manufacturers Urge Tariff Protection," *WDM* 2 (January 19, 1916): 3, 8. During Hesse's testimony, the committee engaged in a brief discussion about German dyes firms making explosives. Hesse, very well informed on most matters related to the German firms, said he had heard conflicting reports and did not know whether the German dyes firms had made explosives. Hesse testimony, House Committee on Ways and Means, *Hearings, To Establish the Manufacture of Dyestuffs* (1916), 39. The German dyes firms, in fact, had undertaken the manufacture of high explosives, to the chagrin of at least Carl Duisberg of Bayer, who bemoaned the "powder keg" under his beloved dyes works. Jeffrey Johnson, quoting Duisberg, in comments at the Chemical Heritage Foundation, August 22, 2001.

16. Charles H. Herty, "The Dyestuff Situation," April 4, 1916, before the American Cotton Manufacturers Association, Atlanta, CHH, box 81; "Cotton Manufacturers Urge Protection on Dyestuffs," *WDM* 2 (April 12, 1916): 14-15.

17. "New Tariff Commission to Consider Dye Situation," *WDM* 2 (March 29, 1916): 15; Herty, Matthews, and Dow testimony in House Committee on Ways and Means, *Hearings, To Establish the Manufacture of Dyestuffs* (1916), 105, 108, 63, 104; "Dye Industry Crippled by the German Embargo," *WDM* 2 (January 5, 1916): 34.

18. John F. Queeny, Monsanto Chemical Works, to Claude Kitchin, House Committee on Ways and Means, January 12, 1916, reprinted in "Monsanto Chemical Works Asks Tariff Protection," *WDM* 2 (January 16, 1916): 4; "Urges Prompt Action on Chemical Tariff Changes," *WDM* 2 (March 1, 1916): 3; Queeny testimony, Senate Committee on Finance, *Hearings, To Increase the Revenue* (1916), 138-43. The subcommittee also received briefs from the Mallinckrodt and Heyden companies advocating the same position as the Monsanto president, 142-44; "Wants Phenol Taxed as Finished Product," *WDM* 2 (July 26, 1916): 6—this was an inaccurate headline. Also, a lawyer representing two large North Carolina textile manufacturers opposed any rate on indigo. Thomas S. Beall testimony, Senate Committee on Finance, *Hearings, To Increase the Revenue* (1916), 146-51.

19. House Committee on Ways and Means, *To Increase the Revenue, and for Other Purposes* (1916), 8-10; "American Dyestuff Manufacture," *JIEC* 8 (October 1916): 952; U.S. Tariff Commission, *Census of Dyes and Coal-Tar Chemicals*, 1918, 11.

20. "Underwood Opposed to Tariff on Dyes," *WDM* 2 (August 30, 1916): 6; "Senate Passes Dyestuffs Tariff with Only Seven Opposing Votes," *WDM* 2 (September 6, 1916): 6; "Says Dyestuff Tariff Is Ineffective," *WDM* 2 (August 30, 1916): 5.

21. "New U.S. Revenue Bill Affords Protection for Coal Tar Dyes," *WDM* 2 (July 5, 1916): 4; Schoellkopf to Herty, July 18, 1916, CHH, box 81; Representative E. J. Hill to Herty, July 8, 1916, CHH, box 81. Herty discusses Beckers's acquiescence in Herty to Hesse, August 18, 1916, CHH, box 81; "E. C. Klipstein Praises Dye Tariff," *DCM* 3 (October 18, 1916): 6.

22. "Tariff Needed for Coal-Tar Medicinals," *DCM* 3 (October 4, 1916): 11-12; "Dye Industry Needs Protection," *WDM* 2 (July 19, 1916): 8, 16; "Indigo and Alizarin Exception Would Paralyze U.S. Dye Industry," *WDM* 2 (August 2, 1916): 6-7.

23. "American Dyestuff Manufacture," *JIEC* 8 (October 1916): 950-52; Hill to *New York Times*, October 1, 1916; Herty to Erwin, February 8, 1917, CHH, box 81.

24. "President Acts to Supervise All Trading with Foe," *New York Times*, October 15, 1917, p. 1; House Committee on Interstate and Foreign Commerce, *Trading with the Enemy* (1917), 4.

25. Executive Order 2729A, "Vesting Power and Authority in Designated Officers and Making Rules and Regulations under Trading with the Enemy Act and Title VII of the Act Approved June 15, 1917," October 12, 1917, reprinted in *Report of the Alien Property Custodian, 1918-1919*, 499-504; "President Acts to Supervise All Trading with Foe," *New York Times*, October 15, 1917, p. 1.

26. House Committee on Interstate and Foreign Commerce, *Hearings, Trading with the Enemy* (1917), 8, 13, 37. Warren, who later became a renowned legal scholar and historian, had spent 1914-1917 as the Department of Justice's point man on enforcing the neutrality laws. "Charles Warren, Law Expert, Dies," *New York Times*, August 17, 1954, p. 21; L. H. Woolsey, "Charles Warren," *American Journal of International Law* 49, no. 1 (January 1955): 50-54.

27. House Committee on Interstate and Foreign Commerce, *Hearings, Trading with the Enemy* (1917), 5-6, 13-14. The law resembled measures established by the belligerent European countries and by the United States in the Civil War. By 1917, the policies of Germany, France, and Great Britain had all evolved to become increasingly punitive. In France, the court system oversaw the initial sequestration of enemy property and ran 12,000 enemy-owned firms by mid-1915. In Germany, national and state officials shared administration of enemy property, and they often adopted policies in retaliation to Allied policies. Already in September 1914, Great Britain passed its own Trading with the Enemy Act that led quickly to registration of enemy property, the appointment of trustees, the trustees' power to sell enemy property, the right to use enemy patents, and a "black list" of firms with which it was unlawful to conduct business. The Board of Trade supervised the administration of enemy property. In 1916, Britain and France consulted and jointly decided to permit liquidation of German property, which they hoped would damage both German morale and its economy, and Germany responded in kind. When the Americans passed their Trading with the Enemy Act in 1917, it most resembled the British law. In deference to the great number of German Americans and in contrast to the European laws, however, the United States did not include law-abiding Germans who resided within the United States. U.S. history also contained precedents for the confiscation of enemy property, and American courts consistently ruled such wartime measures legal, even if they countered international law. Gathings, *International Law and American Treatment of Alien Enemy Property*, 46-59, 65-66.

28. Liebenau, *Medical Science and Medical Industry*, 113-19; Schamberg testimony in Senate Committee on the Judiciary, Subcommittee on the Judiciary, *Hearings, Alleged Dye Monopoly* (1922), 982-1006; "Ask for Permit to Make Salvarsan in America," *WDM* 2 (October 20, 1915): 12-13; House Committee on Interstate and Foreign Commerce, *Hearings, Trading with the Enemy* (1917), 88-91; "P. A. B. Widener, Capitalist, Dies," *New York Times*, November 7, 1915, p. 5.

29. Liebenau, *Medical Science and Medical Industry*, 109-18; Metz testimony in *USA v. The Chemical Foundation, Inc.*, 3:1873-1907; Schamberg testimony in *Hearings, Alleged Dye Monopoly* (1922): 979-97.

30. On the medical profession's evolution in attitude toward patents, see Gabriel, "A Thing Patented Is a Thing Divulged"; Mayo Clinic letter in Senate Committee on Patents,

Hearings, Salvarsan (1917), 5; Schamberg testimony in Senate Committee on the Judiciary, *Hearings, Alleged Dye Monopoly*, 918.

31. Metz testimony in Senate Committee on Patents, *Hearings, Salvarsan* (1917), 20–51, quotation on 22; "H. A. Metz's Offer to the Government," *Pharmaceutical Era* 50 (July 1917): 238.

32. Frank Eldred to Herty, August 9, 1917, CHH, box 73; "The Question of Using German Patents," *DCM* 3 (June 27, 1917): 31.

33. "Trading with the Enemy Act," Public Law 91, 65th Cong., 1st sess. (October 6, 1917); President Woodrow Wilson, "Executive Order Vesting Power and Authority in Designated Officers and Making Rules and Regulations under Trading with the Enemy Act and Title VII of the Act Approved June 15, 1917," Executive Order No. 2729A, October 12, 1917, both printed in *Report of the Alien Property Custodian, 1918-1919*, 483–504.

34. Before the war, Salvarsan had been recognized as a dangerous, if effective, treatment. Over 100 deaths had been attributed to Salvarsan prior to the war, a statistic attributed in part to its difficulty to administer. Ward, "The American Reception of Salvarsan," 59–60; "Use of Foreign Patents in United States," *DCM* 4 (March 13, 1918): 11–12; Metz testimony in Senate Committee on Patents, *Hearings, Salvarsan* (1917), 29–30. Besides Stieglitz, the committee included George W. McCoy of the Public Health Service and Edward S. Rogers, a Chicago patent attorney. Stieglitz testimony in *USA v. The Chemical Foundation, Inc.*, 5:3609; several letters between Stieglitz and Metz, recorded in Senate Committee on the Judiciary, *Hearings, Alleged Dye Monopoly* (1922), 828–35, and *USA v. The Chemical Foundation, Inc.*, Defendant's Exhibits Nos. 168–74, 374, vol. 12 (exhibits vol. 5): 2997–3006, 3489–90, and Plaintiff's Exhibits Nos. 669–78, vol. 10 (exhibits vol. 3): 1549–51, 1721–31. In his laboratory, Stieglitz had the aid of Julius Kahn, who joined the Mallinckrodt firm in the postwar years.

35. Coben, *A. Mitchell Palmer*, 7, 8–15, 127–54, 149; Palmer testimony in *USA v. Chemical Foundation, Inc.*, 4:2506; Higham, *Strangers in the Land*, 229–30; "Tribute from Cummings," *New York Times*, May 12, 1936, p. 26.

36. *Report of the Alien Property Custodian, 1918-1919*, 156; "How Seized German Millions Fight Germany," *New York Times*, January 27, 1918, p. 63.

37. Palmer testimony in Senate Committee on Appropriations, *Urgent Deficiency Appropriation Bill* (1918), 7–17; Palmer testimony, *USA v. The Chemical Foundation, Inc.*, 4:2511, 2536–37.

38. Palmer testimony in Senate Committee on Appropriations, *Urgent Deficiency Appropriation Bill* (1918), 6, 18; *Report of the Alien Property Custodian, 1918-1919*, 14, 25; "Palmer Is Biggest Big Business Man," *New York Times*, December 15, 1918, p. 27.

39. *Report of the Alien Property Custodian, 1918-1919*, especially chaps. 5, 7; "German Insurance Companies Liquidate," *New York Times*, January 21, 1918, p. 14; "Bosch Plants Taken Over," *New York Times*, April 20, 1918, p. 17.

40. House Committee on Ways and Means, *Hearings, Dyestuffs* (1919), 246–47 (excerpting a passage from Palmer's testimony in an earlier appropriations hearing). At the end of the trial, a short but illuminating exchange took place between Garvan and Irénée du Pont, president of Du Pont, over the Chemical Foundation's bill at the Hotel Du Pont in Wilmington, where both sides lodged their lawyers and witnesses. "You surely are pugnacious," du Pont wrote to Garvan after Garvan informed him that he, du Pont, would be paying the bill. Du Pont won that small skirmish, but not without exposing the tensions between the allies and revealing the kind of pressure Garvan could apply. Irénée

du Pont to Francis P. Garvan, October 18, 1923, and Garvan to du Pont, October 18, 1923, Hagley Museum and Library, Accession 1662, box 17. It is a measure of chemists' respect for Garvan that in 1929 the American Chemical Society made him the only nonchemist to win the ACS's Priestley medal. The origin of the Francis P. Garvan Award, given to prominent women chemists since 1937, is discussed in Rossiter, *Women Scientists in America*, 308. Garvan's wife was Mabel Brady Garvan, eventual patron and donor of art to Yale University and a member of the Brady family that made their fortune on Boston utilities. The wealth of Garvan's parents is noted in Tempest, "Francis P. Garvan." Whalen also discusses them in "American Decorative Art Studies at Yale and Winterthur: The Politics of Gender, Gentility, and Academia." In one of his later speeches Garvan suggested, rather dramatically, that "internationalism is like brotherly love taken to the point of homosexuality." Untitled, undated speech, FPG, box 12.

41. "Memorandum In re: Trust No. 6781 (The Bayer Co.)," March 14, 1918; Hardy to F. B. Lynch, December 18, 1917, NARA, RG 131, entry 134, box 3.

42. "Memorandum, The Bayer Company, Trusts 11698-6781, Extracts from Letter of F. B. Lynch," memorandum is undated; Lynch's letter is March 4, 1918; Covington to H. A. Dunn, APC Division of Audits Chief, July 13, 1918; Dunn to Lybrand, Ross Brothers, and Montgomery, July 25, 1918, NARA, RG 131, entry 134, box 3.

43. Hardy to Williams & Crowell, July 17, 1918; "Examination of Dr. Rudolph Hutz by Special Agent Muccino of the Providence Office of the U.S. Department of Justice," n.d. (but likely August 1918), NARA, RG 131, entry 7, box 7.

44. "Deposition of H. C. A. Seebohm (continued)," July 13, 1918; "Report on Williams & Crowell Company, Inc.," July 15, 1918; "Matter of Williams & Crowell Color Co., Inc.," July 15, 1918; "Deposition of Mr. Charles J. Hardy before Francis P. Garvan, Esq.," July 12, 1918, NARA, RG 131, entry 7, box 7.

45. House Committee on Ways and Means, *Hearings, Dyestuffs* (1919), 251. The specific quote came from 1919; the sentiment is expressed by mid-1918.

46. "Statement of Mr. H. C. A. Seebohm Before Louis F. Muccino, Special Agent United States Department of Justice," July 30, 1918, quotation on p. 18, NARA, RG 131, entry 157, box 1.

47. Senate Committee on the Judiciary, *Brewing and Liquor Interests and German and Bolshevik Propaganda*, 2:2026-27; Seebohm part (p. 8) of "Memorandum of Statement made by Dr. Arthur Franz Felix Mothwurf, August 28, 1918," NARA, RG 131, entry 157, box 1; Senate Committee on Finance, *Hearings, Dyestuffs* (1920), 511; Senate, *Hearings, Alleged Dye Monopoly* (1922), 223.

48. J. Harry Covington to Alien Property Custodian, December 13, 1918, and December 11, 1918, NARA, RG 131, entry 155, box 35; "German Plot for After-War Trade Bared by Palmer," *New York Times*, August 22, 1918, p. 1.

49. Lee Bradley testimony, Senate Committee on Appropriations, *Hearings on First Deficiency Appropriation Bill, 1919*, 82, 85; "An Act Making Appropriations to Supply Deficiencies in Appropriations for the Fiscal Year Ending June 30, 1919," Public Law 233, 65th Cong., 2nd sess., November 4, 1918.

50. J. Harry Covington to Alien Property Custodian, December 11, 1918, NARA, RG 131, entry 155, box 35; *Report of the Alien Property Custodian, 1918-1919*, 46.

51. The figures are estimated replacement costs in 1918. Ford, Bacon, and Davis, "The Bayer Company," November 7, 1918, NARA, RG 131, entry 155, box 34; Alien Property Custodian, "Return of Sale of Corporation," April 8, 1919; "Report of Investigation of Purchaser,"

January 7, 1919, NARA, RG 131, entry 155, box 35; Haynes, *American Chemical Industry*, 3:259, 6:175–76; Mann and Plummer, *The Aspirin Wars*, 50–51.

52. "McCombs Will Quit as Committee Head," *New York Times*, April 25, 1916, p. 4; Coben, *A. Mitchell Palmer*, 57, 132–33, 140–41, 247, 298. Palmer's personal assistant since 1912, Joseph Guffey, became the APC's director of sales. In December 1922, Guffey was indicted for embezzling; he used APC trust funds as personal loans. Coben writes in his footnote: "Guffey's indictment was later quashed under suspicious circumstances," and he served in the U.S. Senate for Pennsylvania from 1935 to 1947.

53. Deposition of George Simon, 1918, NARA, RG 131, entry 7, box 3; *Hodgskin and Simon v. USA*; Palmer and Garvan, *Aims and Purposes of the Chemical Foundation*, 45; Ryan purchase: Haynes, *American Chemical Industry*, 3:324.

54. Thomas Merriless, Department of Justice accountant, to Frank Burke, Bureau of Investigation, Department of Justice, May 21, 1920; Royal H. Weller, lawyer, to Francis P. Garvan, May 28, 1920; *Hodgskin and Simon v. USA*, typescript version, pp. 13 and 17, NARA, RG 131, entry 7, box 3.

55. Metz testimony in Senate Committee on the Judiciary, *Hearings, Alleged Dye Monopoly*, 821–25, 847–57; Wolfe, "Preliminary Report on Farbwerke-Hoechst Co. of New York," in *USA v. The Chemical Foundation, Inc.*, n.d. (but late 1918 or early 1919), Defendant's Exhibit No. 52, 11 (exhibits vol. 4): 2332–39.

56. Greeley and Giles, accountants, report on Merck & Co., January 1, 1914 to September 30, 1918, submitted February 21, 1919, NARA, RG 131, entry 134, box 27; Galambos and Sturchio, "The Origins of an Innovative Organization: Merck & Co., Inc., 1891–1960."

57. George Merck to Alien Property Custodian, April 3, 1918, petition to transfer and deliver 8,000 shares of stock; Edward Green, Merck & Company lawyer, 1918 memorandum, reproduced in Elizabeth G. Hayward, *History of Merck, 1908-1927*, vol. 5, 1952–53, appendices 8 and 6, Merck & Company Archive, Whitehouse, New Jersey. In the 1950s, Hayward collected material on the history of Merck, and the archive contains several manuscript volumes of her compilations and summaries. In his April 3, 1918, petition, George Merck did not concede German ownership of the 8,000 shares, because E. Merck owed him money and saw the 8,000 shares as collateral against his assets in Germany. Greeley and Giles, accountants, report on Merck & Co., January 1, 1914 to September 30, 1918, submitted February 21, 1919, NARA, RG 131, entry 134, box 27. This story is well told in Galambos and Sturchio, "Transnational Investment: The Merck Experience, 1891–1925," 238–42.

58. Francis P. Garvan to Advisory Sales Committee, Office of Alien Property, June 5, 1919; "Trust Agreement between George Merck, Goldman, Sachs & Co., and Lehman Brothers, parties of the first part, and James N. Wallace, Frederick J. Horne, and Frank L. Crocker, parties of the second part," July 23, 1919, NARA, RG 131, entry 155, box 187; APC public notice of sale, *New York Times*, April 23, 1919, p. 12; Francis P. Garvan to Advisory Sales Committee, Office of Alien Property, June 5, 1919; "Trust Agreement between George Merck, Goldman, Sachs & Co., and Lehman Brothers, parties of the first part, and James N. Wallace, Frederick J. Horne, and Frank L. Crocker, parties of the second part," July 23, 1919, NARA, RG 131, entry 155, box 187; *Report of the Alien Property Custodian, 1918-1919*, 59; Hayward, *History of Merck, 1908-1927*, 5:31–54, appendices 6 and 8; Greeley and Giles, report on Merck & Co., February 21, 1919. Merck continued lawsuits under TWEA's Section 9, *Annual Report of the Alien Property Custodian for the Year 1924*, 36, 84, 90, 138.

Chapter Six

1. Many works have addressed the theoretical justifications for patent systems. See, e.g., Kaufer, *The Economics of the Patent System;* Machlup, *The Political Economy of Monopoly,* 280-86; and Machlup, *An Economic Review of the Patent System.* On the issue of corporations, see Scherer et al., *Patents and the Corporation,* 56-66.

2. Palmer also ran an unsuccessful presidential campaign in 1920. Coben, *A. Mitchell Palmer,* 136-38, 149, 201-8.

3. Garvan testimony in *USA v. The Chemical Foundation, Inc.,* 4:2790-2803; Haynes, *American Chemical Industry,* 3:260; C. C. Cooper, Du Pont's Development Department, to R. R. M. Carpenter, Du Pont vice president, November 25, 1918, especially pp. 10-11, Hagley Museum and Library, Accession 2091, box 14.

4. Garvan testimony in *USA v. The Chemical Foundation, Inc.,* 5:2976-3005; "Conference of Dye Manufacturers Held at the Office of the Alien Property Custodian," January 27, 1919, recorded in *USA v. The Chemical Foundation, Inc.,* Plaintiff's Exhibit No. 28, 8 (exhibits vol. 1): 85-109. Although research has not revealed a direct connection between the Chemical Foundation and the Research Corporation, the organizers of the Chemical Foundation may have had the model of the Research Corporation in mind. Frederick G. Cottrell, a mining engineer, founded the Research Corporation in 1912 to manage patents and royalties for the purpose of funding research. Initially, Cottrell's own patents formed the core of the organization, but he and his advisors designed the corporation with the hope and expectation that other scientists and engineers would turn over their patents to it. The Research Corporation would manage the patents through licensing, creating a steady stream of income from successful patents to support a variety of research projects. On the Research Corporation see Cameron, *Cottrell: Samaritan of Science;* Bush, "Frederick Gardner Cottrell, 1877-1948"; Joseph Warren Barker, *Research Corporation (1912-1952);* Servos, *Physical Chemistry from Ostwald to Pauling,* 242; Cornell, *Establishing Research Corporation;* Mowery, Nelson, Sampat, and Ziedonis, *Ivory Tower and Industrial Innovation,* 58-84; and LeCain, *Mass Destruction,* 88-89. For a list of stockholders and administrators of the Research Corporation, which included Arthur D. Little and T. Coleman du Pont, see Frederick G. Cottrell, "The Research Corporation, An Experiment in Public Administration of Patent Rights," *JIEC* 4 (December 1912): 866.

5. Will T. Gordon testimony in *USA v. The Chemical Foundation, Inc.,* 2:1050-70; Haynes, *American Chemical Industry,* 3:260; "The Chemical Foundation, Inc.: Schedule of All Licenses Issued and Amount of Royalties Paid on Said Licenses from Date of Incorporation February 19, 1919 to December 31, 1936," FPG, box 17. On total sales, see John W. H. Crim to Henry M. Daugherty, June 29, 1922, NARA, RG 60, Central Files, entry 114 Correspondence, Case 9-17-10-12, box 748.

6. Garvan, "The German Menace," in *Aims and Purposes of the Chemical Foundation,* 61-63; Gordon testimony in *USA v. The Chemical Foundation, Inc.,* 2:1061-62; Garvan testimony in *USA v. The Chemical Foundation, Inc.,* 4:2698-2715, 5:3069-81. A listing of the patents turned over to the Chemical Foundation and Sterling Products is in *Report of the Alien Property Custodian* (1922), as is a listing of the many other properties in the APC's trusts.

7. For the 1921 figures, see "The Chemical Foundation, Inc.: Statement of Costs of Properties Acquired from Alien Property Custodian," recorded in *USA v. The Chemical Foundation, Inc.,* Defendant's Exhibit No. 92, 11 (exhibits vol. 4): 2666; for the licensees, see

"Licensees of the Chemical Foundation and What They Manufacture," December 31, 1922, recorded in *USA v. The Chemical Foundation, Inc.*, Defendant's Exhibit No. 89, 11 (exhibits vol. 4): 2653–54; on royalties collected, see "Comparison of Value of Dyes, Domestic and Imported, Controlled by the Patents Owned by the Chemical Foundation," recorded in *USA v. The Chemical Foundation, Inc.*, Defendant's Exhibit No. 117, 12 (exhibits vol. 5): 2703–4.

8. "Statement of Income and Expenses by Yearly Periods from Beginning of Operation to December 31, 1922," recorded in *USA v. The Chemical Foundation, Inc.*, Plaintiff's Exhibit No. 639, 10 (exhibits vol. 3): 1547–48; *Eighth Annual Report of the United States Tariff Commission, 1924*, 11; Garvan testimony in *USA v. The Chemical Foundation, Inc.*, 5:2873–2906. The numbers reflected the American industry's production of dyes in which the patents had already expired.

9. Palmer and Garvan, *Aims and Purposes of the Chemical Foundation*. Shortly after the Office of Alien Property and the American Dyes Institute formulated the plan for the Chemical Foundation, A. Mitchell Palmer submitted a final report on the activities of the Custodian's office during his tenure. His report, warning of German spies "permeating" the chemical industry, emphasized America's continuing dependence on Germany unless the Custodian dealt wisely with the dyes and pharmaceuticals patents. The published report, bound with Garvan's "The German Menace" in *Aims and Purposes*, circulated widely as a promotional tool. Historian David J. Rhees has discussed the significance of the Chemical Foundation in its support of professional chemists' attempts to popularize chemistry and raise the status of the profession in chapter 6 of "The Chemists' Crusade: The Rise of an Industrial Science in Modern America, 1907–1922," 286–306. "Literature Distributed by The Chemical Foundation, Inc.," recorded in *USA v. The Chemical Foundation, Inc.*, Defendant's Exhibit No. 318, 12 (exhibits vol. 5): 3421–36; "Address of Francis P. Garvan Delivered before the Joint Session of the Society of Chemical Industry and the American Chemical Society at Columbia University," September 7, 1921, uncited copy in FPG, box 14.

10. Coben, *A. Mitchell Palmer*, 131–32. Creel's Committee on Public Information was the leading wartime propaganda agency in the government, which, among other things, published accusations of disloyalty and recruited academic scholars to "educate" the public about the war. Murphy, *World War I and the Origin of Civil Liberties in the United States*, 107–16. On expenses, see "Statement of Income and Expenses by Yearly Periods from Beginning of Operation to December 31, 1922," recorded in *USA v. The Chemical Foundation, Inc.*, Plaintiff's Exhibit No. 639, 10 (exhibits vol. 3): 1547–48. See also "Literature Distributed by the Chemical Foundation, Inc.," recorded in *USA v. The Chemical Foundation, Inc.*, Defendant's Exhibit No. 318, 12 (exhibits vol. 5): 3421–36; and "Address of Francis P. Garvan Delivered before the Joint Session of the Society of Chemical Industry and the American Chemical Society at Columbia University," September 7, 1921, copy in FPG, box 14. For operating costs see "Statement of Salaries and Wages" and "Statement of Educational Literature, Salaries, and Wages," recorded in *USA v. The Chemical Foundation, Inc.*, Plaintiff's Exhibit Nos. 642–43, 10 (exhibits vol. 3): 1553–56.

11. Stieglitz to Rogers, June 2, 1919; Stieglitz to Garvan, July 3, 1919; Stieglitz to S. A. Tucker (Chemical Foundation), October 1, 1919, FPG, box 64, Edward S. Rogers folder; John F. Oberlin and Mark Putnam (Dow Chemical Company) to Chemical Foundation, September 9, 1919; H. E. Bateman (Chemical Foundation) to Douglas McKay (Chemical Foundation), September 9, 1919; Herbert Dow to McKay, August 15, 1919, and Dow to J. K. L. Snyder (Chemical Foundation), October 10, 1919, FPG, box 94, Dow Chemical Company folder.

12. Hawley, "Herbert Hoover, the Commerce Secretariat, and the Vision of an 'Associative State,' 1921-1928"; Hawley, *The Great War and the Search for a Modern Order*.

13. Treaty of Versailles, Article 236 and Annex VI. The treaty contained provisions for other industries as well; the chemical industry was important but not entirely unique.

14. Marks, "Smoke and Mirrors: In Smoke-Filled Rooms and the Galerie des Glaces," 337-38; Kent, *The Spoils of War*, 1-13; Patton, "The German Chemical Industry and Reparations in Kind (1919-1924)." Historian Adam Tooze highlights the slipperiness of the numbers used to measure the German economy in "Trouble with Numbers: Statistics, Politics, and History in the Construction of Weimar's Trade Balance, 1918-24."

15. Leonard A. Yerkes to Baruch, April 1, 1919, copied in Yerkes to Du Pont, April 4, 1919, FPG, box 5. Lawrence E. Gelfand notes the size of the American delegation in "The American Mission to Negotiate Peace: An Historian Looks Back," 189; Yerkes to Du Pont, April 4, 1919, FPG, box 5.

16. Yerkes to Du Pont, April 4, 1919, FPG, box 5; Frank L. Polk to Department of State, September 12, 1919, CHH, box 85; Theodore W. Sill, "The Dyestuff Plants and Their War Activities," *JIEC* 11 (June 1919): 509-12. In 1919, Amos Fries testified that he "sent one of my best engineers into the occupied territory to find out everything we could about chemicals being manufactured there," and he may have been referring to Sill. Fries testimony in House Committee on Ways and Means, *Hearings, Dyestuffs* (1919), 38.

17. Yerkes to Bernard Baruch, April 1, 1919, copied in Yerkes to Du Pont, April 4, 1919, FPG, box 5. Garvan obtained a summary of the detailed British report, "The Chemical Industries of German Rhineland," March 1920, FPG, box 77. Also see Hartley et al., "Report of the British Mission Appointed to Visit Enemy Chemical Factories in the Occupied Zone Engaged in the Production of Munitions of War," recorded in *USA v. The Chemical Foundation, Inc.*, Defendant's Exhibit No. 321, 12 (exhibits vol. 5): 3437-48. The British excursion into the German plants is discussed in Jeffrey Johnson and Roy MacLeod, "The War the Victors Lost: The Dilemmas of Chemical Disarmament, 1919-1926," 224-27; also see Haber, *The Poisonous Cloud*, 1-5; Bureau of Foreign and Domestic Commerce to William Corwine, ADI, August 16, 1919, NARA, RG 151, 232 Dyes (Germany), box 1036.

18. William J. Matheson to William C. Redfield, November 12, 1918; Matheson to Redfield, November 13, 1918; Chief of Bureau to Matheson, November 25, 1918, RG 151, 232 Dyestuffs (Germany), box 1035; Grosvenor M. Jones, BFDC, to Redfield, December 21, 1918, paraphrased in "Department of Commerce File," NARA, RG 60, entry 114 (Correspondence), box 749. For a short biographical sketch of Wigglesworth, who was born in Ireland in 1866, see "Henry Wigglesworth," *DCM* 18 (January 6, 1926): 10; Yerkes to Du Pont, April 4, 1919, FPG, box 5; Kaufman, *Efficiency and Expansion: Foreign Trade Organization in the Wilson Administration, 1913-1921*, 77-80, 219.

19. Yerkes to Du Pont, April 4, 1919 and April 8, 1919, FPG, box 5; Wigglesworth to Redfield, July 2, 1919, RG 151, 232 Dyes (Germany), box 1036.

20. Andrew Imbrie, U.S. Finishing Company, to Charles H. Herty, June 20, 1919; Cluett, Peabody & Company and four other shirt manufacturers to Clarence M. Woolley, WTB, June 28, 1919; C. S. Hawes, WTB, to Herty, July 8, 1919, CHH, box 85.

21. Clarence M. Woolley to Redfield, April 1, 1919; Baruch to George N. Peek, March 18, 1919; Wigglesworth to Department of Commerce, March 19, 1919; Wigglesworth to Burwell S. Cutler, Bureau of Foreign and Domestic Commerce, April 21, 1919, NARA, RG 151, 232 Dyestuffs (Germany), box 1036.

22. Clarence M. Woolley to Redfield, April 1, 1919; Baruch to George N. Peek, March 18, 1919; Redfield to Joseph H. Choate Jr., April 3, 1919; Choate and William T. Miller, April 4, 1919, NARA, RG 151, 232 Dyestuffs (Germany), box 1036.

23. Garvan to Baruch and Bradley Palmer, Alien Property Custodian representative in Paris, April 8, 1919; Garvan to Bradley Palmer, April 9, 1919; Baruch and Charles H. MacDowell to George N. Peek, March 27, 1919; Woolley to Vance C. McCormick, April 11, 1919, NARA, RG 151, 232 Dyes (Germany), box 1036.

24. Textile Alliance to Patterson, April 11, 1919, FPG, box 5; Garvan to Baruch and Bradley Palmer, Alien Property Custodian representative in Paris, April 8, 1919; Garvan to Bradley Palmer, April 9, 1919; Woolley to McCormick, April 11, 1919, NARA, RG 151, 232 Dyes (Germany), box 1036; McCormick to Woolley, April 18, 1919, copied in "The Files of the State Department and the War Trade Board," RG 60, Classified Subject File, entry 114 (Correspondence), box 749. Yerkes was more diplomatic than Wigglesworth and generally fostered a more congenial relationship with American officials, but he still sent periodic memos to them complaining that U.S. interests would suffer if the diplomats continued to exclude technical experts from the American dyes industry. Along with Yerkes, Du Pont sent Elmer Bolton and Eric Kunz to Paris. Yerkes to Baruch, April 1, 1919, copied in Yerkes to Du Pont, April 4, 1919; Yerkes to Du Pont, c. April 14, 1919; Yerkes to Charles H. MacDowell, April 22, 1919; A. M. Patterson to Clarence M. Woolley, April 21, 1919, FPG, box 5.

25. War Trade Board, "Extract of Proceedings of May 8, 1919," NARA, RG 151, 232 Dyes (Germany), box 1036.

26. *The Treaty of Versailles and After*, 380, 460, 516–18. A useful chronology of developments in reparations is U.S. Department of Commerce, Bureau of Foreign and Domestic Commerce, "The Reparation Problem, 1918–1924" (1924).

27. Frank Polk to Secretary of State, August 10, 1919, FPG, box 5.

28. David H. Miller, Department of State to Garvan, August 15, 1919; Miller, memo, August 18, 1919, FPG, box 5; Herty testimony in Senate Committee on the Judiciary, *Hearings, Alleged Dye Monopoly* (1922): 561–64.

29. C. E. Herring, Department of Commerce to P. Kennedy, December 9, 1919, NARA, RG 151, 232 Dyes (Germany), box 1036; War Trade Board to Herty, September 29, 1919, FPG, box 5; Kunz to W. Carpenter, Du Pont, October 2, 1919, FPG, box 5; Herty to Kunz, January 6, 1920, CHH, box 86; "Report of Edward S. Chapin, Paris representative of the Textile Alliance," September 29, 1920, CHH, box 87.

30. Kunz to W. Carpenter, Du Pont, October 2, 1919, FPG, box 5; Herty to Kunz, January 6, 1920, CHH, box 86; Herty testimony in Senate Committee on the Judiciary, *Hearings, Alleged Dye Monopoly* (1922), 561–64.

31. Albert M. Patterson testimony in *USA v. The Chemical Foundation, Inc.*, 2:950, 998; Haynes, *American Chemical Industry*, 3:264. During the war, the Textile Alliance supervised the importation of wool and other products from British territories. The British desired assurances that the products would not be resold and end up benefiting Germany. The Textile Alliance, as an organization of textile manufacturing associations, provided the assurance. Patterson testimony in Senate Committee on the Judiciary, *Hearings, Alleged Dye Monopoly (1922)*, 307; "Report of Edward S. Chapin, Paris representative of the Textile Alliance," September 29, 1920, CHH, box 87; Patterson testimony in *USA v. The Chemical Foundation, Inc.*, 2:964, 978–79; Patterson testimony in Senate Committee on the Judiciary, *Hearings, Alleged Dye Monopoly* (1922), 329–32; Patterson to Horace B. Cheney, March 1, 1928, CHH, box 87; William C. Dickson, Special Accountant, Department

of Justice, "Report of Investigation of the Textile Alliance Incorporated of New York City in Re. Reparation Dyes," NARA, RG 60, entry 233, box 1. The thieves included members of the Widder Dye and Chemical Company of Brooklyn and Chicago. "List of Stolen German Dyes," *DCM* 7 (July 21, 1920): 125; "Seventeen Arrests Follow Seizure of Stolen Textile Alliance Dyes," *DCM* 7 (September 15, 1920): 551; "No Conviction in 'Orange Cat' Cases," *DCM* 8 (June 15, 1921): 1317.

32. William Phillips, Department of State, to American Mission, September 8, 1919; Department of State to American Mission, October 8, 1919, FPG, box 5. Herty irritated the diplomats. In January 1920, officials in the War Trade Board section of the Department of State found Herty's editorials entirely too inflammatory for a member of its advisory committee. One official complained that despite requests to Herty that he "desist from making ridiculous public comments," Herty continued to engage in "vilifications and vituperations" that generated criticism of the War Trade Board. The War Trade Board officials considered asking Herty to resign from the advisory committee and also considered abolishing the committee simply to be rid of Herty. In the end, however, the officials decided that Herty "had too much influence with the dyes manufacturers." He and the committee stayed. St. John Perret to Van S. Merle-Smith, January 5, 1920, and margin notes from Merle-Smith and Van Dyke on the letter, copied in "The Files of the State Department and the War Trade Board," NARA, RG 60, Classified Subject File, entry 114 (Correspondence), box 749; "Why Are Dye Imports a Secret?" *DCM* 7 (July 7, 1920): 8; "Names of German Dye Importers and Consumers Held Confidential," *DCM* 7 (July 7, 1920): 11; Phillips to American Mission, September 8, 1919; Phillips to American Mission, September 27, 1919, FPG, box 5.

33. Herty to Garvan, October 6, 1919, FPG, box 5.

34. The treaty stipulated that the Allies could claim up to 25 percent of "normal" German production for fear that Germany would reduce output to minimize their obligation. The Allies subsequently debated the definition of "normal," declaring only that it did not mean "prewar." Through 1922, "normal" production in practice meant the actual production of the Germans. Joseph H. Choate Jr., to American Dyes Institute, November 15, 1920, recorded in Senate Committee on the Judiciary, *Hearings, Alleged Dye Monopoly* (1922), 533; "Report of Edward S. Chapin, Paris representative of the Textile Alliance," September 29, 1920, CHH, box 87. The percentages of dyes received are discussed in Chapin and also in "U.S. Share in Reparation Dyes $1,123,167," *DCM* 13 (October 17, 1923): 1027.

35. Choate to American Dyes Institute, November 9, 1920, November 15, 1920, November 19, 1920, November 26, 1920, and December 6, 1920, reprinted in Senate Committee on the Judiciary, *Hearings, Alleged Dye Monopoly* (1922), 530-37; "Report of Edward S. Chapin, Paris representative of the Textile Alliance," September 22, 1921, CHH, box 87.

36. The efficiency was not immediate. Bureaucratic challenges, difficulties in the German firms, and a dockworkers strike in Rotterdam kept the reparations dyes from arriving before early 1920. In addition, not all of the dyes arrived in workable quality, and the Textile Alliance had to accept returns or refunds. Of the indigo shipped to China, the Chinese received a refund of nearly 10 percent for poor quality dyes. St. John Perret to Van S. Merle-Smith, January 5, 1920, copied in "The Files of the State Department and the War Trade Board," NARA, RG 60, Classified Subject File, entry 114 (Correspondence), box 749; Textile Alliance Board of Directors, Excerpt of Minutes, December 8, 1921; Herty to Chapin, December 17, 1921, CHH, box 87. Kent, *The Spoils of War*, 373; E. A. Macon to Herty, April 6, 1922; Macon to Herty, June 7, 1922, CHH, box 87; A. C. Imbrie to Textile Alliance, January

28, 1922; E. A. Macon to Herty, April 7, 1922; Macon to Herty, June 7, 1922; Minutes of the Textile Alliance dyes committee, December 13, 1922, CHH, box 87; Reparations Commission, "1753. Sales of Dyestuffs to the United States of America," January 27, 1922, Hoover Papers, Department of Commerce, box 515. *The Treaty of Versailles and After*, 518.

37. "Imports by Messrs. Kuttroff, Pickhardt & Co.," *OPDR*, uncited table recorded in Senate Committee on the Judiciary, *Hearings, Alleged Dye Monopoly* (1922), 340; Macon to Herty, April 7, 1922, CHH, box 87; Patterson testimony in Senate Committee on the Judiciary, *Hearings, Alleged Dye Monopoly* (1922), 347.

38. The profit came from the dyes that the Textile Alliance purchased without specific advance orders from the United States; those with orders were sold at cost. Patterson testimony in *USA v. The Chemical Foundation, Inc.*, 2:964, 978-79; Patterson testimony in Senate Committee on the Judiciary, *Hearings, Alleged Dye Monopoly* (1922), 329; Patterson to Horace B. Cheney, March 1, 1928, CHH, box 87; "Minutes of the Second Meeting to Dispose of the Surplus," December 7, 1927; "The Textile Foundation," *Boston Herald* (June 21, 1930): 10; Herty to Frank A. Fleisch, April 22, 1930, CHH, box 87; Arthur N. Young, David E. Finley, and Edward Pickard, to the Secretaries of State, Treasury, and Commerce, "Disposition of the Surplus Monies of the Textile Alliance, Incorporated," April 13, 1927, Hoover Papers, Department of Commerce, box 598, folder Textiles, 1922-28; "Textile Alliance Foundation Established by Congress," *Chemical Markets* 27 (July 1930): 67; Dillon, *History, Textile Research Institute, 1930-1970*, 1-5; District Court of the United States, Southern District of New York, *U.S.A. v. Textile Alliance, Inc.*, June 4, 1930, NARA, RG 60, entry 233, box 1. This court case was a "friendly" one, with each side agreeing a judge's decree was necessary to clarify the legalities of the original agreement, primarily tax obligations. These records in RG 60, the Department of Justice, are quite complete in their review of the Textile Alliance and the subsequent distribution of money. As of this court case in 1930, the lawyers for the Textile Alliance, Coudert Brothers of New York City, still held the organization's papers.

39. U.S. Tariff Commission, *Census of Dyes and Other Synthetic Organic Chemicals, 1923*, 164-67; "Not to Import German Chemicals," *DCM* 7 (September 8, 1920): 484.

40. Marks, "Smoke and Mirrors: In Smoke-Filled Rooms and the Galerie des Glaces," 339-40. Elisabeth Glaser notes the exceptional place chemicals held in the American agenda in Paris in "The Making of the Economic Peace," 383.

41. The Longworth bill is printed in House Committee on Ways and Means, *Hearings, Dyestuffs* (1919), 5-7. On the lobbying plan, see Poucher and Choate testimony in Senate Committee on the Judiciary, *Hearings, Alleged Dye Monopoly* (1922), 504-7, 528.

42. Haber, *The Chemical Industry, 1900-1930*, 235-36, 242-45; Reader, *Imperial Chemical Industries*, 1:427-31; Breithut, "British Dyestuffs Industry," (1924), 3-11; Charles H. Herty, "The Race Is Not Always to the Swift," *JIEC* 13 (February 1921): 109; Choate testimony in Senate Committee on Finance, *Hearings, Tariff Act of 1921*, 1:683.

43. J. Harry Covington (Chemical Foundation lawyer) testimony in House Committee on Ways and Means, *Hearings, Dyestuffs* (1919), 7; J. H. Burns testimony, p. 37, and William Sibert, p. 11, in Senate Committee on Finance, *Hearings, Dyestuffs* (1920); Fries testimony in Senate Committee on Finance, *Hearings, Tariff Act of 1921* (1922), 375; Ralph Earle testimony in Senate Committee on Finance, *Hearings, Dyestuffs* (1920), 14; Sibert testimony in House Committee on Ways and Means, *Hearings, Dyestuffs* (1919), 32; Fries testimony in Senate Committee on Finance, *Hearings, Tariff Act of 1921* (1922), 380. Fries quoted from the British Hartley report.

44. Fries testimony in Senate Committee on Finance, *Hearings, Dyestuffs* (1920), 26. Herty and a few other leaders in the ACS also maintained close correspondence with Fries and Sibert in relation to the CWS's campaign within the Department of War to remain a separate entity within the military. CHH, boxes 96-97; Jones, "The Role of Chemists." Fries testimony in House Committee on Ways and Means, *Hearings, Dyestuffs* (1919), 41.

45. Sibert and O. M. Hustvedt testimony in House Committee on Ways and Means, *Hearings, Dyestuffs* (1919), 33, 41-42.

46. Herty to Longworth, June 23, 1919, CHH, box 82; Herty testimony in Senate Committee on Finance, *Hearings, Dyestuffs* (1920), 190. The French reportedly wanted to dismantle much of the German chemical industry. Patton, "The German Chemical Industry and Reparations in Kind (1919-1924)," 7. Garvan testimony in House Committee on Ways and Means, *Hearings, Dyestuffs* (1919), 297; Klipstein testimony in Senate Committee on Finance, *Hearings, Dyestuffs* (1920), 177.

47. Theodore W. Sill, "The Dyestuff Plants and Their War Activities," and Frederick Pope, "Condition of Chemical Plants in Germany," both in *JIEC* 11 (June 1919): 509-12 and 512; Sill testimony in Senate Committee on Finance, *Hearings, Dyestuffs* (1920), 154. Choate testimony in House Committee on Ways and Means, *Hearings, Dyestuffs* (1919), 500; Garvan's testimony in House Committee on Ways and Means, *Hearings, Dyestuffs* (1919), 222-300, 327-68, 354 (final quotation); Garvan's testimony in Senate Committee on Finance, *Hearings, Dyestuffs* (1920), 489-551.

48. Witnesses among the manufacturers included Henry Howard (Merrimac Chemical Company), Albert H. Hooker (Hooker Electrochemical Company), James T. Pardee (Dow Chemical Company), Henry Wigglesworth (National Aniline & Chemical Company), Henry B. Rust (Koppers Company), Irénée du Pont (E. I. du Pont de Nemours & Company), E. H. Killheffer (Newport Company), E. C. Klipstein (E. C. Klipstein & Sons), Samuel Isermann (Van Dyk & Company), and Herman A. Metz (Consolidated Color & Chemical Company) in Senate Committee on Finance, *Hearings, Dyestuffs* (1920), 61-141, 155-83, 207-16, 281-326.

49. Cheney testimony and Wilson testimony in House Committee on Ways and Means, *Hearings, Dyestuffs* (1919), 47-49, 43-46.

50. Frederick S. Dickson testimony in Senate Committee on the Judiciary, *Hearings, Alleged Dye Monopoly* (1922), 625-47; John P. Wood, Pequea Mills Company of Philadelphia, testimony in Senate Committee on Finance, *Hearings, Dyestuffs* (1920), 249-51; Arthur Land, Alexander Smith & Sons Carpet Company of Yonkers, New York, testimony in Senate Committee on Finance, *Hearings, Dyestuffs* (1920), 333. Philip Scranton discusses the textile industry's strategies for handling the frequent changes in textile demand in *Figured Tapestry: Production, Markets, and Power in Philadelphia Textiles, 1885-1941*.

51. Wood testimony in Senate Committee on Finance, *Hearings, Dyestuffs* (1920), 222.

52. Sykes testimony in ibid., 407-67; MacKinney testimony in House Committee on Ways and Means, *Hearings, Dyestuffs* (1919), 398.

53. Fordney commentary in House Committee on Ways and Means, *Hearings, Dyestuffs* (1919), 50-53, 426; Demming testimony in Senate Committee on Finance, *Hearings, Dyestuffs* (1920), 375.

54. Herty, "From the Senate Gallery," *JIEC*, 12 (June 1920): 522-24; "Chronology," submitted in Poucher testimony in Senate Committee on the Judiciary, *Hearings, Alleged Dye Monopoly* (1922), 507-12; Dickson testimony in Senate Committee on the Judiciary, *Hearings, Alleged Dye Monopoly* (1922), 625-27.

55. Compared to $7,689 in 1919 and $28,727 in 1920 (educational campaign). *USA v. The Chemical Foundation, Inc.*, Plaintiff's Exhibit No. 639, 10 (exhibits vol. 3): 1547-48. Among other things, the ADI sent hundreds of reprints of Herty's editorial, "From the Senate Gallery," to senators and textile interests. Corwine to Herty, May 20, 1920; Poucher to Herty, June 2, 1920; June 4, 1920, CHH, box 84; Corwine to Herty, July 28, 1920; August 5, 1920; August 12, 1920; Fannie L. Sornborger, New York State Federation of Women's Clubs, to American Dyes Institute, September 3, 1921, CHH, box 84. The significance of home economists as consumer advisors is discussed in Goldstein, *Creating Consumers: Home Economists and Twentieth-Century America*. E. F. Smith to Herty, November 7, 1921, November 8, 1921, CHH, box 83.

56. "Tariff Rates of New Law Compared with Old," *OPDR* 102 (September 25, 1922): 19-20; "Dye-Chemical Section Ceases Its Function," *OPDR* 102 (October 2, 1922): 19.

57. *Dictionary of American Biography*, s.v. "La Follette, Robert Marion." Frear referred to Du Pont and Allied Chemical, a new merger, which will be discussed in chapter 8. Frear speech, *Congressional Record*, July 11, 1921, copy in CHH, box 83.

58. King opening statement in *Hearings, Alleged Dye Monopoly* (1922), 1-4, 22; *National Cyclopedia of American Biography*, 34:88-89, s.v. "King, William Henry."

59. King opening statement in Senate Committee on the Judiciary, *Hearings, Alleged Dye Monopoly* (1922), 176-77. The chart is opposite p. 22. With partial funding from the American Dyes Institute and input from the Chemical Warfare Service, the National Research Council built two exhibits to highlight the "interdependence" of chemistry, the chemical industry, and national defense, explained Harrison Howe, who supervised the exhibits. He argued that the purpose was to educate, not to disperse propaganda, although presumably one exhibit's appearance in the House Office Building's rotunda aimed for a particular audience, which industry advocates especially hoped to educate. This exhibit was later given to the Smithsonian Institution. Howe testimony in Senate Committee on the Judiciary, *Hearings, Alleged Dye Monopoly* (1922), 670-84. Du Pont testimony in Senate Committee on Finance, *Hearings, Dyestuffs* (1920), 163-64; King opening statement in Senate Committee on the Judiciary, *Hearings, Alleged Dye Monopoly* (1922), 13-19, 47-71, 97-100, 122-62.

60. King opening statement in Senate Committee on the Judiciary, *Hearings, Alleged Dye Monopoly* (1922), 169-70.

61. King opening statement and Frelinghuysen statement in ibid., 6, 23-24.

62. Garvan testimony in ibid., 203-95, 355-450. For the Department of Justice investigation, see chapter 7.

Chapter Seven

1. Taussig, "The Tariff Act of 1922," (1922) 15, 17, 20. Taussig explores the retreat of free trade ideas and the rise of protectionist theories in more depth in *Free Trade, the Tariff, and Reciprocity*. This book contains his main critique of theories supporting the equalization of costs of production, pp. 134-36, which appeared in the decade before World War I. Endorsement for the flexible provision came from the editors of *Chemical and Metallurgical Engineering*, who hoped the provision was the "first step" toward "a tariff controlled and administered by an impartial, non-political agency." 27 (September 27, 1922): 625.

2. Haynes, *American Chemical Industry*, 3:255, 272; Harry E. Danner testimony in *USA v. The Chemical Foundation, Inc.*, 2:1003-8.

3. Haynes, *American Chemical Industry*, 3:272-73.

4. Two letters from Herty to Edgar F. Smith, October 31, 1921, CHH, box 74. One letter was Herty's official resignation as the editor of *JIEC*. The second letter, more personal than the first, explained in greater detail Herty's discussions with SOCMA's organizers. The Herty papers collectively demonstrate the extent of Herty's contacts.

5. Woodford received $5,000. SOCMA, "Schedule 1 (a): Payroll," c. July 1926, CHH, box 74. The median wage for chemists in the industry was between $2,500 and $4,000 per year in 1922. U.S. Tariff Commission, *Census of Dyes and Other Synthetic Organic Chemicals, 1922*, 65. As with the other parts of this story involving Herty, a more complete discussion of his actions is available in Reed, *Crusading for Chemistry*. Her chapter 7 covers the years with SOCMA.

6. SOCMA Board of Governors, "Minutes of Meeting," November 18, 1921, and March 30, 1922; SOCMA, "1926 Assessment List," c. July 1, 1926, CHH, box 74; "Synthetic Organic Chemical Manufacturers Association of the United States," *JIEC* 13 (December 1921): 1168.

7. *Eighth Annual Report of the U.S. Tariff Commission, 1924*, 12-13; Goldstein, *Ideas, Interests, and American Trade Policy*, 123; "Extract from [SOCMA] Counsel Report," c. May 10, 1923, CHH, box 75.

8. F. J. H. Krack, Appraiser, to Herty, September 22, 1922, in SOCMA Minutes of the Board of Governors, September 25, 1922, CHH, box 74.

9. Krack to Senator James W. Wadsworth Jr., March 24, 1924, CHH, box 75; untitled summary of meeting between Deputy Appraiser John J. Donnelly and importing representatives, December 1, 1922, CHH, box 83; Donnelly memorandum to appraiser, December 4, 1922, CHH, box 83. The rumors of cheating appeared in "The Pot and the Kettle," *DCM* 12 (January 10, 1923): 73.

10. "Extract from [SOCMA] Counsel Report," c. May 10, 1923; Krack to Wadsworth, March 24, 1924, CHH, box 75.

11. Harrison F. Wilmot, SOCMA, "Annual Report of the Technical Advisor," c. 1924, CHH, box 77; Herty, SOCMA, "President's Annual Report," December 11, 1925, CHH, box 74; SOCMA Dyestuffs Section, "Minutes of Meeting," January 11, 1924, CHH, box 75; SOCMA, "Bulletin No. 29," July 17, 1925; SOCMA, "Special Bulletin No. 5," January, 31, 1923; SOCMA, "Special Bulletin No. 6," n.d. (c. February-June, 1923); SOCMA, "Special Bulletin No. 9," October 17, 1923; SOCMA, "Special Bulletin No. 10," November 14, 1923; SOCMA, "Special Bulletin No. 16," July 2, 1924; SOCMA, "Special Bulletin No. 18," January 5, 1925, CHH, box 75. The dyes standards list put out by the U.S. Tariff Commission appeared regularly in *DCM*; for example, "Dye Standard List No. 12 Issued," *DCM* 15 (November 12, 1924): 1459.

12. Prior to joining SOCMA, Wilmot was affiliated with several firms, including S. R. David and Company, Kalle, A. Klipstein and Company, and Tinc Tura Laboratories. Haynes, *American Chemical Industry*, 4:239-40. In the SOCMA minutes, his affiliation with importers was overlooked, although certainly it was that experience that made him qualified to assist the manufacturers with the customs office. SOCMA Dyestuffs Section, "Minutes of Meeting," December 14, 1922, CHH, box 75; Marion DeVries was formerly a judge on the U.S. Court of Customs Appeals. His partner, Thomas L. Doherty, and another lawyer in the firm, George F. Lamb, were the two primary lawyers serving SOCMA on customs issues. SOCMA, "Bulletin No. 5," December 2, 1922; SOCMA, "Special Bulletin No. 18," January 5, 1925, CHH, box 75.

13. Wilmot, SOCMA, "Annual Report of the Technical Advisor," c. 1924, CHH, box 77.

14. SOCMA, "Total Disbursements [November 1921–June 1926]," c. July 1926, CHH, box 74.

15. The most concise list of the importers' complaints was their brief filed with the House Committee on Ways and Means in 1929 when the Smoot-Hawley tariff was under consideration. House Committee on Ways and Means, *Hearings, Tariff Readjustment, 1929,* 1:430–45.

16. "Analysis of Coal Tar Dye and Color Imports for 1928," compiled from the 1928 Department of Commerce Monthly Import Bulletins by SOCMA, c. 1929, CHH, box 77.

17. House Committee on Ways and Means, *Hearings, Tariff Readjustment, 1929,* 1:424–26.

18. Glassie, a member of the U.S. Tariff Commission, "Some Legal Aspects of the Flexible Tariff, Parts I and II"; Welch, "The Flexible Provisions of the Tariff Act."

19. *Seventh Annual Report of the United States Tariff Commission, 1923,* 38–39; *Eleventh Annual Report of the United States Tariff Commission, 1927,* 7.

20. "Europe Dislikes Chemical Envoys, Says Bobst in Talk to Salesmen," *DCM* 13 (November 14, 1923): 1314, 1325; "Cost Inquiries Resented," *DCM* 16 (April 22, 1925): 1171.

21. Edouard Neron, "America's Customs Inquisition," *DCM* 18 (February 11, 1926): 391–92; Stowell, "Tariff Relations with France."

22. The observer continued: the Americans aggravated their offense in France by sending agents "bearing a German name and speaking the language with a decided German accent." "Agents of U.S. Tariff Commission Not Welcome in French Factories," *DCM* 14 (March 19, 1924): 913.

23. "U.S. Tariff Agents Seeking British Costs of Production Meet Rebuffs," *DCM* 14 (April 16, 1924): 1177.

24. "Europe Dislikes Chemical Envoys Says Bobst in Talk to Salesmen," *DCM* 13 (November 14, 1923): 1314, 1325.

25. No title, but short paragraph begins, "The U.S. Tariff Commission's agents," *DCM* 17 (August 19, 1925): 535; "Protection or Espionage," *DCM* 18 (February 11, 1926): 387.

26. The importers for the Norwegian firm pursued a backdoor strategy in their case. In protesting American officials' demands to see the Norwegian costs of production, the importers demanded to see the cost of production information the tariff officials had gathered from a U.S. producer. Ironically, the Supreme Court ruled that production costs were justifiably deemed trade secrets and should not be revealed. The Norwegians won their point but lost the case. *Norwegian Nitrogen Products Co. v. United States*, 288 U.S. 294 (1933), *Supreme Court Reporter*, October term, 1932, pp. 350–62; "Flexible Tariff," *Chemical Markets* 20 (April 7, 1927): 508.

27. The classic discussion of the Department of Commerce and its relationship to trade associations is Hawley, "Herbert Hoover, the Commerce Secretariat, and the Vision of an 'Associative State,' 1921–1928."

28. Hoover, "Synthetic Organic Chemicals Manufacturers Association, Washington, D.C.—Address," October 28, 1921, Hoover Papers, speech no. 182.

29. H. E. Howe, minutes of the Chemical Advisory Committee meeting, June 7, 1926, Hoover Papers, Department of Commerce, box 129, folder 02232; Howe, minutes of the Chemical Advisory Committee meeting, December 12, 1927, Hagley Museum and Library, Acc. 1662, box 17.

30. *Tenth Annual Report of the Secretary of Commerce* (1922), 96–97; *Eleventh Annual Report of the Secretary of Commerce* (1923), 103.

31. Hawley, *The Great War and the Search for a Modern Order*, 59. The secretary's ideas about appropriate activities for trade associations sometimes clashed with the Department of Justice's quest to enforce antitrust laws, and the chemical manufacturers, in warding off charges of antitrust, looked to support from Hoover. Hawley, *The Great War*, 70; Herty to Hoover, March 31, 1925, Hoover Papers, Department of Commerce, box 592, SOCMA folder.

32. Otis L. Graham discusses the significance of continuity among federal agency officials "to sustain the policy effort" in *Losing Time*, 258; "Charles C. Concannon," *Chemical and Engineering News* 35 (August 26, 1957): 48; "Charles Concannon, U.S. Ex-Aide, Dead; Led Chemical Unit in Commerce Dept.," *New York Times*, August 11, 1957, p. 80.

33. Herty to Concannon, April 2, 1923, NARA, RG 151, 461.1 Imports, Chemicals, box 2226.

34. For example, U.S. Tariff Commission, *Census of Dyes and Other Synthetic Organic Chemicals, 1923*; SOCMA, "Bulletin No. 11," May 4, 1923, CHH, box 75.

35. Edward Clifford, Assistant Secretary of the Treasury, January 10, 1923; Concannon to Herty, February 8, 1923; Concannon to Herty, April 7, 1923, NARA, RG 151, 461.1 Imports, Chemicals, box 2226.

36. Herty to Julius Klein, Bureau of Foreign and Domestic Commerce, July 2, 1923; Herty to Concannon, January 22, 1924; Thomas Delahanty to Concannon, April 21, 1925; Herty to Klein, April 25, 1925; Concannon to Warren N. Watson, U.S. Tariff Commission, July 22, 1925, NARA, RG 151, 461.1 Imports, Chemicals, box 2226. As they had during the war when the Department of Commerce published Thomas Norton's census, the importers protested publication of the statistics as a violation of confidentiality, but the objections received little consideration and the service continued into the mid-1930s. By 1925, officials in the Department of Commerce believed that importers had devised ways to throw off the accuracy of the monthly reports, primarily by "recalling" invoices repeatedly. The officials believed the inaccuracies were negligible, however. Delahanty to Concannon, April 15, 1925; Concannon to Herty, March 13, 1924; Senator George H. Moses to Klein, January 18, 1924; O. P. Hopkins to Moses, February 16, 1924; Concannon to Eugene Pickrell, June 16, 1925; E. K. Halbach to Robert Lamont, Secretary of Commerce, February 3, 1931; Delahanty to J. J. O'Hara, March 28, 1931; Earl Tupper to Mr. Dunn, August 19, 1935; Tupper to Mr. Moore, October 22, 1935, RG 151, 461.1 Imports, Chemicals, box 2226.

37. As mentioned in chapter 6, the Departments of Commerce and State sometimes sparred over which should have jurisdiction over foreign commerce. In this case, the departments cooperated. A decade later, the Department of State took over matters related to foreign commerce, and the Bureau of Foreign and Domestic Commerce shriveled. Kaufman, *Efficiency and Expansion*, 77–80, 218–21; Brandes, *Herbert Hoover and Economic Diplomacy*, 4–5; Delahanty to National Aniline and Chemical Company, April 8, 1924; Herbert C. Hengstler, Department of State, to American Consular Officers, April 11, 1924, NARA, RG 151, 232 Dyes—General, box 1034. William Becker has pointed out that small businesses tended to benefit more than large firms from the Department of Commerce's efforts to promote foreign trade because larger firms had the resources to pursue markets without government help. In the case of gathering foreign trade information on the state of the chemical industries and markets around the world, however, American firms of all sizes supported the department's work. Becker, *The Dynamics of Business-Government Relations*.

38. *Twelfth Annual Report of the Secretary of Commerce* (1924), 100-101; Concannon to S. H. Day, Philadelphia District Office, December 8, 1924, NARA, RG 151, 232 Dyes—General, box 1034.

39. For example, Breithut, "The German Coal-Tar Chemical Industry" (1923); Breithut, "British Dyestuffs Industry" (1924); Delahanty, "The German Dyestuffs Industry" (1924). Breithut was a professor of chemistry at the City College of New York from 1903 to 1930. "Dr. F. E. Breithut, Ex-Professor, 61," *New York Times*, May 13, 1962, p. 88 (the newspaper erred in its math; Breithut was eighty-one at his death).

40. Concannon to Leighton W. Rogers, commercial attaché in Warsaw, May 4, 1925; B. R. Price, National Aniline, to Concannon, May 3, 1926; Price to Concannon, May 21, 1926; Delahanty to Price, May 29, 1926; S. H. Day to Concannon, December 3, 1924; Concannon to Day, December 8, 1924; Price to Concannon, March 8, 1929; Concannon to Price, March 13, 1929, NARA, RG 151, 232 Dyes—General, box 1034; Eric Kunz, Du Pont, to Concannon, June 25, 1924; F. R. Eldridge to Wilbur J. Carr, Department of State, September 12, 1923; Concannon to New York District Office, January 23, 1925, RG 151, 232 Dyes—Germany, box 1036; *Thirteenth Annual Report of the Secretary of Commerce* (1925), 107-8; *Fifteenth Annual Report of the Secretary of Commerce* (1927), 98-100.

41. The U.S. diplomats believed that the Treaty of Versailles explicitly condoned Allied attempts to establish domestic synthetic organic chemicals industries and that revealing German production statistics was therefore legal. However, to facilitate relations with the Germans, the officials recommended that publication, though not distribution, be limited to avoid further confrontation. James A. Logan Jr., Paris, to Secretary of State, April 28, 1924; Reginald Norris, American Unofficial Delegation, April 17, 1924, NARA, RG 151, 232 Dyes—Germany, box 1036.

42. Alan G. Goldsmith, Western European Division, to Charles E. Herring, commercial attaché in Berlin, December 22, 1922; Donald L. Breed, assistant commercial attaché in Berlin, to BFDC, March 13, 1923; Herring to Frederick Breithut, Paris, August 8, 1923; Concannon to Edgar M. Queeny, December 20, 1924, NARA, RG 151, 232 Dyes—Germany, box 1036.

43. Concannon to Herty, August 11, 1923; Chester Lloyd Jones, Paris, to BFDC, c. August 13, 1923; Concannon to Jones and Breithut, August 16, 1923; Breithut to BFDC, August 22, 1923, NARA, RG 151, 232 Dyes—Germany, box 1036; SOCMA Board of Governors, "Minutes of Meeting," August 16, 1923, CHH, box 74; Chemical Division, "Special Circular No. 24—Anglo-German Dye Agreement," January 30, 1924; Concannon to Breithut, February 2, 1924; Breithut to Concannon, February 15, 1924; Breithut to Concannon, February 18, 1924; Concannon to Hugh D. Butler, London, October 14, 1924, RG 151, 232 Dyes—Germany, box 1036; for an account of the German and British negotiations, see Reader, *Imperial Chemical Industries*, vol. 1, chap. 18.

44. I have written about the court case elsewhere. Steen, "Patents, Patriotism, and 'Skilled in the Art.'"

45. On the new administration see Murray, *The Harding Era*, 296-98, and Daugherty and Dixon, *The Inside Story of the Harding Tragedy*, 103. On the audit, see Daugherty to Harding, October 16, 1922, MIC 3 Warren G. Harding papers [microform], roll 170, frames 308-17, Ohio Historical Society. On the Chemical Foundation, see Thomas Miller to Harding, April 21, 1922; Harding to Daugherty, April 22, 1922; quotation from Daugherty to Harding, April 25, 1922, NARA, RG 60, Central Files, entry 114 Correspondence, Case 9-17-10-12, box 748.

46. "Files Chemical Suit," *Washington Post*, September 9, 1922, pp. 1, 11.

47. Herman J. Galloway, Special Assistant to the Attorney General, to Attorney General Daugherty, July 11, 1922; and Galloway to United States Attorney, Pittsburgh, August 31, 1922, NARA, RG 60, Central Files, entry 114 Correspondence, Case 9-17-10-12, box 748.

48. Galloway to Daugherty, July 11, 1922; Galloway to United States Attorney, Pittsburgh, August 31, 1922; Galloway to Col. Henry W. Anderson, Special Assistant to the Attorney General, September 2, 1922; and Galloway to A. C. Vandiver, lawyer for Metz, January 9, 1923, NARA, RG 60, Central Files, entry 114 Correspondence, Case 9-17-10-12, box 748.

49. On Morris, see Hoffecker, *Federal Justice in the First State*, 97-98; on Anderson, see *National Cyclopaedia of American Biography* 44:10-11, s.v. "Anderson, Henry Watkins"; on Kresel, see *Who Was Who in America* 3:490, s.v. "Kresel, Isidor Jacob." Documenting Kresel's employment by the APC is Daugherty to Harding, October 16, 1922, MIC 3 Warren G. Harding papers, roll 170, frames 308-17, Ohio Historical Society. The final published record is *USA v. The Chemical Foundation, Incorporated, Record on Final Hearing*, District Court of the United States for the District of Delaware, No. 502 in Equity. The Hagley Museum and Library possesses the complete record of the case. All volumes are in the library; the district court ruling by Judge Morris, however, is in the manuscripts division (Acc. 1662, box 17).

50. Anderson statement, *USA v. The Chemical Foundation, Inc.*, 2:1082-87; 3:1380-85, 1390-93, 1715-37; 5:2403. The plea on behalf of principle comes from Anderson's opening statement, ibid., 2:684-86.

51. Ibid., 2:1534-37, 1005-8. Will T. Gordon testimony, ibid., 2:1054-69; and Anderson opening statement, ibid., 2:599-627.

52. In December 1919, before the Senate Finance Committee, Garvan described World War I as "an industrial war, brought on by industrial Germany in her lust-made haste to capture the markets of the world." Germany's "arrogance and pride," he asserted, grew in part from the country's "chemical supremacy." Garvan testimony, Senate Committee on Finance, *Hearings, Dyestuffs* (1919-1920), 537-38. Garvan testimony, *USA v. The Chemical Foundation, Inc.*, 4:2714-90.

53. One such book was Victor Lefebure's *The Riddle of the Rhine* (1923), published and distributed by the Chemical Foundation. A retired British officer, Lefebure linked chemical warfare and military strength to a strong synthetic organic chemicals industry and argued that Germany remained a threat to peace in the future because its chemical firms survived the war and the Treaty of Versailles largely intact. Judge Morris in *USA v. The Chemical Foundation, Inc.*, read and was persuaded by Lefebure's argument that a synthetic organic chemicals industry was vital to national defense; see Morris, ibid., 5:3041-50.

54. Kresel's opening statement, *USA v. The Chemical Foundation, Inc.*, 2:699-702; and Frank L. Polk testimony, ibid., 2:1181-82, 1187-1200, 1241, 1258-59.

55. Harry E. Danner testimony, ibid., 2:1008; Kresel opening statement, ibid., 2:689-95; and Reply Brief on Behalf of the United States, ibid., 3-4.

56. Holdermann was one of the German representatives helping to coordinate the Section 10(f) suits to recover royalties from the patents licensed under the FTC. See ibid., 4:2421-24; and Holdermann testimony, ibid., 4:2355-63, 2381, 2405, 2412.

57. George W. Storck, accountant, testimony, ibid., 3:1481-82; and Anderson statements, ibid., 3:1478-80. Some FTC licensees preferred to continue the arrangement with the FTC even after the sale of patents to the Chemical Foundation, hence the income from

royalties through 1922. In its Section 10(f) suits, the Chemical Foundation claimed those FTC royalties.

58. Metz testimony, ibid., 3:1756, 1762, 1771-73, 1779-80.

59. On ties to the ambassador and on Salvarsan, Kresel statements, ibid., 3:1792-93, 1796-1807, 1903; 6:3678; and "Col. Metz Defends His Products," *DCM* 13 (July 18, 1923): 157. On profiteering and on instigating the lawsuit, see Kresel statements in *USA v. The Chemical Foundation, Inc.*, 3:1860-71, 1845.

60. Anderson statements, *USA v. The Chemical Foundation, Inc.*, 3:2050.

61. Kresel statements, ibid., 6:3678.

62. William J. Hale reported to his boss, Herbert Dow, that most people in the industry believed the Chemical Foundation would win because the "government has not been able to secure the services of any chemists willing to testify in their behalf." Hale to Dow, June 13, 1923, DHC, folder 230059. Intriguingly, none of the chemists who testified in *USA v. The Chemical Foundation, Inc.*, against the workability of German patents and in support of the Chemical Foundation held positions at National Aniline. Refusing to permit National Aniline's chemists from testifying for either side would have been consistent with Orlando Weber's preference for secrecy. "Allied Dye and Chemical," *Fortune* (June 1930): 81.

63. Elmer K. Bolton testimony, *USA v. The Chemical Foundation, Inc.*, 6:4070; Volwiler testimony, ibid., 5:3263; and Alfred H. White and Bolton testimonies, ibid., 6:3970, 4077.

64. Volwiler testimony, ibid., 5:3243-44, 3255-59, 3282, 3292-97; and Bolton testimony, ibid., 6:4030, 4070, 4077.

65. Anderson and Morris statements, ibid., 6:4177-83.

66. Julius Stieglitz testimony, ibid., 5:3575, 3616; 6:3626, 3638-39, 3647, 3651-52. On Stieglitz, see *Dictionary of American Biography*, s.v. "Stieglitz, Julius." Anderson and the government lawyers missed an opportunity to trip up Stieglitz. In a congressional hearing in 1919 Stieglitz had testified that Salvarsan "must be extremely pure, because it is injected directly into the veins and any impurity is likely to lead to death" very quickly. The question is whether Stieglitz or any other chemist needed to be told in a patent description that a potent product destined for intravenous consumption must be pure. Stieglitz testimony, House Committee on Ways and Means, *Hearings, Dyestuffs* (1919), 17.

67. Morris statement in *USA v. The Chemical Foundation, Inc.*, 6:4268-69 (on laws governing patents). In essence, Anderson argued that the Americans lacked the "tacit knowledge," knowledge that cannot be articulated, which the German industrial chemists had acquired after decades of experience. The concept comes from Polanyi, *The Tacit Dimension* (1966). For a discussion of the concept applied to business organizations, see Nelson and Winter, *An Evolutionary Theory of Economic Change*, 76-82. For Bolton's admission, see Bolton testimony, *USA v. The Chemical Foundation, Inc.*, 6:4087-88. For other chemists' reports of their experience, see, e.g., Ernest H. Klipstein of E. C. Klipstein and Sons, Moses Crossley of Calco, and Volwiler testimonies, ibid., 6:4169-70, 4225-28; 5:3243-44, 3312-16.

68. Walter G. Christianson testimony, *USA v. The Chemical Foundation, Inc.*, 7:4462-68, 4481.

69. Christianson testimony, ibid., 7:4509-33, 4537-38, 4558, 4568-82, 4591-92.

70. Harry D. Gibbs testimony, ibid., 7:4641-42, 4647-49, 4662, 4664. Garvan subsequently denied that Gibbs told him that he had produced Salvarsan using the patent. Garvan testimony, ibid., 7:4906.

71. Freedman testimony, ibid., 7:4595-96, 4604, 4607-16, 4688-94; on the uses of cinchophen, see Volwiler testimony, ibid., 5:3282. On the increasing number of American-trained

Ph.D. chemists, see Thackray, Sturchio, Carroll, and Bud, *Chemistry in America, 1876-1976*, 43, 46, 265.

72. Freedman testimony, *USA v. The Chemical Foundation, Inc.*, 7:4595-96, 4604; and Morris statements and Freedman testimony, ibid., 7:4607-16, 4688-94.

73. Morris statements and Freedman testimony, ibid., 7:4607-16, 4688-94, 4724-36.

74. Freedman, Gibbs, Bolton, Crossley, and Gellert Alleman testimonies, ibid., 7:4738, 4744-51, 4761-69, 4774-75, 4780-82.

75. Gibbs, Freedman, Bolton, and Crossley testimonies, ibid., 7:4740-42, 4726-36, 4759, 4761-69.

76. Closing Arguments on Behalf of the Defendant, ibid., 223; and Anderson and H. E. Knight statements, ibid., 5:3370-72, 6:3975-77.

77. Closing Arguments on Behalf of the Defendant, ibid., 95, 210-23, 250-51; Brief on Behalf of the Defendant, ibid.; and Reply Brief on Behalf of the United States of America, ibid., 351-69.

78. Midway through the trial, the internal correspondence of the Department of Justice indicates that the lawyers believed Judge Morris had treated them fairly thus far. See Galloway to Daugherty, June 28, 1923, NARA, RG 60, Classified Subject Files, entry 114 Correspondence, Case 9-17-10-12, box 748. The industrialists, on the other hand, believed that Morris "leans very strongly to the side of the Chemical Foundation." Hale to Dow, June 13, 1923, DHC, folder 230059.

79. E. S. Rochester (?) to Arthur Brisbane, *New York America*, July 5, 1923, Harry M. Daugherty Papers, MSS 271, box 1, folder 2, Ohio Historical Society; for the false story, see "Metz Admits He Paid German Spy in Senate Probe," *Wilmington Morning News*, June 26, 1923, p. 1ff.; for the retraction, see Charles E. Gray, managing editor for the News Journal Company, "Col. H. A. Metz Paid No Spy in Senate Probe," *Wilmington Morning News*, June 27, 1923, p. 1; on the government attorneys' response, see undated, unsigned memo beginning "Here is the situation," c. late June 1923, NARA, RG 60, Classified Subject Files, entry 114 Correspondence, Case 9-17-10-12, box 748.

80. Opinion of Court, *USA v. The Chemical Foundation, Inc.*, U.S. District Court, 5-7, 10-15, 17-23, 51-52; Opinion of Court, *USA v. The Chemical Foundation, Inc.*, U.S. Circuit Court of Appeals for the Third Circuit, No. 3160, October term, 1924; and Opinion of Court, *USA v. The Chemical Foundation, Inc.*, Supreme Court of the United States, No. 127, October term, 1926. As the opinions of the district and appellate court showed, the government lawyers were unable to counter the excesses of the anti-German rhetoric at a time when memory of the war was still fresh. Although the court record only rarely mentions the scandals of the Harding administration, it also seems likely that the lawyers from the Department of Justice, which was wracked with suspicions and revelations of misconduct under Harry Daugherty, lacked the credibility to convince the courts that members of the previous administration had committed illegal acts. At one point Kresel had tried to link Metz to Gaston Means, who, as Kresel must have known, was one of the slipperiest among the shady characters in the Harding administration. By 1923 Means's career included a stint as a German spy, an acquittal in a murder trial, and two years as an FBI agent, a position he used to become wealthy by collecting bribes and by bootlegging. Kresel statements, *USA v. The Chemical Foundation, Inc.*, 3:1937. See Murray, *The Harding Era*, 477-78.

81. Gathings, *International Law and American Treatment of Alien Enemy Property*, 1; the same view is expressed throughout.

82. The Chemical Foundation, Inc., "Schedule of All Licenses Issued and Amount of Royalties Paid on Said Licenses from Date of Incorporation, February 19, 1919, to December 31, 1936," FPG, box 17.

83. The Chemical Foundation, Inc., "Statement of Income and Expenditures for the Period from Incorporation, Feb. 19, 1919, to Dec. 31, 1941," FPG, box 17.

84. Typically, a less developed country benefits from a weak patent system because its firms can more readily copy the successes of advanced rivals. See Haber, *Chemical Industry during the Nineteenth Century*, 202–4; Beer, *Emergence of the German Dye Industry*, 54–56; and Kaufer, *The Economics of the Patent System*, 9–10.

85. Not included in the $265 million were additional assets related to merchant ships and a radio station. McHugh, "Settlement of War Claims Act of 1928," 195; "The Trading with the Enemy Act," (1926): 352 (signed by E.M.B.). Jerald A. Combs discusses the relatively quick change in the Americans' attitude toward the war and their analyses of the war's causes. His focus is on historians' and public intellectuals' accounts of the war in the 1920s, but they reflect a broader change in attitude. See Combs, *American Diplomatic History*, 132–52.

86. Manfred Jonas, *The United States and Germany*, 151–93; Schwabe, "America's Contribution to the Stabilization of the Early Weimar Republic," 21–28; and Hans W. Gatzke, *Germany and the United States*, 75–102. Joan Hoff Wilson considers the influence of the divided business community on U.S. foreign policy in *American Business and Foreign Policy, 1920–1933*, 123–56. On ships, see *Congressional Record*, 69th Cong., 2nd sess. (1926–1927), vol. 68, part 1, 598. A similar set of negotiations through the Tripartite Commission addressed claims between the United States and Austria-Hungary. McHugh, "Settlement of War Claims Act of 1928," 194–95.

87. *Congressional Record*, 69th Cong., 2nd sess. (1926–1927), vol. 68, part 1 (December 16, 1926), 597.

88. Ibid., 597, 611.

89. Rep. W. R. Green mentions earlier legislative attempts in *Congressional Record*, 69th Cong., 2nd sess. (1926–1927), vol. 68, part 1, 594; Fehr, "Paying Our Claims against Germany," 408.

90. Under the newly negotiated Dawes Plan, Germany had been given more than sixty years to pay off the debts owed to the United States. Therefore, Congress decided to pay private American citizens the amounts they were due from Germany and then let the U.S. Treasury collect the German payments over the next several decades. *Congressional Record*, 69th Cong., 2nd sess. (1926–1927), vol. 68, part 1, 594–96; McHugh, "Settlement of War Claims Act of 1928," 193. Germany paid 6,574 of 7,025 American claims established by the Mixed Claims Commission but suspended payments to the United States in 1931; in a 1953 agreement, in the midst of sorting claims from World War II, West Germany agreed to pay the remainder of the American claims. "Rights of Germans Upheld by Court," *Washington Post*, February 18, 1936, p. X5; House Committee on Interstate and Foreign Commerce, *Hearings, German Special Deposit Account* (1947). The German Special Deposit Account held the 20 percent of German assets retained under the War Claims Act until Germany paid American claims. "Germany Debt Pacts to be Signed Today," *New York Times*, February 27, 1953, p. 31; *Annual Report of the Secretary of the Treasury, 1939*, 68–76.

91. Holtzoff, "Enemy Patents in the United States," 273–75; Parker, "War Claims Arbiter, Administrative Decision No. 1," 196; "Minutes of the War Claims Arbiter," June 7, 1928, NARA, RG 60, entry 209, box 1. Edwin B. Parker died in the fall of 1929 after having

established most of the procedures by which claims were settled under the act. He was replaced by James W. Remick, who had been a judge on New Hampshire's supreme court. Holtzoff, "Enemy Patents," 277; "Memorandum for Counsel and the Press," January 13, 1930, RG 60, entry 209, box 1.

92. "Minutes of the War Claims Arbiter," September 10-11, 1928, NARA, RG 60, entry 209, box 1; "Hearing before the Arbiter under the Settlement of War Claims Act of 1928," September 10-11, 1928, RG 60, entry 228, box 6.

93. "Hearing before the Arbiter," 43ff.

94. Ibid., 13ff. and 180ff.

95. Parker, "War Claims Arbiter, Administrative Decision No. 1," 213-14.

96. Ibid., 209-13, 221-25.

97. Ibid., 211-14. Although the original TWEA would have permitted patent royalty claims throughout the entire war, the War Claims Act amended that provision to exclude the period when the United States actively participated in the war. "Memorandum for the War Claims Arbiter," October 5, 1928, NARA, RG 60, entry 228, box 3.

98. Generally, this was a smooth relationship, although the lawyers in the War Claims Arbiter's office expressed frustration that the Department of War could not produce their reports on patent use more quickly. Alexander Holtzoff, "Memorandum for Assistant Attorney General [Herman] Galloway," April 16, 1929, NARA, RG 60, entry 228, box 2.

99. The archival records for the War Claims Arbiter (NARA, RG 60, entries 209-28) do not reveal which patents belonged to the secret Group B, but my best guess is that they were the Haber-Bosch patents related to the fixation of nitrogen.

100. Lt. Col. Joseph I. McMullen to James Remick, War Claims Arbiter, May 19, 1930, NARA, RG 60, entry 228, box 2.

101. Alexander Holtzoff, "Memorandum for Assistant Attorney General [Herman] Galloway," May 2, 1929, NARA, RG 60, entry 228, box 3.

102. "Stenographic Notes Taken at Hearing Held in Arbiter's Office, November 26, 1930, in Dockets 1 and 2," NARA, RG 60, entry 228, box 2.

103. Almuth C. Vandiver, lawyer for I.G. Farben, "Before the War Claims Arbiter: Claim of I.G. Farbenindustrie, A.G. (Docket No. 1)," January 8, 1931, NARA, RG 60, entry 228, box 2.

104. "Motion, I.G. Farbenindustrie, A.G., v. the United States, before the War Claims Arbiter," December 9, 1931; undated memoranda and working notes, NARA, RG 60, entry 228, box 4, folder on Group A patents.

105. I have not found documentation for payments beyond the first, but the United States came close to paying out the 80 percent of all awards before transatlantic payments ceased. "Awards entered January 23, 1931," NARA, RG 60, entry 228, box 3. The $739,175.89 was the final, base award due as of December 31, 1928, and whatever portion of the sum retained in the U.S. Treasury would continue to collect 5 percent interest. Ogden L. Mills, Undersecretary of the U.S. Treasury, to I.G. Farbenindustrie A.G., May 26, 1931, Amerika Freigabe Bill (4.10.1928-6.4.1936), 9/A.1.1, Bayer-Archiv, Leverkusen, Germany.

President Herbert Hoover implemented a moratorium on reparations payments, which was not supposed to apply to Germany's payments toward the private claims arranged under the Mixed Claims Commission. However, Germany did not have the wherewithal to pay even these claims, which, under the War Claims Act, would be used to pay the remainder due I.G. Farben and the other private German claims. When the German payments to the United States ceased, the American payment of German private claims ceased. "Confidential memorandum," February 5, 1932, and "Observation on the question why the German government should be prepared to continue payments under the debt settlement between the United States and Germany," n.d., Amerika Freigabe Bill (4.10.1928-6.4.1936), 9/A.1.1, Bayer-Archiv.

106. Seiforde M. Stellwagen to Francis P. Garvan, February 18, 1927, and Stellwagen to Garvan, March 5, 1927, FPG, box 103, folder on Du Pont/Section 10(f) suits. The number of test cases doubled to eight when the APC, Howard Sutherland, argued that his office temporarily owned the property between the seizure and sale and had a claim to it. "Rules on Alien Patents," *New York Times*, November 20, 1928, p. 16; "Arguments Ended in German Patents Royalties Title," *New York Daily News Record*, March 11, 1931, clipping in FPG, box 103, folder on Du Pont/Section 10(f) suits; and Brief for Respondent E. I. du Pont de Nemours and Company, for eight cases, including Case No. 179: *Farbwerke vormals Meister Lucius und Bruning, Deutsche Gold und Silber Scheide Anstalt vorm. Roessler, and Badische-Anilin und Soda-Fabrik v. Chemical Foundation, Inc., E. I. du Pont de Nemours and Company, and Walter O. Woods, as Treasurer of the United States*, and Case Nos. 180, 181, 182, 271, 272, 273, and 274, Supreme Court of the United States, October 1930, pp. 2-11. See also McHugh, "Settlement of War Claims Act of 1928," 193-95, enclosed in Brief for E. I. du Pont de Nemours and Company, District Court of the United States, District of Delaware, 1928, Cases No. 492, 494, and 495. The collection of briefs for this series of cases is in the library of the Hagley Museum and Library.

107. The Chemical Foundation, however, chose not to collect any royalties prior to April 10, 1919, when it purchased most of its patents from the APC. Consequently, by law, the royalties would return to the American companies which had licensed the patents and paid the royalties and which were, not coincidentally, stockholders in the Chemical Foundation. Thomas E. Rhodes, Special Assistant to the Attorney General, "Memorandum for Assistant Attorney General Farnum," February 28, 1929, NARA, RG 60, Central Files, entry 114 Correspondence, Case 9-17-10-12, box 750.

108. Rhodes, "Memorandum for Assistant Attorney General Farnum," February 28, 1929; "Rules on Alien Patents," *New York Times*, November 20, 1928, p. 16; several Section 10(f) cases reached the Supreme Court together in 1930, including Cases 179-182 and 271-74 (see note 106).

Unlike the German firms, some of the American importers for German chemical companies were able to collect seized assets under Section 10(f). "Receipt and Release of the H. A. Metz Laboratories, Inc.," June 16, 1931, NARA, RG 60, entry 114, box 750, case 9-17-10-12.

109. Hart, "Corporate Technological Capabilities and the State."

110. Ibid., especially 175.

Chapter Eight

1. E. H. Kilheffer testimony and SOCMA brief in House Committee on Ways and Means, *Tariff Readjustment* (1929), 1:68-87. The quotation is from the written brief on p. 85.

2. U.S. Tariff Commission, *Census of Dyes and Coal-Tar Chemicals, 1927*, 129; König, "Adolf Hitler vs. Henry Ford: The Volkswagen, the Role of America as a Model, and the Failure of a Nazi Consumer Society."

3. Plumpe, *Die I.G. Farbenindustrie A.G.*; Marsch, "Strategies for Success: Research Organization in German Chemical Companies and I.G. Farben until 1936"; Hughes,

"Technological Momentum in History: Hydrogenation in Germany, 1898-1933"; Stokes, "From the I.G. Farben Fusion to the Establishment of BASF AG, 1925-1952."

4. Schröter, "Cartels as a Form of Concentration in Industry: The Example of the International Dyestuffs Cartel from 1927 to 1939," 123; Haber, *The Chemical Industry, 1900-1930*, 275; Reader, *Imperial Chemical Industries*, 1:442; "French Buy German Dye Secrets," *DCM* 10 (April 19, 1922): 917; Frank B. Gorin, "French Dyestuffs Industry" (1924), 9, quotation on 13; "French Dye Merger Completed," *DCM* 13 (November 7, 1923): 1259.

5. Reader, *Imperial Chemical Industries*, 1:425-49; Frederick E. Breithut discussed the British firms and licensing system in detail in a Department of Commerce report. The British initially instituted the licensing system by an executive order, which a judge overruled at the end of 1919. Parliament replaced the system with a new law in early 1921, as discussed by both Reader and Breithut (the report reprints the law verbatim). Breithut, "British Dyestuffs Industry," (1924). One example of American attitudes to the discussions in Britain: "Britain's Queer Dye Policy," *DCM* 17 (November 18, 1925): 1387-88.

6. Garvan helped to press charges against Kuttroff, Pickhardt & Company for violating the WTB importing regulations, and the WTB held confidential hearings to evaluate the charges. The files do not include a verdict. "Proceedings of a Hearing Accorded by the War Trade Board Section of the Department of State to the Firm of Kuttroff, Pickhardt & Company, Inc.," January 26-28 and February 17, 1920, FPG, box 44; Textile Alliance to E. S. Chapin, February 11, 1921, CHH, box 87. For a short biographical sketch of Ludwig, see "Ludwig Elected National Aniline Head," *Chemical Markets* 27 (August 1930): 167; "Besuch des herrn Carl v. Weinberg mit herrn Ludwig von der National Aniline Comp., New York . . . den 24 juli 1920," Bayer Archives, USA Entwicklung der amerikanischen chemischen Industrie, 1915-1949, #81/2.2. The letter describes a meeting between Ludwig and Weinberg, but the letterhead is Duisberg's. Thanks to department colleague Christian Hunold for help with the German handwriting. Irénée du Pont to Francis P. Garvan, December 12, 1923, Hagley Museum and Library, Acc. 1662, box 17, Chemical Foundation folder; Hounshell and Smith, *Science and Corporate Strategy*, 205.

7. "Sterling Products, Inc.: Resumption of German Relations," a summary of Sterling-Bayer correspondence prepared by the Department of Justice in World War II for an antitrust suit against the pharmaceuticals industry, NARA, RG 60, Central Files, Case 60-21-56, Enclosures, box 1329. See also Verena Schröter, who discusses the relationships based on extant German records, "Participation in Market Control through Foreign Investment: I.G. Farbenindustrie A.G. in the United States, 1920-1938." She also considers the relationships in other branches of chemical products.

8. "Sterling Products, Inc.: Resumption of German Relations" and "Sterling Products, Inc.: The I.G.-Sterling Partnership," NARA, RG 60, Central Files, Case 60-21-56, Enclosures, box 1329.

9. Ernst Moeller to Weiss, July 31, 1920, quoted in "Sterling Products, Inc.: Resumption of German Relations"; "Sterling Products, Inc.: Resumption of German Relations," NARA, RG 60, Central Files, Case 60-21-56, Enclosures, box 1329.

10. Rudolf Mann to Duisberg, April 28, 1922, USA Schriftwechsel, Besprechungen, Historische Entwicklung, 1909-1954, Bayer-Archiv, Bd. 3, 9/A.1. More negotiations with Weiss can be found in FFB und I.G. Farben Warenzeichen, "Weiss-Verträge"/Vorarbeiten, Besprechungen, usw., Bayer-Archiv, #4/C.34.6. The agreement in 1922 with another firm is discussed in "Sterling Products, Inc.: Resumption of German Relations," NARA, RG 60, Central Files, Case 60-21-56, Enclosures, box 1329. The products under the agreement

consisted primarily of pharmaceuticals, perfumes, agricultural chemicals, and photochemicals. It excluded dyes and other chemicals. As part of another accord in 1923, Sterling and Bayer agreed to establish jointly a new corporation, Bayer Company, Ltd., in Britain and avoid direct competition of the rival "Bayer" products coming from the United States and Germany. As in the United States, German ownership was prohibited. Consequently, the American firm established and owned the British company, but promised to transfer half ownership to the German firm at a more "politic" later date (1926, it turned out). In the meantime, the Germans would provide one of four directors, supply the chemicals, and receive half the profits. "The Sterling Products, Inc.: Resumption of German Relations," NARA, RG 60, Central File, Classified Subject File, Case 60-21-56, Enclosures, box 1329. Plumpe has discussed these negotiations and also noted the similarity in the terms of the agreements the Germans offered to the French, British, and Americans. Plumpe, *Die I.G. Farbenindustrie A.G.*, 128-29.

11. Haynes, *American Chemical Industry*, 6:175, 183-84; Garvan to Thomas S. Grasselli, July 24, 1923, FPG, box 18; Rudolf Mann to Carl Duisberg, April 18, 1923; and Mann to Grasselli Company, April 18, 1923, Bayer-Archiv, USA Schriftwechsel, Besprechungen, Historische Entwicklung, 1909-1954, Bd. 3, 9/A.1; Glaser-Schmidt, "Foreign Trade Strategies of I.G. Farben after World War I," 202-3; Minutes, Board of Governors, Synthetic Organic Chemical Manufacturers Association, October 8, 1925, CHH, box 75. Sterling Products owned the pharmaceuticals section of the Rensselaer plant.

12. Haynes, *American Chemical Industry*, 6:271; Galambos and Sturchio, "Transnational Investment: The Merck Experience, 1891-1925." Powers-Weightman-Rosengarten, an amalgamation of firms founded in Philadelphia and dating to as early as 1818, manufactured pharmaceuticals such as iodine, mercury, and morphine products. After World War I, the firm began to produce Salvarsan and related arsenicals. "The Powers-Weightman-Rosengarten Company," *IEC* 17 (January 1925): 99-100. *Census of Dyes and of Other Synthetic Organic Chemicals, 1930*, 60-61. The corporate reorganization in the late 1920s, which coincided with the purchase of Powers-Weightman-Rosengarten, created room on Merck's organization chart for a "pure research" program, but only in 1930 did the firm hire Randolph T. Major to direct the research program and begin to fill out the organization. Galambos and Sturchio, "The Origins of an Innovative Organization: Merck & Co., Inc., 1891-1960," 14-17; Galambos et al., *Values and Visions: A Merck Century*, 66-68. Merck's research ties in the 1930s to Alfred N. Richards of the University of Pennsylvania are discussed in Swann, *Academic Scientists and the Pharmaceutical Industry*, 65-86.

13. Haynes, *American Chemical Industry*, 6:207; *Census of Dyes and of Other Synthetic Organic Chemicals, 1930*, 60-61.

14. Haynes, *American Chemical Industry*, 4:233-35; Metz quotation found in "Metz Explains Grasselli Agreement," *DCM* 17 (November 18, 1925): 1395; "Kuttroff, Pickhardt Joins General Dyestuff Combination," *DCM* 17 (December 30, 1925): 1803; Ludwig's technical specialty was the application of dyes. "Ludwig Elected National Aniline Head," *Chemical Markets* 27 (August 1930): 167.

15. "The Sterling Products, Inc.: The I.G.-Sterling Partnership," NARA, RG 60, Central File, Classified Subject File, Case 60-21-56, Enclosures, box 1329.

16. "American I.G. Chemical Corporation," March 17, 1936, draft report in FPG, box 76; "American I.G. Chemical Corporation," *Moody's Manual of Investments, 1930* (New York: Moody's Investors Service, 1930), 2149-50. In 1928, the German Agfa formed Agfa-Ansco with the purchase of Ansco, the successor firm to American photographic suppliers and

manufacturers since the 1840s. Jenkins, *Images and Enterprise: Technology and the American Photographic Industry, 1839 to 1925*, 20-28, 337. "The Sterling Products, Inc.: The I.G.-Sterling Partnership," NARA, RG 60, Central File, Classified Subject File, Case 60-21-56, Enclosures, box 1329; Schröter, "Participation in Market Control," 179; Plumpe, *Die I.G. Farbenindustrie A.G.*, 127; U.S. Tariff Commission, *Census of Dyes and Other Synthetic Organic Chemicals, 1925*, 54; SOCMA, cross index to U.S. Tariff Commission censuses, 1924 and 1930, CHH, box 76. The General Aniline Works supplied one-quarter of the U.S. dyes market by value, not by pounds. Of dyes produced in the United States, the German firm supplied 20 percent in sales. Stocking and Watkins, *Cartels in Action*, 472, 508. Francis Garvan and others despised and feared the American I.G., but Herbert Dow believed the threat was overstated. "New developments in regard to the American I.G. do not seem to indicate that they will be as aggressive as their first announcement might lead one to expect. Dr. Hale stated that the $30 million they have raised was shipped to Germany for use in Germany before the announcement of the offering of stock was made public in the United States." Dow to G. E. Collings, May 3, 1929, DHC, folder 290044.

17. It is not actually clear from the records whether Garvan sent his letter to the American I.G. directors. Garvan to Charles E. Mitchell, W. C. Teagle, Edsel Ford, Paul M. Warburg, May 1929; "Confidential Memo," October 7, 1926; FPG, box 76, German Developments.

18. For one overview of the variety of cartel agreements in Europe, see Schröter, "Cartels as a Form of Concentration in Industry." Stocking and Watson, *Cartels in Action*, 377, 406-16. Other historians use similar political metaphors to discuss the cartel style of economic decision making. For example, see Reader, *Imperial Chemical Industries*, 1:468-71. In 1929, after three years of negotiating, the German, Swiss, and French dyes manufacturers formed the "Three Party Cartel," which divided markets and set sales quotas. Schröter, "The International Dyestuffs Cartel, 1927-1939."

19. John Newlean testimony in Senate Committee on the Judiciary, *Hearings, Alleged Dye Monopoly* (1922), 600-602. At the hearings in the spring of 1922, Newlean gave an average of 10 percent return in the "last four years," which would have included 1918, a war year. The rate of profits had declined over the period, although Newlean does not indicate by how much. "National Aniline & Chemical Company," *Moody's Manual of Railroads and Corporation Securities*, 21st ed., industrial section (New York: Poor's Publishing Company, 1920), 2783-84. In a pamphlet of 1919, the company reports the total investment of National Aniline & Chemical Company to have been about $45 million, although it does not specify over what time period or whether that includes investments of its constituent companies prior to merger. "Exhibit of the National Aniline & Chemical Company" (National Aniline & Chemical Company, 1919), 3, 5, FPG, box 107; *The General Chemical Company after Twenty Years, 1899-1919*, 51. This publication cites the market share of 60 percent. "Reminiscences of Eugene Meyer" (February 11, 1952), on page 196 in the CCOHC; quotations from "Output of National Aniline Co.," *DCM* 5 (January 8, 1919): 30; S. A. Tucker to Douglas McKay, November 7, 1919, FPG, box 109.

20. "Reminiscences of Eugene Meyer" (February 11, 1952), on pages 194, 202-3 in the CCOHC; *National Cyclopedia of American Biography*, 36:102, s.v. "Weber, Orlando Franklin"; *Dictionary of American Biography*, supplement 6, s.v. "Meyer, Eugene Isaac"; *National Cyclopedia of American History*, vol. G, 460-61, s.v. "Meyer, Eugene Isaac"; "Allied Chemical & Dye," *Fortune* (June 1930): 132. In his interview cited above, Meyer comments on Weber's troubled relationship with *Fortune* magazine; Meyer suspected Weber might have saved himself some difficulty if he would have advertised in the magazine (p. 203).

21. "Reminiscences of Eugene Meyer" (February 11, 1952), on pages 194-97 in the CCOHC. A brief notice of the change of leadership appeared in "William J. Matheson Resigns," *DCM* 5 (July 2, 1919): 19.

22. *Dictionary of American Biography*, supplement 6, s.v. "Meyer, Eugene Isaac"; "Allied Chemical & Dye," *Fortune* (June 1930): 130; Reader, *Imperial Chemical Industries*, 1:361; Spencer Trask & Co., "A Brief Account of the Achievements and Progress of Allied Chemical & Dye Corporation Since Its Formation in 1920, with a Twelve-Year Record of Income and Resources," October 16, 1933, FPG, box 50; Haynes, *American Chemical Industry*, 6:9. Chandler describes the Allied Chemical merger in 1920 as a failed strategy, a collection of firms making disparate chemicals with "no basic related technology or raw materials" and modeled on the Union Carbide merger of 1917. He describes National Aniline as "a small, specialized dye producer." *Shaping the Industrial Century*, 78-79. I disagree with this characterization. The Allied merger grew out of commercial alliances among its constituent parts during the war, partly embodied in the creation of National Aniline. The German firms were more likely the model than Union Carbide. While National Aniline might have been relatively small compared to other firms, it was the largest American dyes firms at the time of the merger in 1920. The organic stocks of Semet-Solvay and Barrett and the acids of General Chemical combined logically for a firm whose founders intended to match the Germans. I do agree, however, that Orlando Weber abandoned the logic of this combination by directing Allied resources toward nitrogen and by skimping on research.

23. "Allied Chemical & Dye," *Fortune* 1, no. 5 (June 1930): 81-82; Reader, *Imperial Chemical Industries*, 1:318-19; "Reminiscences of Eugene Meyer" (February 11, 1952), on page 203 in the CCOHC; "Industrial Notes," *JIEC* 13 (April 1921): 367; Haynes, *American Chemical Industry*, 3:346 note; Derick, "A Foundation Stone of Our Chemical Industry: Schoellkopf Research Laboratory," 52.

24. Derick to Herty, November 13, 1921, CHH, box 74; W. H. Van Winckel to Dow, February 14, 1919, DHC, folder 190128; Hale to Dow, August 1, 1921, folder 210090. After conversing with a former National Aniline employee, Hale reported that National Aniline's indigo operations were insufficient to meet its obligations and that the company had purchased indigo from Du Pont to fill contracts. Dow to C. B. Hall, December 6, 1920, DHC, folder 200080.

25. Much of this news reached Herbert Dow through his son-in-law and chemist William Hale and through his sales agents. Pardee to Dow, April 22, 1921, and November 5, 1921; Hale to Dow, August 1, 1921; Hale to Dow, April 8, 1925, DHC, folder 250052, and April 15, 1925, folder 250053; Hale to Willard Dow, December 8, 1925, folder 250053.

26. Haynes, *American Chemical Industry*, 6:294-95; Earle S. Corey, "A Chronicle of National Aniline Division," *[National Aniline's] Dyestuffs* 41 (June 1956): 117-24; Dow to A. E. Convers, May 5, 1923, DHC, folder 230007.

27. Hale to Dow, April 15, 1925, DHC, folder 250053.

28. "Allied Chemical and Dye," *Fortune* 1, no. 5 (June 1930): 83; Plumpe, *Die I.G. Farbenindustrie A.G.*, 188-89.

29. A rich literature on Du Pont exists, including Alfred Chandler's works and *Science and Corporate Strategy* by David A. Hounshell and John K. Smith. On Du Pont's organization, see Chandler, *Strategy and Structure*, chap. 2; Chandler, *Scale and Scope*, 181-90; Smith, "National Goals, Industry Structure, and Corporate Strategies," 130-61.

30. W. H. Van Winckel to Dow, February 8, 1922, DHC, folder 220072; W. S. Calcott, "History of the Du Pont Organic Chemicals Department and of the Dye Works and Associated

Laboratories," n.d. (but librarian B. M. Walsh dated the paper January 1, 1959), Hagley Museum and Library, Acc. 2227, series II, box 3, History File; Irénée du Pont testimony, Senate Committee on Finance, *Hearings, Dyestuffs* (1920), 156; Hounshell and Smith, *Science and Corporate Strategy*, 88.

31. Hounshell and Smith, *Science and Corporate Strategy*, 92-95; "German Warrants for Du Pont Chemists," *DCM* 8 (February 23, 1921): 411. At least one of the Germans, Max Engelmann, was an expert in pharmaceuticals. Calcott, "The History of the Organic Chemicals Department," July 17, 1945, Hagley Museum and Library, Acc. 2227, series II, box 3, History File.

32. "The Road to Demoralization," *JIEC* 13 (February 1921): 108; "German Warrants for Du Pont Chemists," *DCM* 8 (February 23, 1921): 411; "Tempest in a German Tea Pot," *DCM* 8 (March 30, 1921): 683-84; "German Criticises [sic] American Dyes," *DCM* 8 (June 8, 1921): 1273.

33. U.S. Tariff Commission, *Census of Dyes and of Other Synthetic Organic Chemicals, 1929*, 52-54; O. M. Bishop and J. H. Sachs, "Progress in the Development and Manufacture of Vat Colors in America," *IEC* 18 (December 1926): 1331-34; M. L. Crossley, "Ten Years of Progress in the Dye and Intermediate Industry," *IEC* 18 (December 1926): 1322-23; U.S. Tariff Commission, *Census of Dyes and of Other Synthetic Organic Chemicals, 1922*, 42; Hounshell and Smith, *Science and Corporate Strategy*, 157-60; Haynes, *American Chemical Industry*, 3:218-19; D. W. Jayne to Fin Sparre, "Newport Chemical Company," September 20, 1930, papers of W. F. Harrington, Hagley Museum and Library, Acc. 1813, box 5. Newport manufactured dyes, numbering over 160, in its primary plant in Carrollville and shipped them to Passaic, New Jersey, where the firm maintained its distribution center. At Passaic, Newport staff mixed and standardized colors, tested dyes in laboratories and a dyehouse, and provided the firm's textile customers with technical assistance. Through the decade, Newport had concentrated primarily on producing dyes, but in 1929 the firm purchased the Rhodia Chemical Company of New Brunswick, New Jersey, which manufactured synthetic organic photochemicals and disinfectants and was affiliated with two French firms. E. H. Killheffer, "The Newport Company," *IEC* 21 (December 1929): 1300-1302; "Coal to Dyestuffs," *Chemical Markets* 27 (December 1930): 590-93. *Chambers Works History*, vol. 2, chap. 3, Hagley Museum and Library, Acc. 1387; D. W. Jayne to Fin Sparre, "Newport Chemical Company," September 20, 1930, W. F. Harrington papers, Hagley Museum and Library, Accession 1813, box 5.

34. U.S. Tariff Commission, *Census of Dyes and Coal-Tar Chemicals, 1919*, 52; U.S. Tariff Commission, *Census of Dyes and Coal-Tar Chemicals, 1920*, 61. Not all firms responded to the census, but the writers of the report generally assumed such firms were primarily engaged in activities other than synthetic dyes production. In 1919, for example, they estimated that 214 firms made at least one dye, but 191 responded to the census. U.S. Tariff Commission, *Census of Dyes and Other Synthetic Organic Chemicals, 1922-1930*. Just considering Du Pont's research expenditures by its dyes department, Du Pont contributed a significant percentage of the national totals, approximately 25 percent of the figures given by the U.S. Tariff Commission. Hounshell and Smith, *Science and Corporate Strategy*, 76, 147-50, 288-91; Chandler, *Scale and Scope*, 175-77.

35. Hounshell and Smith, *Science and Corporate Strategy*, 98-110, quotation on p. 108; Chandler, *Strategy and Structure*, 99-112.

36. Charles M. A. Stine, "Coordination of Laboratory and Plant Effort," *IEC* 24 (February 1932): 191-92; Hounshell and Smith, *Science and Corporate Strategy*, 51, 104.

37. Ernest B. Benger, "The Organization of Industrial Research," *IEC* 22 (June 1930): 572–77; Hounshell and Smith, *Science and Corporate Strategy*, 300–312.

38. "Du Pont Chemical Fellowships," *IEC* 14 (March 1922): 185; Hounshell and Smith, *Science and Corporate Strategy*, 287–93; Robert E. Rose, "Dye Research," *JIEC* 12 (August 1920): 805; Robert E. Rose, "The Education of the Research Chemist," *IEC* 12 (October, 1920): 947–51.

39. Hounshell and Smith, *Science and Corporate Strategy*, 297–300; Conant, "A Review of Some of the Interesting Papers Appearing between October, 1929 and April, 1930," n.d., and Conant to Bolton, January 29, 1930; Bolton to Conant, April 26, 1930; Bolton to Conant, January 21, 1930, Harvard University Archives, UAI 15.898, Papers of James B. Conant.

40. The other firm, which had their methanol on the market before Du Pont, was Commercial Solvents Corporation, with plants in Terre Haute, Indiana, and Peoria, Illinois. They manufactured synthetic methyl alcohol in 1927 in a high-pressure works originally designed to produce ammonia. Haynes, *American Chemical Industry*, 4:169–77, 6:86–87. With the French developers of an ammonia process, Du Pont formed the Lazote, Inc., holding company, which produced the methyl alcohol. Hounshell and Smith, *Science and Corporate Strategy*, 186. U.S. Tariff Commission, *Census of Dyes and of Other Synthetic Organic Chemicals, 1927*, 132.

41. "Duco, Oval Shotgun Powder, Low-Freezing Dynamite, Rubber Accelerators, Dyes," *Journal of Chemical Education* 2 (December 1925): 1162; George Oenslager, "Organic Accelerators," *IEC* 25 (February 1933): 236. Oenslager suggests his firm didn't file a patent. Hounshell and Smith, *Science and Corporate Strategy*, 251; W. S. Calcott, "The History of the Organic Chemicals Department," July 17, 1945, Hagley Museum and Library, Acc. 2227, series II, box 3, History File.

42. "Du Pont on Motor Growth," *New York Times*, January 15, 1922, p. 87; "Du Pont Co. Earned $41,113,938 in Year," *New York Times*, January 28, 1928, p. 21 ($47 million in total income; $41 million in net income); Chandler and Salsbury, *Pierre S. du Pont and the Making of the Modern Corporation*, 429–36, 583; Stocking, "The Du Pont–General Motors Case and the Sherman Act"; Hounshell and Smith, *Science and Corporate Strategy*, 155–57.

43. Silas Bent, "Tetraethyl Lead Fatal to Makers," *New York Times*, June 22, 1925, p. 3; Hounshell and Smith, *Science and Corporate Strategy*, 150–55; Stuart W. Leslie, *Boss Kettering*; Haynes, *American Chemical Industry*, 4:399–404; *Census of Dyes and of Other Synthetic Organic Chemicals, 1930*, 81. Understanding the airborne hazards of lead gasoline combustion came much later.

44. Hounshell and Smith, *Science and Corporate Strategy*, 221–29; Hounshell, "The Evolution of Industrial Research in the United States"; Lipartito, "Rethinking the Invention Factory."

45. Furukawa, *Inventing Polymer Chemistry*, 93–144; Hounshell and Smith, *Science and Corporate Strategy*, 228–33.

46. Hounshell and Smith, *Science and Corporate Strategy*, 234–36; Morris, "Transatlantic Transfer of Buna S Synthetic Rubber Technology, 1932–1945"; Haynes, *American Chemical Industry*, 4:409–12.

47. Hounshell and Smith, *Science and Corporate Strategy*, 223–74.

48. Ibid., 190–209.

49. Clarke, Lamoreaux, and Usselman, "Introduction," in *The Challenge of Remaining Innovative*, 1–35; Hounshell and Smith, *Science and Corporate Strategy*, 147.

50. Dow, report to stockholders, July 26, 1929, DHC, folder 290086-00176; Dow to J. S. Crider, July 26, 1926, folder 260034; Dow to A. E. Convers, January 10, 1925, folder 250030.

51. Undated chart, c. 1924, DHC, folder 240023; "Union Carbide Earnings," *New York Times*, March 30, 1926, p. 36; Dow to Byron E. Helman, July 8, 1930, DHC, folder 300033; Dow to J. S. Crider, July 26, 1926, folder 260034.

52. Earl W. Bennett to Dow, December 22, 1923, DHC, folder 230025; Dow to A. W. Smith, March 13, 1922, folder 220003; G. L. Camp to Dow, December 22, 1923, folder 230055. Dow also notes consistent cost reductions from 1918 to 1923 in a letter to A. E. Convers, May 12, 1923, folder 230008; G. L. Camp to Dow, April 28, 1924, folder 240043; Camp to Dow, April 8, 1925, 250044; Dow memo, April 21, 1928, folder 280021. See below for more on the arrangement with the Swiss. When the president of Calco first learned of Dow's plans, he almost pleaded with Dow to show a little "cooperation." At the time, Calco's aniline plant was operating below capacity. Robert C. Jeffcott, Calco, to Dow, September 29, 1925; Jeffcott to Dow, November 18, 1925; Dow to Jeffcott, December 1, 1925; Jeffcott to Dow, December 26, 1925, DHC, folder 250091; Dow to A. F. Lichtenstein, August 26, 1926, folder 260106; Dow to Lichtenstein, October 6, 1926, folder 260108; Dow to A. E. Convers, October 28, 1926, folder 260033; Dow to Convers, March 15, 1929, folder 290042.

53. Haber, *The Chemical Industry, 1900-1930*, 308-9; O. P. Hopkins, "The Chemical Industry and Trade of Switzerland," *JIEC* 13 (April 1921): 285-86. A chemist formerly employed with Ault & Wiborg, Sidney M. Hull, resented the owners' decision in which "they played traitor and sold out to the Swiss." Hull to Herty, October 19, 1921, CHH, box 83. Importers of Swiss chemicals felt the pressure during congressional testimony. P. R. MacKinney testimony in House Committee on Ways and Means, *Hearings, Dyestuffs* (1919), 375-405. Robert Jeffcott, president of Calco, believed that the Swiss competed "fair" when they purchased the Cincinnati plant because they would have the same expenses as American manufacturers. Jeffcott testimony in Senate Committee on the Judiciary, *Hearings, Alleged Dye Monopoly* (1922), 469.

54. Dow to J. S. Crider, January 9, 1920, DHC, folder 200010; J. F. Oberlin to Dow, January 27, 1920, folder 200085; Dow to Henry Mohn, July 22, 1920, folder 200084; G. L. Camp to A. F. Lichtenstein, November 12, 1920, folder 200010; Camp to G. E. Collings, September 27, 1921, folder 210010; Haber, *The Chemical Industry, 1900-1930*, 307-9; SOCMA, cross index to U.S. Tariff Commission censuses, 1924 and 1930, CHH, box 76; Stocking and Watson, *Cartels in Action*, 508. Dow to A. F. Lichtenstein, May 7, 1926, DHC, folder 260107.

55. Jeffcott to Dow, December 26, 1925; Jeffcott to Dow, November 18, 1925; Dow to Jeffcott, December 1, 1925, DHC, folder 250091; Dow to J. S. Crider, April 3, 1925, DHC, folder 250031. Haynes, *American Chemical Industry*, 4:43-46, 231, 6:21-22; "American Cyanamid Company," *IEC* 22 (March 1930): 301-2; SOCMA, cross index to U.S. Tariff Commission censuses, 1924 and 1930, CHH, box 76. For Calco's history more generally, see Travis, *Dyes Made in America, 1915-1980*.

56. Dow to J. S. Crider, March 29, 1924, DHC, folder 240023; Dow to James T. Pardee, January 14, 1924, folder 24001. Only in 1925 did the Dow Company receive the right to produce phenol under the patent. Dow to A. W. Smith, November 13, 1925, DHC, folder 250051. The patent had been filed by J. W. Aylsworth, who had developed a high-pressure process to manufacture phenol intended for Thomas Edison's phonographs. "Bakelite Corporation, 1910-1935," *Bakelite Review* 7 (1936): 7; Vanderbilt, *Thomas Edison, Chemist*, 238-45; Campbell and Hatton, *Herbert H. Dow*, 135; Haynes, *American Chemical Industry*, 4:215-16. On similar research conducted by General Electric, see Kline, *Steinmetz*, 154-56. Hale described the phenol and aniline production in "New Processes for Phenol and Aniline," Haynes, *American Chemical Industry*, 4:534-36 (Appendix 36). Dow to Crider, July

26, 1926, DHC, folder 260034; Thomas Griswold memo, December 16, 1929, folder 290049; George Baekeland to Russell Manufacturing Company, August 14 and August 15, 1929, folder 290060.

57. Dow to J. S. Crider, July 26, 1926, DHC, folder 260034; Dow to Crider, March 29, 1924, folder 240023; George Ashworth to G. Lee Camp, April 14, 1924, folder 240043.

58. Dow to Norman Clark, November 29, 1926, DHC, folder 260008; Dow to J. E. Rudolph, October 22, 1928, folder 28007; Dow to Neil E. Gordon, October 3, 1928, folder 280009; Dow to A. E. Convers, May 1, 1928, folder 280013.

59. Dow to Neil E. Gordon, Johns Hopkins University, October 3, 1928, DHC, folder 280009; Dow to J. E. Rudolph, October 22, 1928, DHC, folder 280007; Dow to R. Perry Shorts, October 21, 1927, folder 270037. For example, Hale, "New Processes for Phenol and Aniline," recorded in Haynes, *American Chemical Industry*, 4:534-35; "Dr. Smith Dies of Heart Trouble," *Case Alumnus* (March 1927): 4, DHC, folder 270040. The obituary notes Smith's primary role in developing Dow's mustard gas process.

60. Dow to A. H. White, University of Michigan, January 7, 1928, DHC, folder 280055; Dow to A. E. Convers, October 28, 1926, DHC, folder 260033; Brandt, *Growth Company: Dow's First Century*, 230.

61. Herbert H. Dow, "Economic Trend in the Chemical Industry," *IEC* 22 (February 1930): 113-16, quote on 116; Thomas Griswold report, December 16, 1929, DHC, folder 290049.

62. "Union Carbide I: The Corporation," "Union Carbide II: Alloys, Gases, and Carbons," and "Carbide and Carbon Chemicals (Union Carbide III)," *Fortune* 23, no. 6 (June 1941): 61ff.; 24, no. 1 (July 1941): 48ff.; 24, no. 3 (September 1941): 56ff. Hereafter "Union Carbide I, II, and III." *Fortune* changed the title format for the third article, but they are a series of three. Quotation from "Union Carbide III," 58.

63. Readers will recognize my debt to Peter H. Spitz throughout the Union Carbide section. Spitz, *Petrochemicals: The Rise of an Industry*, 69-82. "C. K. G. Billings, Noted Sportsman," *New York Times*, May 7, 1937, p. 25; "Luncheon in a Stable," *New York Times*, March 30, 1903, p. 14.

64. "Union Carbide I," 124; Stief, *A History of Union Carbide Corporation*; T. L. Willson and J. J. Suckert, "The Carbides and Acetylene Commercially Considered," *Journal of the Franklin Institute* 139, no. 5 (May 1895): 321-41. Willson, the Canadian, was trying to make aluminum following Paul Héroult's process; Willson noted that one advantage of acetylene in lighting lay in its notable smell: people learned of leaks quickly.

65. Dienel, *Linde: History of a Technology Corporation, 1879-2004*, 80-83; Bowden and Smith, *American Chemical Enterprise*, 28.

66. The firm first used the name National Carbon Company after reorganization in 1886; it reorganized again in 1899. Charles Brush had ties also to National Carbon Company going back to its founding in 1881. Stapleton, "The Rise of Industrial Research in Cleveland," 234-35; "Union Carbide II," 49, 96; "Union Carbide I," 124.

67. Marcinkowsky and Keller, "Ethylene and Its Derivatives," 300; Spitz, *Petrochemicals*, 70; "Prest-O-Lite Co. v. Davis, et al.," *Trademark Reporter* 4 (1919): 92.

68. The Mellon Institute was part of the University of Pittsburgh from 1914 to 1928. Servos, "Changing Partners: The Mellon Institute, Private Industry, and the Federal Patron," 231-33. Duncan first launched his industrial fellowships at the University of Kansas. Duncan, *Some Chemical Problems of Today* (1911), 224-41.

69. Kinzel, "George Oliver Curme, Jr., 1888-1976," 123-24; G. O. Curme, "Industry's Toolmaker," *IEC* 27 (February 1935): 223; "Union Carbide III," 60-61.

70. Quoting H. L. McBain, Columbia University dean, "Chandler Lecture," *IEC* 25 (May 1933): 582. When Curme won the Perkin Medal two years later, Weidlein liked McBain's words enough to use them verbatim (and unattributed): E. R. Weidlein, "Accomplishments of the Medalist," *IEC* 27 (February 1935): 221-22.

71. For a time, his ethylene investigations contributed to government mustard gas research. Shortly after the Union Carbide merger, the chemical warfare program pulled in Curme and his research program. His work on acetylene, ethylene, and carbon monoxide all fit into war gas research, but the war ended before his group's research left the laboratory. Unlike most researchers and manufacturers involved in gas research, Curme believed his war work spilled over into the commercial production that Union Carbide launched after the war. Curme, "Industry's Toolmaker," *IEC* 27 (February 1935): 224; Spitz, *Petrochemicals*, 71-74.

72. Curme, "Industry's Toolmaker," *IEC* 27 (February 1935): 223-25; "Union Carbide III," 60-61; Kinzel, "George Oliver Curme, Jr.," 124-27; Spitz, *Petrochemicals*, 75-79. With the Prohibition Act in 1917 and prohibition amendment to the Constitution in 1920, the manufacture and consumption of industrial ethyl alcohol became highly regulated as enforcement agencies tried to prevent the diversion of ethyl alcohol to illegal uses. Haynes, *American Chemical Industry*, 4:151-60.

73. "Union Carbide III," 59-62; Spitz, *Petrochemicals*, 78-79; Paul S. Greer, oral interview by Peter J. T. Morris, November 13, 1985, Chemical Heritage Foundation.

74. Curme, "Industry's Toolmaker," *IEC* 27 (February 1935): 225. Two engineering firms, rather than Germany's chemical firms, pioneered the electric arc furnace (Siemens) and refrigeration and gas separation (Linde).

75. Curme, "Memorandum of Products and Processes Pertaining to Work of Organic Synthesis Fellowship for Attention of Mr. G. C. Furness and Mr. W. F. Barret [sic]," July 16, 1919, Union Carbide Corporation collection, Subsidiaries, Union Carbide and Carbon Corporation, 1924, West Virginia Division of Culture and History, Archives and History Section, Union Carbide Corporation collection (Ms 2002-064), box 3, folder 67.

76. Curme, "Industry's Toolmaker," *IEC* 27 (February 1935): 226. Several historians have noted the importance of the automobile industry to the chemical industry; Hounshell and Smith, *Science and Corporate Strategy*, 199, and chaps. 6-7.

77. "Union Carbide III," 61-64; William H. Rinkenbach, "Glycol Dinitrate in Dynamite Manufacture," *Chemical and Metallurgical Engineering* 34 (May 1927): 296-98.

78. "Union Carbide III," 64-65; Haynes, *American Chemical Industry*, 4:151-60, 166-67. E. R. Weidlein, "Accomplishments of the Medalist," *IEC* 27 (February 1935): 221-22; Spitz, *Petrochemicals*, 80-81.

79. Curme, "Industry's Toolmaker," *IEC* 27 (February 1935): 27, 223; "Union Carbide I," 143.

80. Reynolds, "Defining Professional Boundaries: Chemical Engineering in the Early twentieth Century," 699, 709-10; "Chemical Engineering Builds a Synthetic Aliphatic Chemicals Industry" and "An Award for Chemical Engineering Achievement," *Chemical and Metallurgical Engineering* 40 (November 1933): 564-70 and 563.

81. "Bakelite Corporation Merges Condensite and Redmanol," *OPDR* (June 19, 1922), reprint in Baekeland papers, Published and Unpublished Writings, box 11, folder B-2, Smithsonian Institution. Friedel distinguishes Bakelite from earlier celluloid plastics in this way. Friedel, *Pioneer Plastic*, 103-9. Also see Bijker, "The Social Construction of Bakelite."

82. Baekeland, "Practical Life as a Complement to University Education," Perkin Medal Award address, *JIEC* 8 (February 1916): quotation on 184. Baekeland published his account of their family car tour in Baekeland, *A Family Motor Tour through Europe* (New York: The Horseless Age, 1907), 3. A useful biographical obituary is Wallace P. Cohoe, "Leo Hendrik Baekeland, 1863-1944," *Year Book of the American Philosophical Society* (Philadelphia: APS, 1944); a typescript obituary, Celine Baekeland, "Dr. L. H. Baekeland," May 1944, Family Correspondence (IV), box 10, Baekeland papers, Smithsonian Institution. "Silver Anniversary, 1910 to 1935," *Bakelite Review* 7, no. 3 (1935): 20-22, Baekeland papers, Published and Unpublished Writings (VI), box 11, Smithsonian Institution.

83. Baekeland, "Practical Life as a Complement to University Education," Perkin Medal Award address, *JIEC* 8 (February 1916): 187; "Bakelite Corporation, 1910-1935," *Bakelite Review* 7 (1936): 12, 18, 30-38; "Bakelite," *Time* 4, no. 12 (September 22, 1924): 20. Jeffrey Meikle discusses efforts to popularize early plastics by the Bakelite Corporation and also others who envisaged a "utopian" world of limitless applications of the material. Meikle, "Materials and Metaphors: Plastics in American Culture"; and Meikle, "Plastic, Material of a Thousand Uses."

84. "Bakelite Corporation Merges Condensite and Redmanol," *OPDR* (June 19, 1922), reprint in Baekeland papers, Published and Unpublished Writings, box 11, folder B-2, Smithsonian Institution.

85. For Baekeland's struggle to defend his patents, see Mercelis, "Leo Baekeland's Transatlantic Struggle for Bakelite"; House Committee on Patents, *Oldfield Revision and Codification of the Patent Statutes, No. 4* (1912), 29-30.

86. Baekeland to E. G. Oberlin, Naval Research Laboratory, May 13, 1926, Baekeland papers, Personal Correspondence, Smithsonian Institution, box 7; *Bakelite Review* 7, no. 3, Silver Anniversary Number, 1910-1935 (New York: Bakelite Corporation, 1935): 14-17, quotation on 14.

87. Haynes, *American Chemical Industry*, 4:215-16. Griswold to Dow, June 5, 1925, DHC, folder 230020; George Baekeland to Russell Manufacturing Company, August 14, 1929, folder 290060. Elko began business in 1927 when entrepreneurs from the Rubber Service Laboratories of Akron, makers of chemical accelerators used in the manufacture of rubber, converted a failing dyestuffs plant into a supplier of phenol and several inorganic chemicals. Haynes, *American Chemical Industry*, 4:215, 224, 414; George Baekeland to Russell Manufacturing Company, August 14, 1929, DHC, folder 290060; *Bakelite Review* 7, no. 3, Silver Anniversary Number, 1910-1935 (New York: Bakelite Corporation, 1935): 18; "Merger Approved by Union Carbide," *New York Times*, August 30, 1939, p. 29.

88. U.S. Tariff Commission, *Census of Dyes and of Other Synthetic Organic Chemicals, 1929*, 60, 136; U.S. Tariff Commission, *Census of Dyes and of Other Synthetic Organic Chemicals, 1927*, 56. National Aniline was one of the manufacturers of the new kind of synthetic resins. Haynes, *American Chemical Industry*, 6:294-95. In 1930, production dropped to 30.8 million pounds. Figures from U.S. Tariff Commission, *Census of Dyes and Other Synthetic Organic Chemicals*, 1922, 1929, 1930. More than most chemical products, Bakelite sales dropped during the Depression, largely because the automobile industry was an important customer. Dow address, annual stockholders' meeting, June 1930, DHC, folder 300057-00182.

89. Ulrich Marsch has compiled a valuable set of numbers on I.G. Farben's research costs and sales for a range of products from 1926 to 1942 in "Strategies for Success," 52, 70.

90. "Says Ford Methods Give Chemists Lead," *New York Times*, August 6, 1928, p. 18.

Conclusion

1. Fritz Merck to Rudolf E. Gruber, March 21, 1922; Gruber to E. Merck, April 21, 1922; Gruber to Fritz Merck, April 26, 1922; E. Merck to Merck & Co., May 26, 1922; Merck & Co. to E. Merck, June 14, 1922, Row 3, 5.7.3, E. Merck correspondence, etc., Merck & Company archives. Fletcher S. Boig and Paul W. Howerton, "History and Development of Chemical Periodicals in the Field of Organic Chemistry, 1877–1949," *Science* 115 (January 11, 1952): 29. Thanks to Peter Murmann for this article. Also, during and shortly after the war, Americans translated many of the classic German reference books in chemistry. "Translating German Chemical Books," *DCM* 5 (January 8, 1919): 7.

2. Hayes, *Industry and Ideology: I.G. Farben in the Nazi Era*. Joseph Borkin served on the team prosecuting I.G. Farben after World War II and later wrote *The Crime and Punishment of I.G. Farben*.

3. Galambos, *The Creative Society and the Price Americans Paid for It*, 54–56.

4. Wheeler testimony, July 26, 1882, excerpted in Bernhard C. Hesse, "Lest We Forget! Who Killed Cock Robin? The U.S. Tariff History of Coal Tar Dyes," *JIEC* 7 (August 1915): 696.

5. Jones, "Chemical Warfare Research during World War I"; Harden, *Inventing the NIH: Federal Biomedical Research Policy, 1887–1937*.

6. Becker, *The Dynamics of Business-Government Relations*; Ndiaye, *Nylon and Bombs: DuPont and the March of Modern America*.

7. Chandler, *Shaping the Industrial Century*; Dow to A. E. Convers, May 14, 1929, DHC, folder 290042. Incidentally, Dow had theories on Great Britain, too. England, he said, was "losing out as a manufacturing nation because the owners are not in touch with the workers and neither one has proper respect or sympathy for the other." Dow to W. R. Veazey, February 22, 1930, DHC, folder 300049.

8. Smith, "National Goals, Industry Structure, and Corporate Strategies: Chemical Cartels between the Wars"; Reader, *Imperial Chemical Industries*; *Annual Report of the Alien Property Custodian for the Period Beginning June 30, 1943*.

Bibliography

Primary Source Periodicals

American Bar Association Journal
American Dyestuff Reporter
American Journal of International Law
Boston Herald
Bureau of Mines Bulletin
Bureau of Standards Journal of Research
Chemical and Engineering News
Chemical and Metallurgical Engineering
Chemical Industries
Chemical Markets
Chemical Week
Chemiker Zeitung
Congressional Record
Drug and Chemical Markets
Drug Markets
Industrial and Engineering Chemistry
Journal of the American Chemical Society
Journal of the American Medical Association
Journal of the American Pharmaceutical Association
Journal of the Franklin Institute
Journal of Industrial and Engineering Chemistry
Los Angeles Times
Metallurgical and Chemical Engineering
New York Commercial
New York Times
Oil, Paint, and Drug Reporter
Pharmaceutical Era
Scientific American
Scientific Papers of the Bureau of Standards
Technologic Papers of the Bureau of Standards
Virginia Law Review
Wall Street Journal
Washington Post
Weekly Drug Markets
Wilmington Morning News

Court Cases

U.S. Circuit Court of Appeals for the Second Circuit. *Thomas Ellett Hodgskin and George Simon v. United States.* January 18, 1922.

U.S. District Court for the District of Delaware. *USA v. The Chemical Foundation, Incorporated, Record of Final Hearing,* No. 502 in Equity, 1923.

U.S. District Court for the Northern District of Illinois. *USA v. Du Pont, General Motors, et al.,* Civil Action No. 49 C 1071, 1949.

U.S. Supreme Court. *Farbwerke vormals Meister Lucius und Bruning, Deutsche Gold und Silber Scheide Anstalt vorm. Roessler, and Badische-Anilin und Soda-Fabrik v. Chemical Foundation, Inc., E. I. du Pont de Nemours and Company, and Walter O. Woods, as Treasurer of the United States.* Case Nos. 180, 181, 182, 271, 272, 273, and 274. October 1930.

Archives

BASF, Unternehmensarchiv, Ludwigshafen, Germany
Bayer-Archiv, Leverkusen, Germany
Chemical Heritage Foundation Archives, Chemical Heritage Foundation, Philadelphia, Pa.
   Dow Historical Collection, H. H. Dow Series (DHC)
Columbia University, Columbia Center for Oral History Collection (CCOHC), New York, N.Y.
   Reminiscences of Eugene Meyer
Columbia University, University Archives and Columbiana Library, New York, N.Y.
   Marston Taylor Bogert Papers
   Biographical Files
   Commencement Programs
   Annual Reports
Cornell University, Division of Rare and Manuscript Collections, Cornell University Library, Ithaca, N.Y.
   Department of Chemistry
   Deceased Alumni Files
   Commencement Programs
   Annual Reports
George Eastman House, Rochester, N.Y.
   George Eastman Papers
Eastman Kodak Company, Rochester, N.Y.
   Central File
Emory University, Robert Woodruff Library, Manuscript, Archives, and Rare Book Library, Atlanta, Ga.
   Charles H. Herty Papers (CHH)
Hagley Museum and Library, Wilmington, Del.
   Chambers Works History, Accession 1387
   E. I. du Pont de Nemours and Company, Administrative Papers, Accession 1662
   Du Pont Executive Committee, Accession 2091
   Lammot du Pont Jr., Papers, Accession 676

W. F. Harrington Papers, Accession 1813
Harvard University, Nathan Pusey Library, Special Collections, Cambridge, Mass.
    James B. Conant Papers, UAI 15.898
    Department of Chemistry, 1924-1924, UAV 275.275
    Elmer Peter Kohler, HUG 1495
    G. P. Baxter, Correspondence About Positions Wanted and Offered, 1912-1928, UAV 275.45, and 1918-1929, UAV 275.42
Herbert Hoover Presidential Library, West Branch, Iowa
    Herbert Hoover Papers
Library of Congress, Manuscripts Division, Washington, D.C.
    Eugene Meyer Papers
Merck and Company, Whitehouse Station, N.J.
    Company Archives
National Academy of Sciences, Washington, D.C.
    National Research Council Papers
National Archives and Records Administration, Washington, D.C., and College Park, Md. (by Record Group [RG])
    RG 39, Department of the Treasury
    RG 40, Department of Commerce
    RG 59, Department of State
    RG 60, Department of Justice
    RG 70, Bureau of Mines
    RG 81, U.S. Tariff Commission
    RG 97, Bureau of Chemistry
    RG 107, Office of the Secretary of War
    RG 131, Office of Alien Property
    RG 151, Bureau of Foreign and Domestic Commerce (BFDC)
    RG 156, Office of the Chief of Ordnance
    RG 175, Chemical Warfare Service
The National Archives of the United Kingdom, Public Records Office, London, U.K.
    Foreign Office Papers
Ohio Historical Society, Columbus, Ohio
    Harry M. Daugherty Papers
    MIC 3 Warren G. Harding Papers [microform]
Smithsonian Institution, National Museum of American History, Archives Center, Washington, D.C.
    Leo H. Baekeland Papers
U.S. Army Heritage and Education Center, Carlisle, Pennsylvania
University of Chicago, Joseph Regenstein Library, Department of Special Collections, Chicago, Ill.
    University Presidents' Papers, 1889-1925 and 1925-1945
    Julius Stieglitz Biographical File
University of Illinois, University Archives, Champaign-Urbana, Ill.
    Roger Adams Papers
    College of Liberal Arts and Sciences: Chemistry and Chemical Engineering
    Presidential Papers

Register of the University of Illinois
Alumni Morgue File
University of Michigan, Bentley Historical Library, Ann Arbor, Mich.
Moses Gomberg Papers
Board of Regents
College of Pharmacy
Edward Henry Kraus Papers
Chemical Laboratory and Department of Chemistry
Necrology Files
University of Virginia, Albert and Shirley Small Special Collections Library, Charlottesville, Va.
Papers of Edward Reilly Stettinius Sr., Accession 2723
University of Wyoming, American Heritage Center, Laramie, Wyo.
Francis P. Garvan Collection
Yale University, Manuscripts and Archives, Yale University Library, New Haven, Conn.
Catalogue of Yale University, 1909–1930
Reports of the President, Bulletin of Yale University
Class Books
Presidential Records

Articles, Books, Theses, and Government Documents

The Abbott Almanac: 100 Years of Commitment to Quality Health Care. Elmsford, N.Y.: Benjamin Company, Inc., 1987.

Abelshauser, Werner, Wolfgang von Hippel, Jeffrey Allan Johnson, and Raymond G. Stokes. *BASF: The History of a Company*. Cambridge: Cambridge University Press, 2004.

Ackerknecht, Erwin H. *A Short History of Medicine*. 1955. Rev. ed., Baltimore: Johns Hopkins University Press, 1982.

Ackerman, Carl W. *George Eastman*. Boston: Houghton Mifflin Company, 1930.

"An Act to Reduce Tariff Duties and to Provide Revenue for the Government, and for Other Purposes." 63rd Cong., 1st sess., October 3, 1913. In *The Statutes at Large of the United States of America*, vol. 38, pt. 1. Washington, D.C.: Government Printing Office, 1915.

Aftalion, Fred. *A History of the International Chemical Industry*. 2nd ed. Philadelphia: Chemical Heritage Foundation Press, 2001.

"Allied Chemical & Dye." *Fortune* 1, no. 5 (June 1930): 81–83, 130, 132, 134.

Allison, David K. *New Eye for the Navy: The Origins of Radar at the Naval Research Laboratory*. Washington, D.C.: Government Printing Office, 1981.

Amatori, Franco, and Geoffrey Jones, eds. *Business History around the World*. New York: Cambridge University Press, 2003.

"American I. G. Chemical Corporation." In *Moody's Manual of Investments, 1930*, 2149–50. New York: Moody's Investors Service, 1930.

Annual Report of the Alien Property Custodian for the Year 1922. 67th Cong., 4th sess., January 24, 1923, H. Doc. 525. Washington, D.C.: Government Printing Office, 1923.

Annual Report of the Alien Property Custodian for the Year 1924. 68th Cong., 2nd sess., January 26, 1925. S. Doc. 203. Washington, D.C.: Government Printing Office, 1925.

Annual Report of the Alien Property Custodian for the Period Beginning June 30, 1943. 79th Cong., 1st sess., May 21, 1945. H. Doc. 184. Washington, D.C.: Government Printing Office, 1945.

Archibugi, Daniele, and Bengt-Åke Lundvall, eds. *The Globalizing Learning Economy.* New York: Oxford University Press, 2001.

Arora, Ashish, Ralph Landau, and Nathan Rosenberg, eds. *Chemicals and Long-Term Economic Growth: Insights from the Chemical Industry.* New York: John Wiley and Sons, in conjunction with the Chemical Heritage Foundation, 1998.

Badash, Lawrence. "British and American Views of the German Menace in World War I." *Notes and Records of the Royal Society of London* 34, no. 1 (July 1979): 91–121.

"Bakelite Corporation, 1910–1935." *Bakelite Review* 7 (1936): 5–18.

Balogh, Brian. *A Government Out of Sight: The Mystery of National Authority in Nineteenth-Century America.* New York: Cambridge University Press, 2009.

———. "Reorganizing the Organizational Synthesis: Federal-Professional Relations in Modern America." *Studies in American Political Development* 5 (Spring 1991): 119–72.

Barker, Joseph Warren. *Research Corporation (1912–1952).* New York: Newcomen Society, 1952.

Bartlett, Howard R. "The Development of Industrial Research in the United States." In *Research—A National Resource*, 19–77. Report of the Science Committee to the National Resources Committee. Washington, D.C.: Government Printing Office, 1938.

Bartrams, Kenneth, Nicolas Coupain, and Ernst Homburg. *Solvay: History of a Multinational Family Firm.* Cambridge: Cambridge University Press, 2013.

Baruch, Bernard M. *American Industry in the War, a Report of the War Industries Board.* New York: Prentice-Hall, 1941.

Bäumler, Ernst. *A Century of Chemistry.* Düsseldorf: Econ Verlag GmbH, 1968.

Baumol, William J. "Productivity Growth, Convergence, and Welfare." *American Economic Review* 76 (1986): 1072–85.

Becker, William H. *The Dynamics of Business-Government Relations: Industry and Exports, 1893–1921.* Chicago: University of Chicago Press, 1982.

Beer, John J. *The Emergence of the German Dye Industry.* Urbana: University of Illinois Press, 1959.

Beiträge zur hundertjährigen Firmengeschichte, 1863–1963. Cologne: Farbenfabriken Bayer A.G., 1964.

Ben-Atar, Doron S. *Trade Secrets: Intellectual Piracy and the Origins of American Industrial Power.* New Haven, Conn.: Yale University Press, 2004.

Bender, Thomas. "Historians, the Nation, and the Plenitude of Narratives." In *Rethinking American History in a Global Age*, 1–21. Berkeley: University of California Press, 2002.

Berk, Gerald. "Communities of Competitors: Open Price Associations and the American State, 1911–1929." *Social Science History* 20 (Autumn 1996): 375–400.

Bijker, Wiebe E. "The Social Construction of Bakelite: Toward a Theory of Invention." In *The Social Construction of Technological Systems: New Directions in the Sociology and History of Technology*, edited by Wiebe E. Bijker, Thomas P. Hughes, and Trevor J. Pinch, 159–87. Cambridge, Mass.: MIT Press, 1987.

Blaszczyk, Regina L. *The Color Revolution.* Cambridge, Mass.: MIT Press, 2012.

———. *Imagining Consumers: Design and Innovation from Wedgwood to Corning.* Baltimore: Johns Hopkins University Press, 2002.

Boemeke, Manfred F., Gerald D. Feldman, and Elisabeth Glaser, eds. *The Treaty of Versailles: A Reassessment after 75 Years.* New York: German Historical Institute and Cambridge University Press, 1998.

Boig, Fletcher S., and Paul W. Howerton. "History and Development of Chemical Periodicals in the Field of Organic Chemistry, 1877-1949." *Science* 115 (January 11, 1952): 25-31.

Borkin, Joseph. *The Crime and Punishment of I.G. Farben.* New York: Free Press, 1978.

Bowden, Mary Ellen, and John Kenly Smith Jr. *American Chemical Enterprise.* Philadelphia: Chemical Heritage Foundation, 1994.

Brandes, Joseph. *Herbert Hoover and Economic Diplomacy: Department of Commerce Policy, 1921-1928.* Pittsburgh: University of Pittsburgh Press, 1962.

Brandt, Allan M. *No Magic Bullet: A Social History of Venereal Disease in the United States since 1880.* New York: Oxford University Press, 1985.

Brandt, E. N. *Growth Company: Dow Chemical's First Century.* East Lansing: Michigan State University Press, 1997.

Braun, Hans-Joachim. "The National Association of German-American Technologists and Technology Transfer between Germany and the United States, 1884-1930." *History of Technology* 8 (1983): 15-38. London: Mansell Publishing, 1984.

Breithut, Frederick E. "British Dyestuffs Industry." *Supplement to Commerce Reports.* Department of Commerce. Bureau of Foreign and Domestic Commerce. Trade Information Bulletin No. 231, May 19, 1924. Washington, D.C.: Government Printing Office, 1924.

———. "The German Coal-Tar Chemical Industry: Production, Export, and Import Statistics." *Supplement to Commerce Reports.* Department of Commerce. Bureau of Foreign and Domestic Commerce. Trade Information Bulletin No. 141, September 3, 1923. Washington, D.C.: Government Printing Office, 1923.

Breithut, Frederick E. and J. Allen Palmer. "The Italian Dyestuffs Industry: Production, Export, and Import Statistics." *Supplement to Commerce Reports.* Department of Commerce. Bureau of Foreign and Domestic Commerce. Trade Information Bulletin No. 234, May 26, 1924. Washington, D.C.: Government Printing Office, 1924.

Brose, Eric Dorn. *A History of the Great War: World War One and the International Crisis of the Early Twentieth Century.* New York: Oxford University Press, 2009.

Burke, Kathleen. *Britain, America, and the Sinews of War, 1914-1918.* Boston: George Allen and Unwin, 1985.

Bush, Vannevar. "Frederick Gardner Cottrell, 1877-1948." *Biographical Memoirs of the Fellows of the National Academy of Sciences* 27 (1952): 1-11.

By-Product Coke and Gas Oven Plants. Pittsburgh, Pa.: Koppers Company, 1919.

Cain, John Cannell, and Jocelyn Field Thorpe. *The Synthetic Dyestuffs.* 5th ed. London: Charles Griffin & Company, 1920.

———. *The Synthetic Dyestuffs and the Intermediate Products from Which They Are Derived.* 2nd ed., rev. London: Charles Griffin and Company, 1913.

Cameron, Frank. *Cottrell: Samaritan of Science.* New York: Doubleday, 1952.

Campbell, Murray, and Harrison Hatton. *Herbert H. Dow: Pioneer in Creative Chemistry.* New York: Appleton-Century-Crofts, 1951.

"Carbide and Carbon Chemicals (Union Carbide III)." *Fortune* 24, no. 3 (September 1941): 56-67, 154.

Carpenter, Daniel P. *The Forging of Bureaucratic Autonomy: Reputations, Networks, and Policy Innovation in Executive Agencies, 1862-1928.* Princeton, N.J.: Princeton University Press, 2001.

Carroll, P. Thomas. "Academic Chemistry in America, 1876-1976: Diversification, Growth, and Change." Ph.D. diss., University of Pennsylvania, 1982.

Chambers, John Whiteclay II. *To Raise an Army: The Draft Comes to Modern America.* New York: Free Press, 1987.

Chandler, Alfred D. Jr. *Scale and Scope.* Cambridge, Mass.: Harvard University Press, 1990.

———. *Shaping the Industrial Century: The Remarkable Story of the Evolution of the Modern Chemical and Pharmaceutical Industries.* Cambridge, Mass.: Harvard University Press, 2009.

———. *Strategy and Structure.* Cambridge, Mass.: MIT Press, 1962.

———. *The Visible Hand: The Managerial Revolution in American Business.* Cambridge, Mass.: Harvard University Press, 1977.

Chandler, Alfred D. Jr., and Stephen Salsbury. *Pierre S. du Pont and the Making of the Modern Corporation.* New York: Harper and Row, 1971.

Chemical Warfare Service, Development Division. *History of the Development Division.* Nela Park, Cleveland, Ohio, March 1919. Available at Western Reserve Historical Society, Cleveland.

———. *Report of Midland Section.* Midland, Mich., 1918. Available at the U.S. Army Military Historical Institute, Carlisle, Pa.

———. *Standard Methods for the Manufacture of New G-34.* Nela Park, Cleveland, Ohio. Available at the U.S. Army Military Historical Institute, Carlisle, Pa.

Childs, William Hamlin. "By-Products Recovered in the Manufacture of Coke." Paper presented at the American Iron and Steel Institute, May 26, 1916.

Clarke, Sally H., Naomi R. Lamoreaux, and Steven W. Usselman, eds. *The Challenge of Remaining Innovative: Insights from Twentieth-Century American Business.* Stanford, Calif.: Stanford University Press, 2009.

Coben, Stanley. *A. Mitchell Palmer: Politician.* New York: Columbia University Press, 1963.

Cohen, Wesley M., and David A. Levinthal. "Absorptive Capacity: A New Perspective on Learning and Innovation." *Administrative Science Quarterly* 35 (1990): 128-52.

Combs, Jerald A. *American Diplomatic History: Two Centuries of Changing Interpretations.* Berkeley: University of California Press, 1983.

Cooper, Carolyn, ed. "Patents and Invention" (special issue). *Technology and Culture* 32 (October 1991).

Copeland, Melvin Thomas. *The Cotton Manufacturing Industry of the United States.* New York, 1917. Reprint, New York: August M. Kelley, 1966.

Corey, Earle S. "A Chronicle of National Aniline Division." *[National Aniline's] Dyestuffs* 41 (June 1956): 117-24.

Cornell, Thomas D. *Establishing Research Corporation.* Tucson, Ariz.: Research Corporation, 2004.

Cottrell, F. G. "Division of Chemistry and Chemical Technology." In *Report of the National Research Council,* 34-37. Washington, D.C.: Government Printing Office, 1923.

Craig, Gordon A. *Germany, 1866-1945.* New York: Oxford University Press, 1978.

Crawford, Elisabeth, Terry Shinn, and Sverker Sörlin, eds. "The Nationalization and Denationalization of the Sciences: An Introductory Essay." In *Denationalizing Science: The Contexts of International Scientific Practice*, 1–37. Dordrecht: Kluwer Academic Publishers, 1993.

Crowell, Benedict. *America's Munitions, 1917–1918*. Washington, D.C.: Government Printing Office, 1919.

Crowell, Benedict, and Robert Forrest Wilson. *The Armies of Industry*. Vols. 1–2 in *How America Went to War*. New Haven, Conn.: Yale University Press, 1921.

Cuff, Robert D. *The War Industries Board: Business-Government Relations during World War I*. Baltimore: Johns Hopkins University Press, 1973.

———. "We Band of Brothers—Woodrow Wilson's War Managers." *Canadian Review of American Studies* 5 (Fall 1974): 113–43.

Daugherty, Harry M., with Thomas Dixon. *The Inside Story of the Harding Tragedy*. New York: Churchill Company, 1932.

Delahanty, Thomas. "The German Dyestuffs Industry." *Miscellaneous Series* no. 126. Department of Commerce. Bureau of Foreign and Domestic Commerce. Washington, D.C.: Government Printing Office, 1924.

Derick, Clarence. "A Foundation Stone of Our Chemical Industry: Schoellkopf Research Laboratory." N.p.: Clarence Derick family, August 15, 1961.

Dewing, Arthur S. *Corporate Promotion and Reorganization*. Cambridge, Mass.: Harvard University Press, 1924.

Dienel, Hans-Liudger. *Linde: History of a Technology Corporation, 1879–2004*. New York: Palgrave Macmillan, 2004.

Dillon, John H. *History, Textile Research Institute, 1930–1970*. 2nd ed. Princeton, N.J.: Textile Research Institute, 1970.

Doerries, Reinhard R. *Imperial Challenge: Ambassador Count Bernstorff and German-American Relations, 1908–1917*. Translated by Christa D. Shannon. Chapel Hill: University of North Carolina Press, 1989.

Dreicer, Gregory. "Building Bridges and Boundaries: The Lattice and the Tube, 1820–1860." *Technology and Culture* 51 (January 2010): 126–63.

Duncan, Robert Kennedy. *Some Chemical Problems of Today*. New York: Harper & Brothers, 1911.

Dunlavy, Colleen A. *Politics and Industrialization: Early Railroads in the United States and Prussia*. Princeton, N.J.: Princeton University Press, 1994.

Dupree, A. Hunter. *Science in the Federal Government: A History of Policies and Activities*. 1957. Reprint, Baltimore: Johns Hopkins University Press, 1987.

Dutton, Harold I. *The Patent System and Inventive Activity during the Industrial Revolution, 1750–1852*. Manchester: Manchester University Press, 1984.

Dyer, Davis, and David B. Sicilia. *Labors of a Modern Hercules: Evolution of a Chemical Company*. Boston: Harvard Business School Press, 1990.

Edgerton, David. *The Shock of the Old: Technology and Global History since 1900*. New York: Oxford University Press, 2007.

The Edgewood Arsenal. Published as an issue in *The Chemical Warfare* 1, no. 5 (March 1919).

Eksteins, Modris. *Rites of Spring: The Great War and the Birth of the Modern Age*. Boston: Houghton Mifflin, 1989.

Estevadeordal, Antoni, Brian Frantz, and Alan M. Taylor. "The Rise and Fall of World Trade, 1870–1939." *Quarterly Journal of Economics* 118, no. 2 (May 2003): 359–407.

Fagerberg, Jan, David C. Mowery, and Richard R. Nelson, eds. *The Oxford Handbook of Innovation.* New York: Oxford University Press, 2005.

Fearon, Peter. *War, Prosperity, and Depression: The U.S. Economy, 1917–1945.* Oxford: Philip Allan Publishers, 1987.

Fehr, Joseph Conrad. "Paying our Claims against Germany." *American Bar Association Journal* 12 (1926): 408–12.

Feldman, Martin. "Asleep at the Swatch: The Non-Emergence of the American Synthetic Dye Industry, 1856–1916." Paper presented to the History Division, American Chemical Society, March 14, 1994, San Diego, Calif.

Ferguson, Niall. *The Pity of War: Explaining World War I.* New York: Basic Books, 1999.

Finlay, Mark. *Growing American Rubber: Strategic Plants and the Politics of National Security.* New Brunswick, N.J.: Rutgers University Press, 2009.

Fleischer, Arndt. *Patentgesetzgebung und chemisch-pharmazeutische Industrie im deutschen Kaiserreich, 1871–1918.* Stuttgart: Deutscher Apotheker Verlag, 1984.

Forman, Paul. "Scientific Internationalism and the Weimar Physicists: The Ideology and its Manipulation in Germany after World War I." *Isis* 64 (1973): 151–80.

Forrestal, Dan J. *Faith, Hope, and $5000: The Story of Monsanto.* New York: Simon and Schuster, 1977.

Fox, M. R. *Dye Makers of Great Britain, 1856–1976.* Manchester: Imperial Chemical Industries, 1987.

Foy, Fred C. *Ovens, Chemicals, and Men! Koppers Company, Inc.* New York: Newcomen Society, 1958.

Freemantle, Michael. *Gas! Gas! Quick, Boys! How Chemistry Changed the First World War.* Stroud, U.K.: History Press, 2012.

French, Michael. *The U.S. Tire Industry: A History.* Boston: Twayne Publishers, 1991.

Friedel, Robert. *Pioneer Plastic: The Making and Selling of Celluloid.* Madison: University of Wisconsin Press, 1983.

Furnas, C. C., ed. *Roger's Industrial Chemistry.* 6th ed. 2 vols. New York: D. Van Nostrand, 1942.

Furukawa, Yasu. *Inventing Polymer Chemistry: Staudinger, Carothers, and the Emergence of Macromolecular Chemistry.* Philadelphia: University of Pennsylvania Press, 1998.

Fussell, Paul. *The Great War and Modern Memory.* New York: Oxford University Press, 1975.

Gabriel, Joseph M. "A Thing Patented Is a Thing Divulged: Francis E. Stewart, George S. Davis, and the Legitimization of Intellectual Property Rights in Pharmaceutical Manufacturing, 1879–1911." *Journal of the History of Medicine and Allied Sciences* 64, no. 2 (April 2009): 135–172.

Galambos, Louis. *Competition and Cooperation: The Emergence of a National Trade Association.* Baltimore: Johns Hopkins University Press, 1966.

———. *The Creative Society and the Price Americans Paid for It.* New York: Cambridge University Press, 2012.

———. "The Emerging Organizational Synthesis in Modern American History." *Business History Review* 44, no. 3 (Autumn 1970): 279–90.

———. "Recasting the Organizational Synthesis: Structure and Process in the Twentieth and Twenty-first Centuries." *Business History Review* 79, no. 1 (Spring 2005): 1–38.

———. "Technology, Political Economy, and Professionalization: Central Themes of the Organizational Synthesis." *Business History Review* 7, no. 4 (Winter 1983): 471–93.

Galambos, Louis, et al. *Values and Visions: A Merck Century*. Rahway, N.J.: Merck and Company, 1991.

Galambos, Louis, Takashi Hikino, and Vera Zamagni, eds. *The Global Chemical Industry in the Age of the Petrochemical Revolution*. New York: Cambridge University Press, 2007.

Galambos, Louis, and Jeffrey L. Sturchio. "The Origins of an Innovative Organization: Merck & Co., Inc., 1891–1960." Manuscript draft presented at the annual meeting of the Society for the History of Technology, Uppsala, Sweden, August 1992.

———. "Transnational Investment: The Merck Experience, 1891–1925." In *Transnational Investment from the Nineteenth Century to the Present*, edited by Hans Pohl, 227–43. Stuttgart: Franz Steiner Verlag, 1994.

Gathings, James. *International Law and American Treatment of Alien Enemy Property*. Washington, D.C.: American Council on Public Affairs, 1940.

Gatzke, Hans W. *Germany and the United States: A "Special Relationship."* Cambridge, Mass.: Harvard University Press, 1980.

Geiger, Roger. *To Advance Knowledge: The Growth of American Research Universities, 1900–1940*. New York: Oxford University Press, 1986.

Gelfand, Lawrence E. "The American Mission to Negotiate Peace: An Historian Looks Back." In *The Treaty of Versailles: A Reassessment after 75 Years*, edited by Manfred F. Boemeke, Gerald D. Feldman, and Elisabeth Glaser, 189–202. New York: German Historical Institute and Cambridge University Press, 1998.

The General Chemical Company after Twenty Years, 1899–1919. New York: General Chemical Company, 1919.

Geroski, Paul. "Markets for Technology." In *Handbook of the Economics of Innovation and Technology Change*, edited by Paul Stoneman, 100–108. Cambridge, Mass.: Blackwell, 1995.

Gerschenkron, Alexander. *Economic Backwardness in Historical Perspective*. Cambridge, Mass.: Harvard University Press, 1962.

Gilchrist, Harry L. *A Comparative Study of World War Casualties from Gas and Other Weapons*. Edgewood Arsenal, Md.: Chemical Warfare School, 1931.

Gispen, Kees. *Poems in Steel: National Socialism and the Politics of Invention from Weimar to Bonn*. New York: Berghahn Books, 2002.

Glaser, Elisabeth. "The Making of the Economic Peace." In *The Treaty of Versailles: A Reassessment after 75 Years*, edited by Manfred F. Boemeke, Gerald D. Feldman, and Elisabeth Glaser, 371–399. New York: German Historical Institute and Cambridge University Press, 1998.

Glaser-Schmidt, Elisabeth. "Foreign Trade Strategies of I.G. Farben after World War I." *Business and Economic History* 23 (Fall 1994): 201–11.

Glassie, Henry H. "Some Legal Aspects of the Flexible Tariff, Parts I and II." *Virginia Law Review* 11 (1925): 329–48, 442–66.

Goldstein, Carolyn. *Creating Consumers: Home Economists and Twentieth-Century America*. Chapel Hill: University of North Carolina, 2012.

Goldstein, Judith. *Ideas, Interests, and American Trade Policy*. Ithaca, N.Y.: Cornell University Press, 1993.

Gorin, Frank B. "French Dyestuffs Industry." *Supplement to Commerce Reports*. Department of Commerce. Bureau of Foreign and Domestic Commerce. Trade Information Bulletin No. 253, August 4, 1924. Washington, D.C.: Government Printing Office, 1924.

Graham, Otis. L. Jr. *Losing Time: The Industrial Policy Debate.* Cambridge, Mass.: Harvard University Press, 1992.

Greer, Paul S. Oral interview by Peter J. T. Morris. Philadelphia: Chemical Heritage Foundation, November 13, 1985.

Gruber, Carol S. *Mars and Minerva: World War I and the Uses of the Higher Learning in America.* Baton Rouge: Louisiana State University Press, 1975.

Haber, Ludwig F. *The Chemical Industry during the Nineteenth Century: A Study of the Economic Aspect of Applied Chemistry in Europe and North America.* Oxford: Clarendon Press, 1958, 1969.

———. *The Chemical Industry, 1900–1930: International Growth and Technological Change.* Oxford: Clarendon Press, 1971.

———. *The Poisonous Cloud.* Oxford: Clarendon Press, 1986.

Hamilton, Alice. *Exploring the Dangerous Trades: The Autobiography of Alice Hamilton, MD.* Boston: Northeastern University Press, 1985.

———. *Industrial Poisoning in Making Coal-Tar Dyes and Dye Intermediates.* Department of Labor. Bureau of Labor Statistics. Industrial Accidents and Hygiene Series No. 280. Washington, D.C.: Government Printing Office, 1921.

Harden, Victoria A. *Inventing the NIH: Federal Biomedical Research Policy, 1887–1937.* Baltimore: Johns Hopkins University Press, 1986.

Hart, David M. "Corporate Technological Capabilities and the State: A Dynamic Historical Interaction." In *Constructing Corporate America: History, Politics, Culture,* edited by Kenneth Lipartito and David B. Sicilia, 168–87. New York: Oxford University Press, 2004.

———. *Forged Consensus: Science, Technology, and Economic Policy in the United States, 1921–1953.* Princeton, N.J.: Princeton University Press, 1998.

———. "Herbert Hoover's Last Laugh: The Enduring Significance of the Associative State in the United States." *Journal of Policy History* 10, no. 4 (October 1998): 419–44.

Hawes, Charles S. "Coal-Tar Dyes for which Import Licenses were Granted during the Fiscal Year 1920." Department of State. War Trade Board Section. Washington, D.C.: Government Printing Office, 1921.

Hawley, Ellis. *The Great War and the Search for a Modern Order: A History of the American People and Their Institutions, 1917–1933.* New York: St. Martin's Press, 1979.

———. "Herbert Hoover, the Commerce Secretariat, and the Vision of an 'Associative State,' 1921–1928." *Journal of American History* 61 (June 1974): 116–40.

Hayes, Peter. *Industry and Ideology: I.G. Farben in the Nazi Era.* New York: Cambridge University Press, 1987.

Haynes, Williams. *American Chemical Industry.* 6 vols. New York: D. Van Nostrand, 1945–1954.

———, ed. *Who's Who in the Chemical and Drug Industry.* New York: Haynes Publications, Inc., 1928.

Headrick, Daniel R. *The Tools of Empire: Technology and European Imperialism in the Nineteenth Century.* New York: Oxford University Press, 1981.

Hershberg, James G. *James B. Conant: Harvard to Hiroshima and the Making of the Nuclear Age.* Stanford, Calif.: Stanford University Press, 1993.

Higham, John. *Strangers in the Land: Patterns of American Nativism, 1860–1925.* 1955. 2nd ed., New Brunswick, N.J.: Rutgers University Press, 1988.

A History of the National Research Council, 1919-1933. Washington, D.C.: National Research Council, 1933. Reprint, Wilmington, Del.: Scholarly Resources, 1974.

Hochheiser, Sheldon. *Rohm and Haas: History of a Chemical Company.* Philadelphia: University of Pennsylvania Press, 1986.

Hoffecker, Carol. *Federal Justice in the First State: A History of the United States District Court for the District of Delaware.* Wilmington: Historical Society for the United States District Court for the District of Delaware, 1992.

Holtzoff, Alexander. "Enemy Patents in the United States." *American Journal of International Law* 26 (April 1932): 272-79.

Homburg, Ernst. "The Emergence of Research Laboratories in the Dyestuffs Industry, 1870-1900." *British Journal for the History of Science* 25 (1992): 91-111.

Horn, Martin. *Britain, France and the Financing of the First World War.* Montreal and Kingston: McGill-Queen's University Press, 2002.

———. "A Private Bank at War: J. P. Morgan & Co. and France, 1914-1918." *Business History Review* 74, no. 1 (Spring 2000): 85-112.

Horne, John. "German Atrocities, 1914: Fact, Fantasy, or Fabrication?" *History Today* 52, no. 4 (April 2002): 47-53.

Hounshell, David A. "The Evolution of Industrial Research in the United States." In *U.S. Research at the End of an Era,* edited by Richard S. Rosenbloom and William J. Spencer, 13-85. Boston: Harvard Business School Press, 1996.

Hounshell, David A., and John K. Smith Jr. *Science and Corporate Strategy: Du Pont R & D, 1902-1980.* New York: Cambridge University Press, 1988.

Hughes, Thomas P. *Elmer Sperry: Inventor and Engineer.* Baltimore: Johns Hopkins University Press, 1971.

———. *Networks of Power: Electrification in Western Society, 1880-1930.* Baltimore: Johns Hopkins University Press, 1983.

———. "Technological Momentum in History: Hydrogenation in Germany, 1898-1933." *Past and Present* 44 (August 1969): 106-32.

Ihde, Aaron J. *The Development of Modern Chemistry.* 1964; reprint, New York: Dover Publications, 1984.

Ivey, Dean B. "Origins of the American Synthetic Dye Industry, 1865-1925, with Special Emphasis upon Government Policy." M.A. thesis, University of Delaware, 1963.

James, Harold. *The End of Globalization: Lessons from the Great Depression.* Cambridge, Mass.: Harvard University Press, 2001.

James, T. H. *A Biography-Autobiography of Charles Edward Kenneth Mees: Pioneer of Industrial Research.* Rochester, N.Y.: Eastman Kodak Company, n.d.

Jenkins, Reese V. *Images and Enterprise: Technology and the American Photographic Industry, 1839 to 1925.* Baltimore: Johns Hopkins University Press, 1975.

Jeremy, David J., ed. *International Technology Transfer: Europe, Japan, and the USA, 1700-1914.* London: Edward Elgar Publishing Company, 1991.

———. *Transatlantic Industrial Revolution: The Diffusion of Textile Technologies Between Britain and American, 1790-1830.* Cambridge, Mass.: MIT Press, 1981.

———, ed. *The Transfer of International Technology: Europe, Japan, and the USA in the Twentieth Century.* London: Edward Elgar Publishing Company, 1992.

John, Richard R. "Elaborations, Revisions, Dissents: Alfred D. Chandler, Jr.'s, 'The Visible Hand' after Twenty Years." *Business History Review* 71, no. 2 (Summer 1997): 151-200.

———. *Network Nation: Inventing American Telecommunications.* Cambridge, Mass.: Harvard University Press, 2010.

———. *Spreading the News: The American Postal System from Franklin to Morse.* Cambridge, Mass.: Harvard University Press, 1998.

Johnson, Jeffrey A. "German Women in Chemistry, 1895-1945" (two parts). *NTM: International Journal of History and Ethics of Natural Sciences, Technology and Medicine* 6, nos. 1 and 2 (1998): 1-21 and 65-90.

———. *The Kaiser's Chemists: Science and Modernization in Imperial Germany.* Chapel Hill: University of North Carolina Press, 1990.

———. "Hierarchy and Creativity in Chemistry, 1871-1914." In *Science in Germany* (special issue), edited by Kathryn M. Olesko. *Osiris* 5, 2nd ser. (1989): 214-40.

Johnson, Jeffrey A. and Roy MacLeod. "The War the Victors Lost: The Dilemmas of Chemical Disarmament, 1919-1926." In *Frontline and Factory: Comparative Perspectives on the Chemical Industry at War, 1914-1924,* edited by Roy MacLeod and Jeffrey A. Johnson, 221-45. Dordrecht: Springer, 2006.

Jonas, Manfred. *The United States and Germany: A Diplomatic History.* Ithaca, N.Y.: Cornell University Press, 1984.

Jones, Daniel P. "American Chemists and the Geneva Protocol." *Isis* 71 (September 1980): 426-40.

———. "Chemical Warfare Research during World War I: A Model of Cooperative Research." In *Chemistry and Modern Society: Historical Essays in Honor of Aaron J. Ihde,* edited by John Parascandola and James C. Whorton, 165-85. Washington, D.C.: American Chemical Society, 1983.

———. "From Military to Civilian Technology: The Introduction of Tear Gas for Civil Riot Control." *Technology and Culture* 19 (April 1978): 151-68.

———. "The Role of Chemists in Research on War Gases in the United States during World War I." Ph.D. diss., University of Wisconsin, 1969.

Jones, Geoffrey. "The End of Nationality? Global Firms and 'Borderless Worlds.'" *Zeitschrift für Unternehmensgeschichte* 51, no. 2 (2006): 149-65.

———. *Multinationals and Global Capitalism: From the Nineteenth to the Twenty-first Century.* New York: Oxford University Press, 2005.

Kahn, Ely J. Jr. *The Problem Solvers: A History of A. D. Little, Inc.* Boston: Little, Brown and Company, 1986.

Kargon, Robert H. *The Rise of Robert Millikan: Portrait of a Life in American Science.* Ithaca, N.Y.: Cornell University Press, 1982.

Kaufer, Erich. *The Economics of the Patent System.* New York: Harwood Academic Press, 1989.

Kaufman, Burton I. *Efficiency and Expansion: Foreign Trade Organization in the Wilson Administration, 1913-1921.* Westport, Conn.: Greenwood Press, 1974.

Kaufmann, Carl B. "Grand Duke, Wizard, and Bohemian: A Biographical Profile of Leo Hendrik Baekeland, 1863-1944." M.A. thesis, University of Delaware, 1968.

Kennedy, David M. *Over Here: The First World War and American Society.* New York: Oxford University Press, 1980.

Kent, Bruce. *The Spoils of War: The Politics, Economics, and Diplomacy of Reparations, 1918-1932.* Oxford: Clarendon Press, 1989.

Kettering, Charles F. "Biographical Memoir of Leo Hendrik Baekeland, 1863-1944." *Biographical Memoirs.* Washington, D.C.: National Academy of Sciences, 1947.

Kevles, Daniel J. *The Physicists: The History of a Scientific Community in Modern America*. Cambridge, Mass.: Harvard University Press, 1971.

Kinzel, Augustus B. "George Oliver Curme, Jr." In *Biographical Memoirs*, 121-37. Washington, D.C.: National Academy of Sciences, 1980.

Kleinman, Daniel Lee. *Impure Cultures: University Biology and the World of Commerce*. Madison: University of Wisconsin Press, 2003.

Kline, Ronald R. *Steinmetz: Engineer and Socialist*. Baltimore: Johns Hopkins University Press, 1992.

Kocka, Jürgen. "Entrepreneurs and Managers in German Industrialization." In *The Industrial Economies: Capital, Labour, and Enterprise*, edited by Peter Mathias and M. M. Postan, vol. 7 of *Cambridge Economic History*, part 1, 492-589. Cambridge: Cambridge University Press, 1978.

Kohler, Robert E. *Partners in Science: Foundations and Natural Scientists, 1900-1945*. Chicago: University of Chicago Press, 1992.

Koistinen, Paul A. C. "The 'Industrial-Military Complex' in Historical Perspective: World War I." *Business History Review* 41 (Winter 1967): 378-403.

———. *Mobilizing for Modern War, 1865-1919*. Lawrence: University Press of Kansas, 1997.

König, Paul. *Voyage of the Deutschland: The First Merchant Submarine*. New York: Hearst's International Library Company, 1916.

König, Wolfgang. "Adolf Hitler vs. Henry Ford: The Volkswagen, the Role of America as a Model, and the Failure of a Nazi Consumer Society." *German Studies Review* 27, no. 2 (May 2004): 249-68.

Krige, John. *American Hegemony and the Postwar Reconstruction of Science in Europe*. Cambridge, Mass.: MIT Press, 2006.

Krige, John, and Kai-Henrik Barth, eds. "Global Power Knowledge: Science and Technology in International Affairs." *Osiris* 21, 2nd ser. (2006).

Lamoreaux, Naomi R. *The Great Merger Movement in American Business, 1895-1904*. New York: Cambridge University Press, 1985.

Landau, Ralph, Basil Achilladelis, and Alexander Scriabine, eds. *Pharmaceutical Innovation: Revolutionizing Human Health*. Philadelphia: Chemical Heritage Press, 1999.

Landes, David S. *The Unbound Prometheus: Technological Change and Industrial Development in Western Europe from 1750 to the Present*. New York: Cambridge University Press, 1969.

LeCain, Timothy J. *Mass Destruction: The Men and Giant Mines That Wired America and Scarred the Planet*. New Brunswick, N.J.: Rutgers University Press, 2009.

Lesch, John E., ed. *The German Chemical Industry in the Twentieth Century*. Dordrecht: Kluwer Academic Publishers, 2000.

Leslie, Stuart W. *Boss Kettering*. New York: Columbia University Press, 1983.

Leuchtenburg, William E. *The Perils of Prosperity, 1914-1932*. 1958. 2nd ed., Chicago: University of Chicago Press, 1993.

Liebenau, Jonathan. *Medical Science and Medical Industry: The Formation of the American Pharmaceutical Industry*. London: Macmillan Press, 1987.

———. "Patents and the Chemical Industry: Tools of Business Strategy." In *The Challenge of New Technology: Innovation in British Business since 1850*, edited by Jonathan Liebenau, 135-50. Aldershot: Gower, 1988.

Link, Arthur S., ed. *The Papers of Woodrow Wilson*. Vols. 31 and 36. Princeton, N.J.: Princeton University Press, 1979.

Lipartito, Kenneth. "Rethinking the Invention Factory: Bell Laboratories in Perspective." In *The Challenge of Remaining Innovative*, edited by Sally Clarke, Naomi Lamoreaux, and Steven Usselman, 132-59. Stanford, Calif.: Stanford University Press, 2009.

Lubar, Steven. "New, Useful, and Nonobvious." *American Heritage of Invention and Technology* 6 (1990): 8-16.

Lyons, Michael J. *World War I: A Short History*. 2nd ed. Upper Saddle River, N.J.: Prentice Hall, 2000.

Machlup, Fritz. *An Economic Review of the Patent System*. Washington, D.C.: Government Printing Office, 1958.

———. *The Political Economy of Monopoly: Business, Labor, and Government Policies*. Baltimore: Johns Hopkins University Press, 1952.

MacLeod, Christine. *Inventing the Industrial Revolution*. New York: Cambridge University Press, 1988.

MacLeod, Roy. "Secrets among Friends: The Research Information Service and the 'Special Relationship' in Allied Scientific Information and Intelligence, 1916-18." *Minerva* 37 (1999): 201-33.

MacLeod, Roy, and Jeffrey A. Johnson, eds. *Frontline and Factory: Comparative Perspectives on the Chemical Industry at War, 1914-1924*. Dordrecht: Springer, 2006.

Madison, James H. *Eli Lilly, 1885-1977*. Indianapolis: Indiana Historical Society, 1989.

Mann, Charles C., and Mark L. Plummer. *The Aspirin Wars*. New York: Alfred A. Knopf, 1991.

Marcinkowsky, Arthur E., and George E. Keller II. "Ethylene and Its Derivatives: Their Chemical Engineering Genesis and Evolution at Union Carbide Corporation." In *A Century of Chemical Engineering*, edited by William F. Furter, 293-352. New York: Plenum Press, 1982.

Marks, Sally. "Smoke and Mirrors: In Smoke-Filled Rooms and the Galerie des Glaces." In *The Treaty of Versailles: A Reassessment after 75 Years*, edited by Manfred F. Boemeke, Gerald D. Feldman, and Elisabeth Glaser, 327-70. New York: German Historical Institute and Cambridge University Press, 1998.

Marsch, Ulrich. "Strategies for Success: Research Organization in German Chemical Companies and I.G. Farben until 1936." *History and Technology* 12 (1994): 23-77.

McGerr, Michael. "The Price of the 'New Transnational History.'" *American Historical Review* 96 (October 1991): 1056-67.

McHugh, Glenn. "Settlement of War Claims Act of 1928." *American Bar Association Journal* 14, no. 4 (April 1928): 193-95.

McMaster, John Bach. *The United States in the World War, 1918-1920*. New York: D. Appleton and Company, 1920.

McTavish, Jan. "What's in a Name? Aspirin and the American Medical Association." *Bulletin of the History of Medicine* 61 (Fall 1987): 343-66.

Merck Index, 12th ed. Whitehouse Station, N.J.: Merck & Company, 1996.

Meikle, Jeffrey L. "Materials and Metaphors: Plastics in American Culture." In *New Perspectives on Technology and American Culture*, edited by Bruce Sinclair, 31-47. Library Publication No. 12. Philadelphia: American Philosophical Society, 1986.

———. "Plastic, Material of a Thousand Uses." In *Imagining Tomorrow: History, Technology, and the American Future*, edited by Joseph J. Corn, 77–96. Cambridge, Mass.: MIT Press, 1986.

Meissner, Carl. "Some Recent Developments in By-Product Coke Ovens." Paper read at the Sixth General Meeting of the American Iron and Steel Institute, New York, May 22, 1914.

Mercelis, Joris. "Leo Baekeland's Transatlantic Struggle for Bakelite: Patenting Inside and Outside of America." *Technology and Culture* 53 (April 2012): 366–400.

Messimer, Dwight R. *The Merchant U-Boat: Adventures of the Deutschland, 1916–1918*. Annapolis, Md.: Naval Institute Press, 1988.

Meyer-Thurow, Georg. "The Industrialization of Invention: A Case Study from the German Chemical Industry." *Isis* 73 (September 1982): 363–81.

Milward, Alan S., and S. B. Saul. *The Economic Development of Continental Europe, 1780–1870*. Totowa, N.J.: Rowman and Littlefield, 1973.

Mizuno, Hiromi. *Science for the Empire: Scientific Nationalism in Modern Japan*. Stanford, Calif.: Stanford University Press, 2009.

Morris, Peter J. T. *The American Synthetic Rubber Research Program*. Philadelphia: University of Pennsylvania Press, 1989.

———. "Transatlantic Transfer of Buna S Synthetic Rubber Technology, 1932–1945." In *The Transfer of International Technology: Europe, Japan, and the U.S.A. in the Twentieth Century*, edited by David J. Jeremy, 57–89. London: Edward Elgar Publishing Company, 1992.

Morris, Peter J. T., and Anthony S. Travis. "A History of the International Dyestuff Industry." *American Dyestuff Reporter* 81 (November 1992): 59–100, 192–95.

Mowery, David C., Richard Nelson, Bhavat Sampat, and Arvids Ziedonis, eds. *Ivory Tower and Industrial Innovation: University-Industry Technology Transfer before and after the Bayh-Dole Act*. Stanford, Calif.: Stanford Business Books, 2004.

Mowery, David C., and Nathan Rosenberg. *Technology and the Pursuit of Economic Growth*. New York: Cambridge University Press, 1989.

Murmann, Johann Peter. *Knowledge and Competitive Advantage: The Coevolution of Firms, Technology, and National Institutions*. New York: Cambridge University Press, 2003.

Murphy, Paul. *World War I and the Origin of Civil Liberties in the United States*. New York: W. W. Norton & Company, 1979.

Murray, Robert K. *The Harding Era: Warren G. Harding and His Administration*. Minneapolis: University of Minnesota Press, 1969.

Nass, David L. "The Rural Experience." In *Minnesota in a Century of Change: The State and Its People since 1900*, edited by Clifford E. Clark Jr., 129–54. St. Paul: Minnesota Historical Society Press, 1989.

"National Aniline & Chemical Company." *Moody's Manual of Railroads and Corporation Securities*. 21st ed. Industrial section, 2783–84. New York: Poor's Publishing Company, 1920.

Ndiaye, Pap A. *Nylon and Bombs: DuPont and the March of Modern America*. Translated by Elborg Forster. Baltimore: Johns Hopkins University Press, 2006.

Nelson, Richard R., and Sidney G. Winter. *An Evolutionary Theory of Economic Change*. Cambridge, Mass.: Harvard University Press, 1982.

Neustadt, Richard E., and Ernest R. May. *Thinking in Time: The Uses of History for Decision Makers*. New York: Free Press, 1986.

Noble, David F. *America by Design: Science, Technology, and the Rise of Corporate Capitalism.* New York: Alfred A. Knopf, 1977.
Norton, Thomas H. *Artificial Dyestuffs Imported into the United States, 1913-1914: Supplemental Statistics to Accompany Special Agents Series No. 121.* Department of Commerce. Bureau of Foreign and Domestic Commerce. Mimeographed circular, 1922.
——. *Artificial Dyestuffs Used in the United States.* Department of Commerce. Bureau of Foreign and Domestic Commerce. Special Agents Series No. 121. Washington, D.C.: Government Printing Office, 1916.
——. *Dyestuff Situation in the United States, November 1915.* Department of Commerce. Bureau of Foreign and Domestic Commerce. Special Agents Series No. 111. Washington, D.C.: Government Printing Office, 1916.
——. *Dyestuffs for American Textile and Other Industries.* Department of Commerce. Bureau of Foreign and Domestic Commerce. Special Agents Series No. 96. Washington, D.C.: Government Printing Office, 1915.
Noyes, Arthur A. "The Supply of Nitrogen Products for the Manufacture of Explosives." In *The New World of Science*, edited by Robert M. Yerkes, 123-33. New York: Century Company, 1920.
Obstfeld, Maurice, and Alan M. Taylor. *Global Capital Markets: Integration, Crisis, and Growth.* New York: Cambridge University Press, 2004.
O'Rourke, Kevin H., and Jeffrey G. Williamson. *Globalization and History: The Evolution of a Nineteenth-Century Atlantic Economy.* Cambridge, Mass.: MIT Press, 1999.
Palazzo, Albert. *Seeking Victory on the Western Front: The British Army and Chemical Warfare in World War I.* Lincoln: University of Nebraska Press, 2000.
Palmer, A. Mitchell, and Francis P. Garvan. *Aims and Purposes of the Chemical Foundation.* New York: De Vinne Press, 1919.
Parascandola, John. "Charles Holmes Herty and the Effort to Establish an Institute for Drug Research in Post World War I America." In *Chemistry and Modern Society*, edited by Parascandola and James C. Whorton, 85-103. Washington, D.C.: American Chemical Society, 1983.
——. *The Development of American Pharmacology: John J. Abel and the Shaping of a Discipline.* Baltimore: Johns Hopkins University Press, 1992.
——. "Industrial Research Comes of Age: The American Pharmaceutical Industry, 1920-1940." *Pharmacy in History* 27 (1985): 12-21.
——. "The Theoretical Basis of Paul Ehrlich's Chemotherapy." *Journal of the History of Medicine* 36 (January 1981): 19-43.
Parascandola, John, and Ronald Jasensky. "Origins of the Receptor Theory of Drug Action." *Bulletin of the History of Medicine* 48 (Summer 1974): 199-220.
Parker, Edwin B. "War Claims Arbiter, Administrative Decision No. 1." *American Journal of International Law* 23 (1929): 193-233.
Patton, Craig D. "The German Chemical Industry and Reparations in Kind (1919-1924)." Paper presented at the annual meeting of the American Historical Association, December 20, 1992, Washington, D.C.
Plumpe, Gottfried. *Die I.G. Farbenindustrie A.G.: Wirtschaft, Technik und Politik, 1904-1945.* Berlin: Duncker and Humblot, 1990.
Polanyi, Michael. *The Tacit Dimension.* Garden City, N.Y.: Doubleday, 1966.
Possehl, Ingunn. *Modern by Tradition: The History of the Chemical-Pharmaceutical Factory E. Merck Darmstadt.* Dreieich, Germany: HMS Druckhaus for E. Merck Darmstadt, 1995.

Post, Robert C. "'Liberalizers' versus 'Scientific Men' in the Antebellum Patent Office." *Technology and Culture* 17 (January 1976): 24-54.

Pusey, Merlo J. *Eugene Meyer.* New York: Alfred A. Knopf, 1974.

Quivik, Fredric L. "Industrial Pollution on the Southwest Pennsylvania Countryside: Beehive Coking in the Connellsville Coke Region, 1860-1920." Paper presented at the Annual Meeting of the American Society for Environmental History, March 1993, Pittsburgh, Pa.

Rae, John B. *The American Automobile.* Chicago: University of Chicago Press, 1965.

Rasmussen, Nicolas. "What Moves When Technologies Migrate? 'Software' and Hardware in the Transfer of Biological Electron Microscopy to Postwar Australia." *Technology and Culture* 40 (1999): 47-73.

Reader, William J. *Imperial Chemical Industries.* 3 vols. London: Oxford University Press, 1970.

Redfield, William C. *The New Industrial Day: A Book for Men Who Employ Men.* New York: The Century Co., 1912.

Reed, Germaine M. *Crusading for Chemistry: The Professional Career of Charles Holmes Herty.* Athens: University of Georgia Press, 1995.

Reimer, Thomas M. "Bayer & Company in the United States: German Dyes, Drugs, and Cartels in the Progressive Era." Ph.D. diss., Syracuse University, 1996.

Reinhardt, Carsten. *Forschung in der chemischen Industrie: Die Entwicklung synthetischer Farbstoffe bei BASF und Hoechst, 1863 bis 1914.* Freiberg: Technische Universität Bergakademie, 1997.

———. "An Instrument of Corporate Strategy: The Central Research Laboratory at BASF, 1868-1890." In *The Chemical Industry in Europe, 1850-1914,* edited by Ernst Homburg, Anthony S. Travis, and Harm G. Schröter, 239-59. Dordrecht: Kluwer Academic Publishers, 1998.

Reinhardt, Carsten, and Anthony S. Travis. *Heinrich Caro and the Creation of Modern Chemical Industry.* Dordrecht: Kluwer Academic Publishers, 2000.

Report of the Alien Property Custodian. 67th Cong., 2nd sess., April 10, 1922, S. Doc. 181. Washington, D.C.: Government Printing Office, 1922.

Report of the Alien Property Custodian, 1917. 65th Cong., 2nd sess., January 18, 1918. H. Doc. 840. Washington, D.C.: Government Printing Office, 1918.

Report of the Alien Property Custodian, 1918-1919. 65th Cong., 3rd sess., March 1, 1919, S. Doc. 435. Washington, D.C.: Government Printing Office, 1919.

Reynolds, Terry S. "Defining Professional Boundaries: Chemical Engineering in the Early Twentieth Century." *Technology and Culture* 27 (October 1986): 694-716.

———. *75 Years of Progress: A History of the American Institute of Chemical Engineers, 1908-1983.* New York: American Institute of Chemical Engineers, 1983.

Rhees, David J. "The Chemists' Crusade: The Rise of an Industrial Science in Modern America, 1907-1922." Ph.D. diss., University of Pennsylvania, 1987.

Rocke, Alan J. *The Quiet Revolution: Hermann Kolbe and the Science of Organic Chemistry.* Berkeley: University of California Press, 1993.

Rockoff, Hugh. *America's Economic Way of War: War and the U.S. Economy from the Spanish-American War to the Persian Gulf War.* New York: Cambridge University Press, 2012.

Rodgers, Daniel T. *Atlantic Crossings: Social Politics in a Progressive Age.* Cambridge, Mass.: Harvard University Press, 1998.

Roggersdorf, Wilhelm, and Badische Anilin- & Soda-Fabrik AG. *In the Realm of Chemistry*. Düsseldorf: Econ-Verlag GmbH, 1965.
Rossiter, Margaret W. *Women Scientists in America: Struggles and Strategies to 1940*. Baltimore: Johns Hopkins University Press, 1982.
Rostow, Walter W. *The Stages of Economic Growth*. Cambridge, Mass.: MIT Press, 1960.
Sabel, Charles, and Jonathan Zeitlin. "Stories, Strategies, Structures: Rethinking Historical Alternatives to Mass Production." In *World of Possibilities: Flexibility and Mass Production in Western Industrialization*, edited by Sabel and Zeitlin, 1-34. Cambridge: Cambridge University Press, 1997.
Scherer, Frederic M., S. E. Herzstein, A. W. Dreyfoos, W. G. Whitney, O. J. Bachman, C. P. Pesek, C. J. Scott, T. G. Kelly, and J. J. Galvin. *Patents and the Corporation*. 2nd ed. Boston: privately printed, 1959.
Schnietz, Karen. "The 1916 Tariff Commission: Democrats' Use of Expert Information to Constrain Republican Tariff Protection." *Business and Economic History* 23, no. 1 (1994): 176-89.
Schröter, Harm. "Cartels as a Form of Concentration in Industry: The Example of the International Dyestuffs Cartel from 1927 to 1939." In *German Yearbook on Business History 1988*, edited by Hans Pohl and Bernd Rudolph, 113-44. Berlin: Springer-Verlag, 1990.
———. "Die Auslandsinvestitionen der deutschen chemischen Industrie, 1870 bis 1930." *Zeitschrift für Unternehmensgeschichte* 35 (1990): 1-22.
———. "The International Dyestuffs Cartel, 1927-1939, with Special Reference to the Developing Areas of Europe and Japan." In *International Cartels in Business History*, edited by Akira Kudo and Terushi Hara, 33-52. The International Conference on Business History 18: Proceedings of the Fuji Conference. Tokyo: University of Tokyo Press, 1992.
Schröter, Verena. *Die deutsche Industrie auf dem Weltmarkt, 1929 bis 1933*. Frankfurt: Verlag Peter Lang, 1984.
———. "Participation in Market Control through Foreign Investment: I.G. Farbenindustrie A.G. in the United States, 1920-1938." In *Multinational Enterprise in Historical Perspective*, edited by Alice Teichova, Maurice Lévy-Leboyer, and Helga Nussbaum, 171-84. Cambridge: Cambridge University Press, 1986.
Schultz, Gustav. *Farbstofftabellen*. Berlin: Weidmann, 1914.
Schwabe, Klaus. "America's Contribution to the Stabilization of the Early Weimar Republic." In *Germany and America: Essays on Problems of International Relations and Immigration*, edited by Hans L. Trefousse, 21-28. New York: Brooklyn College Press, 1980.
Scranton, Philip. *Endless Novelty: Specialty Production and American Industrialization, 1865-1925*. Princeton, N.J.: Princeton University Press, 1997.
———. *Figured Tapestry: Production, Markets, and Power in Philadelphia Textiles, 1885-1941*. New York: Cambridge University Press, 1989.
Seely, Bruce E. "Historical Patterns in the Scholarship of Technology Transfer." *Comparative Technology Transfer and Society* 1 (April 2003): 7-48.
Servos, John W. "Changing Partners: The Mellon Institute, Private Industry, and the Federal Patron." *Technology and Culture* 35 (April 1994): 221-57.
———. *Physical Chemistry from Ostwald to Pauling: The Making of a Science in America*. Princeton, N.J.: Princeton University Press, 1990.

Sharrer, G. Terry. "Naval Stores, 1781-1881." In *Material Culture of the Wooden Age*, edited by Brooke Hindle, 241-70. Tarrytown, N.Y.: Sleepy Hollow Press, 1981.

Sherwood, Morgan. "The Origins and Development of the American Patent System." *American Scientist* 71 (1983): 500-506.

Sicherman, Barbara. *Alice Hamilton: A Life in Letters*. Cambridge, Mass.: Harvard University Press, 1984.

Skolnik, Herman, and Kenneth M. Reese. *A Century of Chemistry*. Washington, D.C.: American Chemical Society, 1976.

Skowronek, Stephen. *Building a New American State: The Expansion of National Administrative Capacities, 1877-1920*. New York: Cambridge University Press, 1982.

Smith, John K. Jr. "National Goals, Industry Structure, and Corporate Strategies: Chemical Cartels between the Wars." In *International Cartels in Business History*, edited by Akira Kudo and Terushi Hara, 130-61. The International Conference on Business History 18: Proceedings of the Fuji Conference. Tokyo: University of Tokyo Press, 1992.

Smith, Merritt Roe. *Harpers Ferry Armory and the New Technology*. Ithaca, N.Y.: Cornell University Press, 1977.

Spitz, Peter. *Petrochemicals: The Rise of an Industry*. New York: John Wiley and Sons, 1988.

Stapleton, Darwin H. "The Rise of Industrial Research in Cleveland, 1870-1930." In *Beyond History of Science: Essays in Honor of Robert E. Schofield*, edited by Elizabeth Garber, 231-45. Bethlehem, Pa.: Lehigh University Press, 1990.

——. *The Transfer of Early Industrial Technologies to America*. Philadelphia: American Philosophical Society, 1987.

Steen, Kathryn. "Confiscated Commerce: American Importers of German Synthetic Organic Chemicals, 1914-1929." *History and Technology* 12 (1995): 261-84.

——. "German Chemicals and American Politics, 1919-1921." In *The German Chemical Industry in the Twentieth Century*, edited by John E. Lesch, 323-46. Dordrecht: Kluwer Academic Publishers, 2000.

——. "Patents, Patriotism, and 'Skilled in the Art': USA v. The Chemical Foundation, Inc., 1923-1926." *Isis* 92 (March 2001): 91-122.

——. "Technical Expertise and U.S. Mobilization, 1917-1918: High Explosives and War Gases." In *Frontline and Factory: Comparative Perspectives on the Chemical Industry at War, 1914-1924*, edited by Roy MacLeod and Jeffrey A. Johnson, 103-22. Dordrecht: Springer, 2006.

Stevenson, Earl P. "'Scatter Acorns That Oaks May Grow': Arthur D. Little, Inc., 1886-1953." New York: Newcomen Society, 1953.

Stief, Robert D. *A History of Union Carbide Corporation from the 1890s to the 1990s*. Danbury, Conn.: Carbide Retiree Corps, 1998.

Stocking, George W. "The Du Pont-General Motors Case and the Sherman Act." *Virginia Law Review* 44, no. 1 (January 1958): 1-40.

Stocking, George W., and Myron W. Watkins. *Cartels in Action*. New York: Twentieth Century Fund, 1946.

Stokes, Raymond G. *Opting for Oil: The Political Economy of Technological Change in the West German Chemical Industry, 1945-1961*. New York: Cambridge University Press, 1994.

The Story of the Development Division, Chemical Warfare Service. Privately printed. Cleveland, Ohio: General Electric, 1920.

Stowell, Ellery C. "Tariff Relations with France." *American Journal of International Law* 24, no. 1 (January 1930): 110–18.

Stranges, Anthony. "Germany's Synthetic Fuel Industry, 1927–1945." In *The German Chemical Industry in the Twentieth Century*, edited by John E. Lesch, 147–216. Dordrecht: Kluwer Academic Publishers, 2000.

Sturchio, Jeffrey L. "Chemists and Industry in Modern America: Studies in the Historical Application of Science Indicators." Ph.D. diss., University of Pennsylvania, 1981.

———. "Experimenting with Research: Kenneth Mees, Eastman Kodak, and the Challenges of Diversification." Paper presented at the Hagley R&D Pioneers Conference, Wilmington, Delaware, October 7, 1985.

Swann, John P. *Academic Scientists and the Pharmaceutical Industry: Cooperative Research in Twentieth-Century America*. Baltimore: Johns Hopkins University Press, 1988.

Tarbell, D. Stanley, and Ann Tracy Tarbell. *Roger Adams: Scientist and Statesman*. Washington, D.C.: American Chemical Society, 1981.

Taussig, Frank W. *Free Trade, the Tariff, and Reciprocity*. New York: Macmillan Company, 1924.

———. "The Tariff Act of 1922." *Quarterly Journal of Economics* 37 (November 1922): 1–28.

Taylor, Graham D., and Patricia E. Sudnik. *Du Pont and the International Chemical Industry*. Boston: Twayne Publishers, 1984.

Tempest, Ronald. "Francis P. Garvan." *Chemical Heritage* 16 (Fall 1998): 15.

Thackray, Arnold. "University-Industry Connections and Chemical Research: An Historical Perspective." In *University-Industry Research Relationships: Selected Studies*, 193–233. Washington, D.C.: National Science Board, 1983.

Thackray, Arnold, Jeffrey L. Sturchio, P. Thomas Carroll, and Robert Bud. *Chemistry in America, 1876–1976: Historical Indicators*. Dordrecht: D. Reidel Publishing Company, 1985.

Tipton, Frank B., and Robert Aldrich. *An Economic and Social History of Europe, 1890–1939*. Baltimore: Johns Hopkins University Press, 1987.

Tooze, Adam. "Trouble with Numbers: Statistics, Politics, and History in the Construction of Weimar's Trade Balance, 1918–24." *American Historical Review* 113, no. 3 (June 2008): 678–700.

"The Trading with the Enemy Act." *Yale Law Journal* 35 (January 1926): 345–57.

Travis, Anthony S. *Dyes Made in America, 1915–1980: The Calco Chemical Company, American Cyanamid, and the Raritan River*. Jerusalem: Hexagon Press for the Sidney M. Edelstein Center at the Hebrew University of Jerusalem, 2004.

———. *The Rainbow Makers: The Origins of the Synthetic Dyestuffs Industry in Western Europe*. Bethlehem, Pa.: Lehigh University Press, 1993.

———. "Science as Receptor of Technology: Paul Ehrlich and the Synthetic Dyestuffs Industry." *Science in Context* 3 (1989): 383–408.

The Treaty of Versailles and After: Annotations of the Text of the Treaty. Washington, D.C.: Government Printing Office, 1947. Reprint, Grosse Pointe, Mich.: Scholarly Press, 1969.

Trescott, Martha Moore. *The Rise of the American Electrochemicals Industry, 1880–1910: Studies in the American Technological Environment*. Westport, Conn.: Greenwood Press, 1981.

Tyrrell, Ian. "American Exceptionalism in an Age of International History." *American Historical Review* 96 (October 1991): 1031–55.

"Union Carbide I: The Corporation." *Fortune* 23, no. 6 (June 1941): 61-68, 123-33.
"Union Carbide II: Alloys, Gases, and Carbons." *Fortune* 24, no. 1 (July 1941): 48-57, 92-100.
"Union Carbide III." See "Carbide and Carbon Chemicals" above.
U.S. Bureau of the Census. *Thirteenth Census of the United States, 1910.* Vol. 10, *Manufactures, Report for Principal Industries.* Washington, D.C.: Government Printing Office, 1913.
U.S. Council of Defense. *Third Annual Report of the United States Council of National Defense, 1919.* 66th Cong., 2nd sess., H. Doc 434. Washington, D.C.: Government Printing Office, 1919.
U.S. Department of Agriculture. *Annual Reports of the Department of Agriculture, 1918.* November 15, 1918. Washington, D.C.: Government Printing Office, 1919.
———. *Annual Reports of the Department of Agriculture, 1919.* November 15, 1919. Washington, D.C.: Government Printing Office, 1920.
———. *Annual Reports of the Department of Agriculture, 1920.* November 15, 1920. Washington, D.C.: Government Printing Office, 1921.
———. *Annual Reports of the Department of Agriculture, 1923.* November 15, 1923. Washington, D.C.: Government Printing Office, 1924.
———. *Report of the Secretary of Agriculture, 1927.* November 3, 1927. Washington, D.C.: Government Printing Office, 1927.
U.S. Department of Commerce. Bureau of Foreign and Domestic Commerce. "The Reparation Problem, 1918-1924." *Supplement to Commerce Reports.* Trade Information Bulletin No. 278, October 20, 1924. Washington, D.C.: Government Printing Office, 1924.
———. *Eleventh Annual Report of the Secretary of Commerce.* November 1, 1923. Washington, D.C.: Government Printing Office, 1923.
———. *Fifteenth Annual Report of the Secretary of Commerce.* October 1, 1927. Washington, D.C.: Government Printing Office, 1927.
———. *Tenth Annual Report of the Secretary of Commerce.* September 20, 1922. Washington, D.C.: Government Printing Office, 1922.
———. *Thirteenth Annual Report of the Secretary of Commerce.* November 2, 1925. Washington, D.C.: Government Printing Office, 1925.
———. *Twelfth Annual Report of the Secretary of Commerce.* November 1, 1924. Washington, D.C.: Government Printing Office, 1924.
U.S. House of Representatives. Committee on Interstate and Foreign Commerce. *Hearings, German Special Deposit Account.* 80th Cong., 1st sess., 1947, H.R. 4043. Washington, D.C.: Government Printing Office, 1947.
———. Committee on Interstate and Foreign Commerce. *Hearings, Trading with the Enemy.* 65th Cong., 1st sess., May 29 and June 4, 1917, H.R. 4704. Washington, D.C.: Government Printing Office, 1917.
———. Committee on Military Affairs. *Hearings, Selective Service Act.* 65th Cong., 2nd sess., August 19, 1918, H.R. 12731. Washington, D.C.: Government Printing Office, 1918.
———. Committee on Patents. *Oldfield Revision and Codification of the Patent Statutes, No. 4.* 62nd Cong., 2nd sess., April 12, 1912, H.R. 23417. Washington, D.C.: Government Printing Office, 1912.
———. Committee of Ways and Means. *Hearings, Chemical and Optical Glassware and Scientific Apparatus.* 69th Cong., 1st sess., June 11-13, 1919, H.R. 3734, 3735, 4386. Washington, D.C.: Government Printing Office, 1919.

———. *Hearings, Dyestuffs*. 66th Cong., 1st sess., 1919, H.R. 2706 and H.R. 649. Washington, D.C.: Government Printing Office, 1919.

———. *Hearings, To Establish the Manufacture of Dyestuffs*. 64th Cong., 1st sess., 1916, H. Rept. 702. Washington, D.C.: Government Printing Office, 1916.

———. *Hearings, To Increase the Revenue, and for Other Purposes*. 64th Cong., 1st sess., 1916, H. Rept. 922. Washington, D.C.: Government Printing Office, 1916.

———. *Hearings, Tariff Readjustment, 1929*. 70th Cong, 2nd sess., 1929.

———. *Tariff Act of 1922*. 67th Cong., 2nd sess., 1922, H. Doc. 393. Washington, D.C.: Government Printing Office, 1922.

———. *Tariff Act of 1930*. 71st Cong., 2nd sess., 1930, H. Doc 476. Washington, D.C.: Government Printing Office, 1930.

U.S. International Trade Commission. *Synthetic Organic Chemicals, U.S. Production and Sales, 1990*. USITC Publication No. 2470. Washington, D.C.: USITC, 1991.

U.S. Senate. Committee on Appropriations. *Hearings on First Deficiency Appropriation Bill, 1919*. 65th Cong., 2nd sess., October 18, 1918. On H.R. 13086. Washington, D.C.: Government Printing Office, 1918.

———. Committee on Appropriations. *Urgent Deficiency Appropriation Bill. Statement of A. Mitchell Palmer: Supplemental Hearing*. 65th Cong., 2nd sess., March 7, 1918. H.R. 9867. Washington, D.C.: Government Printing Office, 1918.

———. Committee on Finance. *Hearings, Dyestuffs*. 66th Cong., 2nd sess., 1920, H.R. 8078. Washington, D.C.: Government Printing Office, 1920.

———. Committee on Finance. *Hearings, Tariff Act of 1921*. 67th Cong., 2nd sess., 1921, S. Doc 108. Washington, D.C.: Government Printing Office, 1921.

———. Committee on Finance. *Hearings, To Increase the Revenue*. 64th Cong., 2nd sess., 1916, H. Rept. 16763. Washington, D.C.: Government Printing Office, 1916.

———. Committee on Finance. Subcommittee on H.R. 7785. *Hearings, Laboratory Glassware and Scientific and Surgical Instruments*. 66th Cong., 2nd sess., December 12–13, 1919, H.R. 7785. Washington, D.C.: Government Printing Office, 1919.

———. Committee on the Judiciary. *Brewing and Liquor Interests and German and Bolshevik Propaganda: Report and Hearings*. 66th Cong., 1st sess., December 1918, S. Doc. 62. Washington, D.C.: Government Printing Office, 1919.

———. Committee on the Judiciary. Subcommittee on S. Res. 77. *Hearings, Alleged Dye Monopoly*. 67th Cong., 2nd sess., 1922, S. Res. 77. Washington, D.C.: Government Printing Office, 1922.

———. Committee on Military Affairs, *Hearings, Amending the Draft Law*. 65th Cong., 2nd sess., vol. 2, August 6–9, 1918. Washington, D.C.: Government Printing Office, 1918.

———. Committee on Patents. *Hearings, Salvarsan*. 65th Cong., 1st sess., June 4, 1917. On S. 2178 and S. 2363. Washington, D.C.: Government Printing Office, 1917.

U.S. Tariff Commission. *Census of Dyes and Coal-Tar Chemicals, 1917*. Tariff Information Series No. 6. Washington, D.C.: Government Printing Office, 1918.

———. *Census of Dyes and Coal-Tar Chemicals, 1918*. Tariff Information Series No. 11. Washington, D.C.: Government Printing Office, 1919.

———. *Census of Dyes and Coal-Tar Chemicals, 1919*. Tariff Information Series No. 22. Washington, D.C.: Government Printing Office, 1921.

———. *Census of Dyes and Coal-Tar Chemicals, 1920*. Tariff Information Series No. 23. Washington, D.C.: Government Printing Office, 1921.

———. *Census of Dyes and Other Synthetic Organic Chemicals, 1921.* Tariff Information Series No. 26. Washington, D.C.: Government Printing Office, 1922.

———. *Census of Dyes and Other Synthetic Organic Chemicals, 1922.* Tariff Information Series No. 31. Washington, D.C.: Government Printing Office, 1923.

———. *Census of Dyes and Other Synthetic Organic Chemicals, 1923.* Tariff Information Series No. 32. Washington, D.C.: Government Printing Office, 1924.

———. *Census of Dyes and Other Synthetic Organic Chemicals, 1924.* Tariff Information Series No. 33. Washington, D.C.: Government Printing Office, 1925.

———. *Census of Dyes and Other Synthetic Organic Chemicals, 1925.* Tariff Information Series No. 34. Washington, D.C.: Government Printing Office, 1926.

———. *Census of Dyes and Other Synthetic Organic Chemicals, 1926.* Tariff Information Series No. 35. Washington, D.C.: Government Printing Office, 1927.

———. *Census of Dyes and of Other Synthetic Organic Chemicals, 1927.* Tariff Information Series No. 37. Washington, D.C.: Government Printing Office, 1929.

———. *Census of Dyes and of Other Synthetic Organic Chemicals, 1928.* Tariff Information Series No. 38. Washington, D.C.: Government Printing Office, 1930.

———. *Census of Dyes and of Other Synthetic Organic Chemicals, 1929.* Tariff Information Series No. 39. Washington, D.C.: Government Printing Office, 1930.

———. *Census of Dyes and of Other Synthetic Organic Chemicals, 1930.* Report No. 19, 2nd series. Washington, D.C.: Government Printing Office, 1931.

———. *Eighth Annual Report of the United States Tariff Commission, 1924.* Washington, D.C.: Government Printing Office, 1924.

———. *Eleventh Annual Report of the United States Tariff Commission, 1927.* Washington, D.C.: Government Printing Office, 1928.

———. *Information Concerning Optical Glass and Chemical Glassware.* Washington, D.C.: Government Printing Office, 1919.

———. *Second Annual Report of the United States Tariff Commission for the Fiscal Year Ended June 30, 1918.* 65th Cong., 3rd sess., H. Doc. 1369. Washington, D.C.: Government Printing Office, 1919.

———. *Seventh Annual Report of the United States Tariff Commission, 1923.* Washington, D.C.: Government Printing Office, 1924.

———. *Third Annual Report of the United States Tariff Commission, 1919.* 66th Cong., 2nd sess., H. Doc. 319. Washington, D.C.: Government Printing Office, 1920.

U.S. Treasury. *Annual Report of the Secretary of the Treasury, 1939.* 76th Cong., 2nd sess. Washington, D.C.: Government Printing Office, 1940.

Usselman, Steven W. *Regulating Railroad Innovation: Business, Technology, and Politics in America, 1840–1920.* New York: Cambridge University Press, 2002.

———. "Unbundling IBM: Antitrust and the Incentives to Innovation in American Computing." In *The Challenge of Remaining Innovative: Insights from Twentieth-Century American Business*, edited by Sally Clarke, Naomi Lamoreaux, and Steven Usselman, 249–79. Stanford, Calif.: Stanford University Press, 2009.

Vanderbilt, Byron M. *Thomas Edison, Chemist.* Washington, D.C.: American Chemical Society, 1971.

Van Gelder, Arthur Pine, and Hugo Schlatter. *History of the Explosives Industry in America.* New York: Columbia University Press, 1927.

Wahl, André. *The Manufacture of Organic Dyestuffs.* Translated by F. W. Atack. London: G. Bell and Sons, 1914.

Ward, Patricia Spain. "The American Reception of Salvarsan." *Journal of the History of Medicine* 36 (January 1981): 44-62.

Warren, Kenneth. *The American Steel Industry, 1850-1970: A Geographical Interpretation.* Oxford: Clarendon Press, 1973.

———. "Technology Transfer in the Origins of the Heavy Chemicals Industry in the United States and the Russian Empire." In *International Technology Transfer*, edited by David J. Jeremy, 153-77. London: Edward Elgar, 1991.

Weems, F. Carrington. *America and Munitions: The Work of Messrs. J. P. Morgan & Co. in the World War.* New York: privately printed, 1923.

Welch, Walter. "The Flexible Provisions of the Tariff Act." *Virginia Law Review* 13 (1927): 206-28.

West, Carolyn V. *1700 Milton Avenue: The Solvay Story, 1881-1981.* Solvay, N.Y.: Allied Chemical, c. 1981.

West, Clarence J. "The Chemical Warfare Service." In *The New World of Science*, edited by Robert M. Yerkes, 148-74. New York: Century Company, 1920.

Whalen, Catherine. "American Decorative Art Studies at Yale and Winterthur: The Politics of Gender, Gentility, and Academia." *Studies in the Decorative Arts* 9 (Fall/Winter 2001-2002): 108-44.

Wilkins, Mira. *The History of Foreign Investment in the United States to 1914.* Cambridge, Mass.: Harvard University Press, 1989.

———. *The History of Foreign Investment in the United States, 1914-1945.* Cambridge, Mass.: Harvard University Press, 2004.

Williams, William B. *History of the Manufacture of Explosives for the World War, 1917-1918.* 1920.

———. *Munitions Manufacture in the Philadelphia Ordnance District.* Philadelphia: n.p., 1921.

Wilson, Joan Hoff. *American Business and Foreign Policy, 1920-1933.* Boston: Beacon, 1971.

Wingate, P. J. *The Colorful Du Pont Company.* Wilmington, Del.: Serendipity Press, 1982.

Winter, Jay, ed. *The Legacy of the Great War: Ninety Years On.* Columbia: University of Missouri Press, 2009.

Wolff, Stefan L. "Physicists in the 'Krieg der Geister': Wilhelm Wien's 'Proclamation.'" *Historical Studies in the Physical and Biological Sciences* 33, no. 2 (2003): 337-68.

Wolman, Paul. *Most Favored Nation: The Republican Revisionists and U.S. Tariff Policy, 1897-1912.* Chapel Hill: University of North Carolina Press, 1992.

Wotiz, John H., and Susanna Rudofsky, "The Unknown Kekulé." In *Essays on the History of Organic Chemistry*, edited by James G. Traynham, 21-34. Baton Rouge: Louisiana State University Press, 1987.

Yerkes, Robert M., ed. *The New World of Science.* New York: Century Company, 1920.

Index

Abbott Laboratories, 135, 178, 222-23
Aberdeen Proving Ground, 99
Accidents. *See* Health and safety hazards
Acetone, 276, 281
Acetphenetidin, 147
Acetylene, 35, 38, 266; and Union Carbide & Chemical Corporation, 274-79, 281
Adams, Roger, 130, 132, 135
Adamson, William C., 151
Administrative state, 12
Aetna Explosives Company, 81-83, 87, 90, 135
Agfa (Aktiengesellschaft für Anilinfabrikation, Germany), 23-24, 169, 354 (n. 16)
Albert, Heinrich, 162
Alexander, William, 242, 253
Alien Property Custodian (APC), 139, 156-71; and Chemical Foundation, Inc., 174-76; Office of Alien Property organization, 150-51, 157; Sterling Products Company, 244; Treaty of Versailles, 179, 183. *See also* Garvan, Francis P.; Palmer, A. Mitchell; Trading with the Enemy Act; *USA v. The Chemical Foundation, Inc.*
Aliphatic synthetic organic chemicals: as "American," 237-39, 288, 292, 303 (n. 35); and Bakelite, 282, 285; definition, 17-18; and Dow, 273; and Du Pont, 255, 262-66; products, 17, 38, 238, 285; raw materials, 18, 238; and Union Carbide, 274-75, 278-79, 281-82
Alizarin dyes, 23, 29; tariff rates, 22, 32, 147-49
Allegheny College, 134
Alleman, Gellert, 226

Allied Dye & Chemical Corporation, 207, 242, 250-52, 254-55, 268, 275, 281. *See also* National Aniline & Chemical Company
Allison, James A., 276
Amatol, 80, 88, 313 (n. 3)
American Aniline, 170
American Chemical Society (ACS), 114, 125-26, 131, 133, 177; Division of Industrial and Engineering Chemistry, 39-40; dyes report (1914), 120-22; early years, 39-40; lobbying, 120-22, 199; and World War I, 40, 119. *See also* Herty, Charles Holmes
American Cotton Manufacturers' Society, 146
American Cyanamid, 270, 323 (n. 11)
American Dyes Institute (ADI), 175, 181, 191, 199-201, 206
American Dyestuff Manufacturers' Association (ADMA), 73-75
American Expeditionary Force (AEF), 95, 319 (n. 42)
American I.G., 249
American Medical Association (AMA), 153-54
American Synthetic Color Company, 104-6
American Synthetic Dyes, Inc., 81
American University Experiment Station, 98. *See also* War gases
American valuation, 199, 202, 207-8, 210-11, 293
Ammonia, 20, 229, 238, 250, 254, 262, 279, 358 (n. 40)

391

Ammonium nitrate, 3, 80–81, 88. *See also* Explosives
Ammonium picrate, 81
Anderson, Henry W., 217–21, 223–25, 227–29
Aniline, 23, 34–35, 50, 59–60, 63, 67, 251, 263, 269–70, 272
Aniline Dyes & Chemicals, 61, 269, 309 (n. 36)
Anthracene, 19–20, 22, 34
Anti-German sentiment, 8–9; and chemists, 10, 114, 118, 127, 146; and government policy, 4, 9, 138, 159, 174, 190, 288; and importers, 51, 65; and manufacturers, 72, 76, 242–43; subsided, 203, 205, 215, 230. *See also* Alien Property Custodian; Trading with the Enemy Act
Antitrust law, 24, 36–37, 80–81, 121, 173–74
Armistice, 171, 172
Aromatic Chemical Company, 59
Aromatic synthetic organic chemicals, 255, 273; definition, 17, 237; as "German," 17, 237, 274, 278, 281; products, 17, 21, 38, 263, 282; raw materials, 19–20, 238, 274
Aronson, Paul, 70
Arsphenamine. *See* Salvarsan
Aspirin, 2, 77, 163, 238, 248; and Bayer, 7, 27–28, 52, 244–45, 248; and Bayer Company, 49–50, 59, 163; and Dow, 59, 61, 271; and Heyden, 50; and Merck & Co., 246; and Monsanto, 57; patent, 33, 49–50, 288; tariff on, 21
The Aspirin Corporation, 50. *See also* Bayer Company
Associative state, 212–14
Atlas Powder Company, 80–81, 88
Atophan (Cinchophen), 58, 222, 225–27
Ault, Lee A., 72
Ault & Wiborg, 58, 61, 72, 269–70
Autarky, 5, 138. *See also* Isolationism
Automobile industry: as consumer of chemicals, 17–18, 38, 238, 263, 274, 276, 280, 283; and General Motors, 263–64
Azo dyes, 23, 29, 66–67, 70

Badische Company, 47, 61, 68. *See also* Kuttroff, Pickhardt & Company

Baekeland, George, 271
Baekeland, Leo H.: and Belgium and war, 116; on dyes, 28–29; early life, 37–38, 282–83; on patents, 284; tariffs, 147–48; Velox photographic paper, 38, 283. *See also* Bakelite Corporation
Baeyer, Adolf von, 25–26, 117
Bakelite Corporation, 37–38; and "American" chemicals, 38, 40, 282, 285; competition, 284–85; foreign subsidiaries, 283; and phenol, 285; products, 37–38, 283–84; research, 284–85; and tariff rates, 148
Baker, Newton D., 85, 129
Barbital, 58, 211. *See also* Veronal
Barrett, William F., 279
Barrett Company, 34–35, 65–66, 252
Baruch, Bernard, 183–84
BASF (Badische Anilin & Soda Fabrik), 19, 122, 232, 238
Baskerville, Charles, 104
Batteries, carbon, 276
Bayer (Farbenfabriken vorm. Friedr. Bayer & Co., Germany), 117, 194, 232; and Grasselli Chemical Company, 245–46; and Sterling Products, Inc., 244–45
Bayer Company (New York), 48–50; and Alien Property Custodian, 159–66, 169–70; foreign markets, 160, 162; patents, 165, 174; sale of, 159, 164–65
Beckers, William, 31, 56–57, 143–45, 148
Beilstein, Friedrich Konrad, 132
Belgium: invasion of, 115; and peace negotiations, 179, 182
Benzene, 19, 25, 34, 82, 111, 145, 237–39, 274
Benzol Products Company, 34–35, 65, 303 (n. 34)
Berlin Aniline Works, 169
Bernstorff, Johann H. G. von, 44, 221
Billings, C. K. G., 275
Bishop, George T., 63–64
Bismarck, Otto von, 24
Blockade. *See* Great Britain: blockade and dye shortages
Board of General Appraisers, 208. *See also* U.S. Customs Service
Bogert, Marston, 119, 129–30; and NRC, 85

Bohn, René, 19
Bolton, Elmer K., 222-24, 226, 261-62, 338 (n. 24)
Bosch, Carl, 241, 286. *See also* Haber-Bosch process
Bosch Magneto Company, 159
Botany Worsted Mills, 166
Bradley, Lee, 164
Brady, Nicholas, 166
Breithut, Frederick E., 213
Bribery, 28, 227, 301 (n. 22)
British Dyestuffs Corporation, 11, 188, 192, 242, 253
Brown, Kirk, 284
Brunner Mond, 88
Brush, Charles, 275, 360 (n. 66)
Brussels Agreement, 183, 185. *See also* Reparations
Bryan, William Jennings, 44, 141
Bureau of Foreign and Domestic Commerce. *See* U.S. Department of Commerce
Bureau of Mines, war gas research by, 89, 96-101, 103, 107, 130-31
Butler, Nicholas Murray, 129-30
Butterworth Judson Company, 81, 83, 89
By-product coke ovens, 34-35, 81, 83, 90, 92, 111-12, 252
By-Products Steel Corporation, 253

Calcium carbide, 240, 275-77
Calco Chemical Company, 59, 71, 136, 222, 225-26, 269-70; and TNA, 90, 94, 193
Calloway, Fuller, 145
Canada, 93-94, 129, 276, 283
Caro, Heinrich, 26
Carothers, Wallace H., 265-66
Cartels, 7, 24, 53, 214, 241, 250, 253, 269, 299 (n. 13); concern about, 28, 218, 284
Case School of Applied Science, 60, 106, 135, 272
Cassella (Leopold Cassella & Company, Germany), 23-24, 51, 53, 66
Cassella Color Company, 46-47, 53, 65-66, 69, 169, 243, 247, 251
Catholic University, 109
Censorship Board, 150

Central Dyestuff & Chemical Company, 31, 52, 70
Century Colors Company, 47-48. *See also* Cassella Color Company
Chemical engineering, 18, 34, 112, 120, 130, 134, 238, 260, 268, 271-73, 281-82, 287, 292
Chemical Foundation, Inc.: employees, 177; and fraud accusations, 200-201, 219; income and expenditures, 176, 229; licensing patents, 177-78, 229; lobbying, 177-78, 191, 199, 206; and monopoly accusations, 200-202; organization of, 6, 175-76; publications, 176-77; and Treaty of Versailles, 184-85; war claims, 232-35. See also *USA v. The Chemical Foundation, Inc.*
Chemical warfare. *See* War gases
Chemical Warfare Service, 98-99, 101, 126, 130, 135, 181, 193-94
Chemische Fabrik von Heyden, 27, 166-68
Chemists, 38-39; and lobbying, 120-22; and national defense, 194; nationalism of, 10, 114; and publicity, 120-21; women, 133. *See also* American Chemical Society; Universities
Chemists' Club, 149
"Chemists' war," 3
Cheney, Frank D., 196
China, 20, 27, 160, 187, 339 (n. 36)
Chlorine, 20, 33, 95, 99, 193, 268
Chlorpicrin, 95, 99, 101, 104-7, 110. *See also* War gases
Choate, Joseph H., Jr, 188, 192-93, 200
Christianson, Walter G., 224-25
CIBA (Society of Chemical Industry), 269-70. *See also* Switzerland
Cinchophen, 58, 222, 225-27
Cincinnati Chemical Works, 269
Civil liberties, 9, 156-57, 159, 200
Coal, 18, 19-20, 34, 62, 238, 241, 274, 279
Coal tar, 17, 19-20, 34-35, 50, 238, 274
College of the City of New York, 104
Collier, J. W., 231
Color Lab, 13
Columbia University, 85, 92, 119-20, 129-30, 136-37

Committee on Public Information, 177
Commodity chemicals, 238, 288, 292
La Compagnie Nationale des Matières Colorantes et des Produites Chimiques, 241
Competitive and non-competitive dyes, 207–10
Conant, James B., 59, 109, 130, 262, 308 (n. 28), 319 (n. 42)
Concannon, Charles C., 213
Condensite Company of America, 284
Cone, Lee Holt, 60–61, 135
Coors brewery, 132
Copyright, 132
Cordite, 81
Cornell University, 136–37
Corning glassware, 132
Corwine, William, 199
Cott-a-Lap Company, 136
Council of National Defense, 84
Court of Customs Appeals, 208. *See also* U.S. Customs Service
Covington, J. Harry, 160–61, 166
Creel, George, 177
Crossley, M. L., 226, 253
Crozier, William, 86
Curme, George O., Jr., 277–81
Customs administration. *See* Synthetic Organic Chemicals Manufacturers' Association; Tariffs; U.S. Customs Service

Daugherty, Henry M., 215–16, 349 (n. 80)
Davis, T. H., 50
Dawes Plan, 231
Demming, George, 198
Derick, Clarence G., 56, 132, 135, 253
Dermatological Research Laboratories, 152–55
Deutschland submarine, 1–2, 44, 47–48, 162, 295 (n. 1)
DeVries and Doherty, 209
Dickson, Frederick S., 199
Diphenylamine, 67, 304 (n. 37)
Dohme, A. R. L., 143
Dow, Herbert H., 107, 133, 143–44, 147, 198
Dow Chemical Company: and aniline, 60, 270, 272; and aspirin, 59, 61, 271; chemists, 133, 135–36, 271–72; engineers, 268, 271, 273–74; and ethylene, 273; and indigo, 60–61, 195, 268; mass production strategy, 59–60, 267; and mustard gas, 102, 106–9; Novocaine, 177–78; phenol, 81–82, 272; polystyrene plastic, 273; Swiss relations, 60–61, 269–70; university relations, 272. *See also* Dow, Herbert H.
Drug and Chemical Markets, 70, 73, 109, 119, 211, 257
Duisberg, Carl, 24, 26, 52, 160–61, 241, 243, 330 (n. 15). *See also* Bayer
du Pont, Irénée, 68, 192, 195, 200–201, 243, 256, 261, 263, 332 (n. 40), 341 (n. 48)
du Pont, Pierre, 200, 258, 261, 263
Du Pont Company (E. I. du Pont de Nemours and Company), 174; aliphatic chemicals, 262–66; and antitrust law, 36–37, 80–81; assets, 268; and Badische Company, 68–69; and Chemical Foundation, 175, 200, 222; chemists, 89, 133, 136, 256–58; and Department of Commerce, 214; dyes production, 67, 256–58; explosives production, 36–37, 80–83, 86, 88–91, 93, 95, 110–11; and General Motors, 263–64; and Germany, 243, 256–57, 266; and Great Britain, 68, 266–67; lobbying, 256; and nitrocellulose, 255; research, 68, 258–61, 265–66; and SOCMA, 207; and tetraethyllead, 264, 273; and Treaty of Versailles, 179, 186; university relations, 261–62; and war gases, 101
Dyes, synthetic: classification, 22–23, 63, 299 (n. 11); indanthrene blue GCD, 19–23; and national defense, 2, 78; and overcapacity, 239, 247; quality problems, 69–70; shortages of, 2, 11, 42–45, 52, 55; tariff rates on, 31, 147–48; before World War I, 30–31. *See also* Tariffs
Dynamite, 36–37, 280, 303 (n. 37), 313 (n. 4)

Eastman, George, 133
Eastman Kodak, 14, 133, 136, 224, 273, 283
Eddy, Arthur J., 75, 311–12 (n. 64)
Edgewood Arsenal, 99–102, 106–7, 109–10, 131. *See also* War gases

Edison, Thomas, 58-59, 85, 120
Education. *See* Universities
Ehrlich, Paul, 21-22, 152
E. I. du Pont de Nemours and Company. *See* Du Pont Company
Eli Lilly and Company, 33, 57, 155
Elko Chemical Company, 285
E. Merck (Germany), 14, 27, 50, 52, 170, 246, 287. *See also* Merck & Company
E. R. Squibb and Sons, 33
Espionage Act, 150
Etablissements Kuhlmann, 242
Ethyl alcohol, synthetic, 280-81, 361 (n. 72)
Ethylene, 17, 107, 238, 273-74, 278
Ethylene chlorhydrin, 98
Ethylene dibromide, 273
Ethylene dichloride, 281
Ethylene glycol (antifreeze), 18, 280
Ethylene oxide, 281
Ethyl Gasoline Company, 273
Evans, William Lloyd, 131
Everly M. Davis (manufacturer), 81
Explosions. *See* Health and safety hazards
Explosives, 2, 15, 35-37, 226-27; accidents, 93; and dyes, 67, 76, 111-12, 145-46, 193-94, 255; mobilization, 3, 15, 79-95; and Treaty of Versailles, 181. *See also* Amatol; Ammonium nitrate; Du Pont Company; Picric acid; TNT
Export Department, J. P. Morgan & Company, 83, 86-88, 315 (n. 21)

Farbenfabriken vormals Friedrich Bayer & Company. *See* Bayer
Farbwerke Hoechst, 168. *See also* Metz, Herman A.
Federal Dyestuff & Chemical Corporation, 62-64
Federal Trade Commission, 150, 216; advisory board, 155, 177-78; licenses, 216, 220
Fieldhouse, W. H., 61
Fischer, Emil, 25, 117, 277
Fisher, Carl G., 276
Flexible provision of Tariff of 1922, 210-11
Florasynth Laboratories, 58

Flurscheim, Bernhard J., 90
Ford, Edsel, 249
Ford, Henry, 286
Fordney, Joseph W., 198-99
Foreign valuation, 207
Formaldehyde, 38, 262, 282-83, 285
France, 23, 211; and chemical industry, 34, 214, 241-42, 250; and explosives, 79, 82-83, 87; and Germany, 241-42; and Treaty of Versailles, 179, 181-82; and war gases, 97-98
Frank Hemingway, Inc., 103-4
Frear, James, 200
Free, Edward E., 105
Freedman, Louis, 225-27
Frelinghuysen, Joseph F., 201
Fries, Amos A., 95, 194
Full-line forcing, 141

Garvan, Francis P., 6, 159, 161-63, 200; Chemical Foundation, Inc., 172-78, 199, 229, 249; lobbying, 192, 194-95, 200-202; Office of Alien property, 159-71 passim; Treaty of Versailles, 183-85; *USA v. The Chemical Foundation Inc.*, 218, 228
Gas masks, 95-96
Gas warfare. *See* War gases
Gathings, James, 228-29
Geigy, 269. *See also* Switzerland
Geisenheimer & Company, 269
General Aniline Works, 248-49, 355 (n. 16)
General Bakelite Company. *See* Bakelite Corporation
General Chemical Company, 34-35, 48, 65-66, 125, 252
General Dyestuff Company, 246-47
General Motors, 263-64
General Motors Research Corporation, 264
Geneva Protocol, 10
Georgetown College, 117
German Americans, 1, 7-9, 138, 161, 163, 173
German chemical industry, 2, 7, 11-12, 19-29, 211; and Alien Property Custodian, 6, 156, 158; and chemists, 26, 114, 136-37; and dyes, 23, 286; global agreements, 241-42; and global markets, 27,

241–42, 245; and I.G. Farben formation, 241; new products, 26–27, 240; and reparations, 182–83, 185–88; research laboratories, 12, 25–26; and spying accusations, 158, 163–64; and U.S. market, 210, 240, 245, 249; and U.S. patents, 24–25, 123, 150, 163–64, 171, 173–74, 178, 214–15, 219–28, 233–34; and U.S. trademarks, 32, 50, 150, 153, 165, 174, 176, 221, 245; and *USA v. The Chemical Foundation, Inc.*, 221; and war claims, 230–35. *See also individual firms*

German Special Deposit Account, 231. *See also* War claims

Germany, 24, 30, 158; industries of, 6–7; Nazis, 14, 288–89; shipping lines, 6, 159, 162; and war gases, 95, 110. *See also* German chemical industry

Gewerbe Institute, 26

Gibbs, Harry D., 225–26

Globalization, 4, 7–9, 13, 290; destruction of, 7–8, 13, 288, 290; and German chemical industry, 7–8, 179, 187

Goldman Sachs, 170

Gomberg, Moses, 98, 319 (n. 42)

Goodrich-Lockhart Company, 81, 93

Gottschalk, V. H., 131

Grasselli, Eugene, 34

Grasselli Chemical Company, 34, 248; Bayer purchase, 165, 174; German relations, 245–46

Grasselli Dyestuff Corporation, 246–48

Great Britain, 129, 155; blockade and dye shortages, 3, 44–45, 51–52, 139, 152; chemists, 10, 14, 103, 114, 117; and dyes industry, 11, 24–25, 32, 35, 47, 211, 242; and explosives concentration, 37, 82, 266–67; intelligence, 163; policies, 11, 142, 173, 175, 179, 185, 191–92, 203, 242; propaganda, 116–17; purchases of war matériel by, 15, 79–84, 146; and Treaty of Versailles, 179–82, 185, 188; and war gases, 97–98, 105; wartime collaboration, 15, 68, 78–79, 82, 87–88

Greer, Paul S., 279

Gruber, Rudolf, 287

Gubelmann, Ivan, 59, 257

Guffey, Joseph, 334 (n. 52)

Gunpowder Neck Reservation, 99

Haber, Fritz, 117, 277

Haber-Bosch process, 17, 118, 219–20, 238, 241, 351 (n. 99)

Hale, William J., 271–72, 348 (n. 62), 356 (n. 25)

H. A. Metz & Company, 168. *See also* Metz, Herman A.

H. A. Metz Laboratories, Inc., 153, 168, 178. *See also* Metz, Herman A.

Hamilton, Alice, 133, 317 (n. 33)

Harding, Warren, 215–16, 219, 230, 349 (n. 80)

Hardy, Charles J., 160–62

Harvard University, 59, 89, 109–10, 118, 130–31, 134, 136–37, 205, 224, 261–62, 265, 277

Health and safety hazards, 61, 70–71, 93–94, 102–3, 105–9, 133, 264

Hebden, John C., 62–63, 309 (n. 39)

Heller & Merz, 31, 67

Hemingway, Frank, 72, 74, 103–4

Hercules Powder Company, 80–81, 83, 86, 93, 95

Herrick, Myron T., 276

Herty, Charles Holmes, 125–26, 131–32, 146–48, 155; lobbying, 194, 200; and reparations (Herty Option), 185–87, 339 (n. 32); and SOCMA, 206, 213

Hess, Charles, 160

Hesse, Bernhard C., 25, 118, 121–26, 131, 142, 146, 148–49, 181; American citizenship and education, 118

Heyden (Chemische Fabrik von Heyden, Germany), 27, 166–68

Heyden Chemical Company (New York), 28, 33, 50, 166–70, 174, 246, 271

High pressure technology, 238, 241, 249, 254, 262, 266, 271, 279, 282, 286

Hill, Ebenezer J., 143–44, 149

H. K. Mulford Company, 33

Hodgskin, T. Ellet, 166–68

Hoechst (Meister, Lucius & Brüning, Germany), 152–53, 168, 232

Hofmann, August W., 223

Holdermann, Karl, 220-21, 227, 347 (n. 56)
Holmes, Edward O., Jr., 110-11
Holt, Lee Cone. *See* Cone, Lee Holt
Hoover, Herbert, 178, 212-13
Hughes, Charles Evan, 230-31
H. W. Jayne Chemical Company, 35
Hyperinflation, 208, 239
Hydroquinone, 133

I.G. Farben A.G., 289; and American I.G., 249, 355 (n. 16); dyes, 286; formation of, 232, 241-42; and synthetic rubber, 266-67, 286; U.S. relations, 246-50, 266-67, 293; and war claims, 232-35
Immigration: of chemists, 14, 38; hostility to, 8-9, 156-57; and importers, 7
Imperial Chemical Industries (ICI), 242, 267, 293
Importers: and bribery, 28; customs cases, 209-10; and dyes shortages, 43-53; and lobbying and tariffs, 197-98; as manufacturers, 46-52; mergers of, 246-48; relations with German firms, 7, 27-28, 41-42, 51-52; and reparations, 187, 189; and spying accusations, 6, 158, 162-63, 218; technical expertise, 22, 27, 61; trade association, 72-73. *See also* Bayer Company; Cassella; Kuttroff, Pickhardt & Company; Merck & Company; Metz, Herman A.
Indanthrene blue GCD, 19-23, 54
Indigo, 19-20, 22, 24, 26, 29, 238, 281, 301 (n. 24); and Dow Chemical, 59-61, 144, 147, 253, 268-70; and Du Pont, 67, 253, 256-57; and National Aniline, 253-54; and reparations dyes, 187-88; and tariff rates, 32, 144, 147-49, 192, 195
Industrial research. *See* Research and development
Inorganic chemicals, 39; British industry, 88, 267; and dye-making, 20, 145; in explosives, 88, 92, 145; and German industry, 29, 241; sulphuric acid, 33-34; and U.S. industry, 29, 33-35, 48, 55, 60, 165, 252, 254, 268, 274, 287
Institute for Experimental Therapeutics, 21
Intellectual property. *See* Copyright; Patents; Trademarks
Intermediates: examples of, 20, 27, 34, 50, 55, 59, 145; and national defense, 172, 194; tariffs on, 147, 192, 199, 203. *See also individual firms*
Internment, 163
Isolationism, 4, 8, 204, 212, 235, 288; and autarky, 5, 138; and chemists, 10, 113-14, 118-19
Italy, 79, 182-83, 185, 214, 283

Jacobson, C. Alfred, 119
Japan, 81, 126, 214, 283, 314 (n. 6)
Jeffcott, Robert C., 59, 74, 136, 270, 359 (n. 53)
Johns Hopkins University, 125, 131, 154
Jones, L. C., 66
Journal of Industrial and Engineering Chemistry, 39, 119-20, 126, 143, 185
J. P. Morgan & Company, Export Department, 83, 86-88, 315 (n. 21)

Kaiser Wilhelm Institutes, 119, 277, 323 (n. 9)
Kalle & Company (Germany), 23-24, 59, 169, 343 (n. 12)
Kaye, Benjamin, 72
Kekulé, August, 25
Kemmerick, W. E., 271
Keohan, William, 177, 228
Kilheffer, Elvin H., 59, 257
King, William H., 200-201
Kitchin, Claude, 143-49, 329 (n. 13)
Klipstein, E. C., 148, 194
Kny, Richard, 167
Kodak. *See* Eastman Kodak
Kohler, Elmer P., 130-31, 325 (n. 32)
Kolbe, Herman, 27
Koppers, 34
Kresel, Isidor J., 217, 219, 221-22, 224-27, 349 (n. 80)
Kunz, Eric, 186, 256, 338 (n. 24)
Kuttroff, Adolf, 22, 47, 247, 249. *See also* Kuttroff, Pickhardt & Company
Kuttroff, Pickhardt & Company, 22, 47, 169, 195, 197, 243, 247, 353 (n. 6)

Labor, 93-94, 105, 140
Laboratory equipment, 132

INDEX / 397

LaFollette, Robert M., 200
Lansing, Robert, 150-51, 156
Lehman Brothers, 170
Lenard, Philipp, 117
Levinstein Ltd., 68
Lewis, Winford Lee, 109
Lewisite, 109
Liberty Bonds, 151, 160
Lichtenstein, Alfred F., 269-70
Liebig, Justus, 117
Linde, Carl, 275-76
Linde Air Products, 275-77, 279
Little, Arthur D., 39
Longworth, Nicholas, 191-92, 194
Loomis, Chauncey C., 59
Lowell, A. Lawrence, 130
Ludwig, Berthold A., 46, 53, 243, 247, 353 (n. 6)
Lunge, Georg, 125
Lynch, Frederick, 160-61, 165-66

MacKinney, P. R., 198
Mallinckrodt, Edward, 89, 330 (n. 18)
Mann, Rudolph, 160
Manning, Van H., 96-97, 100
Marden, Orth & Hastings, 72
Market information, 14-15; government role in, 289, 291, 345 (n. 36)
Mass production, 12, 16, 18, 59, 267, 274, 280, 282, 286, 292; and chemical engineering, 34
Matheson, William J., 47, 53, 65-66, 181-82, 251
Matthews, J. Merritt, 147
Maxwell Automobile Company, 251
Mayo Clinic, 154
McKerrow, H. Gardner, 72
McPherson, William, 101
Mees, C. E. K., 14, 133
Meister, Lucius & Brüning (Hoechst, Germany), 152-53, 168, 232
Mellon Institute, 277, 279, 281
"Merchants of death," 292
Merck, George, 14, 50, 128, 169-71, 246, 334 (n. 57)
Merck, George W., 246
Merck & Company (New York), 50-51; and Alien Property Custodian, 169-71; chemists, 134-35, 287, 326 (n. 39); confiscation and sale of, 170-71, 174; German relations, 14, 50-51, 246, 287; and military conscription, 128. See also E. Merck
Methanol (methyl alcohol), 17, 211, 238, 249, 262, 281, 358 (n. 40)
Methylene blue, 58
Metol, 133
Metz, Gustave, 52, 153
Metz, Herman A., 51-52, 146; and ACS dyes report (1914), 121; and Alien Property Custodian, 168-69; as Democrat, 200; and General Dyestuffs Company, 246-48; and imports through blockade, 45; and lobbying and tariffs, 195, 197, 202; and Salvarsan, 152-56; and USA v. The Chemical Foundation, Inc., 220-21. See also H. A. Metz Laboratories, Inc.
Meyer, Eugene I., 56, 66, 141, 251-52, 307 (n. 22), 355 (n. 20)
Military conscription, 61, 77, 114, 128
Military-industrial complex, 291
Miller, William T., 66
MIT (Massachusetts Institute of Technology), 105, 124, 134, 190, 261
Mixed Claims Commission, 231-32, 350 (n. 90), 351 (n. 105)
Mobilization, 79, 83-86
Monopoly. See Antitrust law; Cartels; Shortridge hearings
Monsanto Chemical Company, 42, 133, 170, 285
Morris, Hugh M., 217-18, 223, 225-28, 235, 347 (n. 53), 349 (n. 78)
Moses, George H., 200-201
Mothwurf, Arthur, 48
Murphy, Winnifred, 108-9
Mustard gas, 2-3, 95, 98-100, 102, 104, 106-11, 273, 319 (n. 42), 321 (n. 61), 322 (n. 69), 361 (n. 71). See also War gases

Naphthalene, 19, 34, 55, 67
National Academy of Science, 85
National Aniline & Chemical Company, 125, 174; Allied Dye & Chemical

Corporation merger, 250, 252; and Chemical Foundation, Inc., 175; and Department of Commerce, 214; dyes production, 66–67, 250–51, 254; formation of, 65–66, 251; and indigo, 253; and mustard gas, 102; and SOCMA, 207; and Treaty of Versailles, 179, 181–82
National Association of Cotton Manufacturers, 196
National Carbon Company, 276
National defense, chemical industry as, 144, 193–94
National Exposition of Chemical Industries, 120
Nationalism. *See* Anti-German sentiment
National Research Council, 85, 87, 96–98, 130
Naval Consulting Board, 85
Nazism, 14, 288–89
Neidich Processing Company, 58
NELA war gas plant (Cleveland), 102, 109
Neoprene, 265–66
Neosalvarsan, 153, 155
Nernst, Walther, 125, 277
Newport Chemical Works, 59, 257–58, 357 (n. 33)
Nichols, William H., 34, 48, 65–66, 118, 252. *See also* General Chemical Company
Nieuwland, Julius, 266
Nitrobenzene, 34, 50, 59, 253
Nitrocellulose, 36, 45, 80, 255, 279
Nitro Chemical Company, 81
Nitrogen, 17, 29, 67, 118, 219, 240–41, 250, 252, 254, 277, 317 (n. 32)
Nobel, Alfred, 37
Northwestern University, 109, 277, 286
Norton, Thomas H., 142–43, 328 (nn. 6–7)
Norvell Chemical Corporation, 246
Norwegian Nitrogen Products Company, 211–12
Novocaine, 20, 27, 52, 58, 177–78, 221–22
Noyes, W. A., 135
Nye, Gerald, 292
Nylon, 17, 265–67, 292

Oenslager, George, 263
Office of Alien Property. *See* Alien Property Custodian
Oldbury Electrochemical Company, 103, 320 (n. 52)
Open price society, 43, 74–75, 206, 218–19, 311–12 (n. 64), 312 (n. 66)
Ordnance Department, 79, 101, 105, 135; and explosives, 86–95; and war gasses, 98–99
Ostwald, Wilhelm, 117
Oxyacetylene, 38, 275–76, 279
Oxygen, 22, 104, 275–77

Palmer, A. Mitchell, 5–7, 139, 156–61, 164–66, 171, 174, 176–77, 218–19, 228, 336 (n. 9)
Paris peace conference. *See* Treaty of Versailles
Parke-Davis and Company, 33, 57
Parker, Edwin B., 232
Parmelee, James, 276
Patents, 32–33; confiscation of, 174–75; for dyes, 123, 223; and explosives agreement, 37; medical, 153–55; pioneer, 227; strategic management of, 26, 223; U.S. patent law, 173, 224; working clause, 123–24, 324 (n. 220). See also *USA v. The Chemical Foundation, Inc.*
Patterson, Albert M., 186
Peace, 174–75. *See also* Treaty of Versailles
Pennock, Stanley B., 59, 308 (n. 28)
Penrose, Boies, 148
People's Gas, Light, and Coke (Chicago), 275
Perfume, 58, 206, 354 (n. 10)
Petroleum, 18, 34, 40, 85, 93, 238, 274, 282, 310 (n. 42), 317 (n. 32)
Pharmaceuticals industry: American, 33, 38, 49–50, 57–58, 246–48; German, 11, 20–22, 26–27, 57, 244–45, 248; and national defense, 193–94; and reparations, 190; and shortages, 44–45; and tariff rates, 21, 147–48. *See also individual drugs and manufacturers*
Phenacetin, 27, 33, 300 (n. 19)
Phenol, 34, 50, 82, 111, 211, 285, 317 (n. 33), 362 (n. 87); and Bakelite, 38, 147, 271, 282–83, 285; and Dow, 61, 71, 81–82, 271–72, 313 (n. 6), 359 (n. 56); and Du Pont, 67; and Edison, 50, 58–59, 120;

and Merck & Co., 50; and Monsanto, 57
Phenolic resin, 284. *See also* Bakelite Corporation
Phenolphthalein, 33, 57, 147
"Phenol plot," 163, 166
Philadelphia School of Textiles, 190
Phosgene, 95, 99, 101–4, 110. *See also* War gases
Photochemicals, 2, 50, 133, 354 (n. 10)
Phthalic anhydride, 57, 254
Physical chemistry, 133
Pickhardt, Carl, 47. *See also* Kuttroff, Pickhardt & Company
Pickhardt, Paul, 22, 197. *See also* Kuttroff, Pickhardt & Company
Picric acid (trinitrophenol), 3, 80–82, 87, 89, 92–94, 105. *See also* Explosives
Polk, Frank L., 219
Pollution, 94, 106, 109
Polymers, 263, 265–66
Pope Toledo Company, 251
Postmaster General, 150
Poucher, Morris R., 68–69, 73–75, 218, 310 (n. 53)
Powers-Weightman-Rosengarten Company, 246
Pratt Institute, 121
Prest-O-Lite, 275–78, 281
Prestone antifreeze, 18, 280
Price fixing, 75, 201–2
Princeton University, 190, 262
Procaine. *See* Novocaine
Progressive Era, 71, 84, 107, 128, 140, 200, 205, 236
Prohibition, 211, 280–81, 361 (n. 72); in Colorado, 132
Promotional state, 4, 289
Propaganda, 115–17, 127, 162–64, 177, 201, 218
Public Health Service, 156, 178, 225
Pyroacemic acid, 226
Pyruvic acid, 226

Queeny, John F., 57, 71, 147. *See also* Monsanto Chemical Company

Rahway Coal Tar Products, 50, 134
Rainey, Henry T., 146, 231
Ramsay, William, 117
Reagent chemicals, 131–33, 136
Rector Chemical Company, 178
Redfield, William C., 140–43, 146, 150–51, 155, 182
Redman, L. V., 284
Redmanol Chemical Products Company, 284
Red Scare (1919-1920), 156–57
Reese, Charles, 37, 261, 304 (n. 37)
Remick, James W., 350–51 (n. 91). *See also* War claims
Remsen, Ira, 39, 125, 131
Reparations, 179, 182–90, 230; Herty option for, 185–87; and pharmaceuticals, 190. *See also* Treaty of Versailles
Reparations Commission, 185–86
Research and development (R&D), 12–13, 25–26, 173, 300 (n. 16)
Resins, synthetic, 37–38, 58–59, 148, 262, 282–86. *See also* Bakelite Corporation
Richards, Theodore, 118
Rockefeller, John D., 141, 327 (n. 4)
Rockefeller Institute, 152, 323 (n. 9)
Roeber, Eugene F., 115–16
Roessler & Hasslacher, 285
Rogers, Allen, 121
Roos, Lester L., 107–9, 321 (n. 63)
Rose, Robert, 261
Rubber, 40, 134; accelerator, 34, 67, 263, 266, 362 (n. 87); synthetic, 17, 265–66, 280, 292
Russia, 27, 79, 81, 314 (n. 6)
Rutgerswerke, 283
Ryan, Allen A., 168

Saccharin, 33, 147
Salvarsan, 19, 21–23, 152–56; and arsphenamine, 9; development, 21–22; hazards, 22, 155–56, 221; manufacturing of, in United States, 52; patents, 152–56, 220–21, 223–25, 233; tariff, 22. *See also* Hoechst; Metz, Herman A.
Sandoz, 269. *See also* Switzerland
Scandinavia, 37, 44, 182, 211–12
Schamberg, Jay Frank, 152–55, 168
Schoellkopf, C. P. Hugo, 31

Schoellkopf, Jacob F., Jr., 31, 143-45, 148
Schoellkopf Aniline & Chemical Works, 42, 55-56, 70, 247, 302 (n. 28); lobbying, 144-45; mergers, 65-67, 253-54; research, 135, 253, 310 (n. 46); before World War I, 28, 31
School of Explosives Manufacture, 92
Schweitzer, Hugo, 49, 53, 162-63
Section 10(f) lawsuits. *See* Trading with the Enemy Act: Section 10(f) lawsuits
Seebohm, Herman C. A., 52, 159-63. *See also* Bayer Company
Semet-Solvay Company, 34-35, 61, 65-66, 81, 83, 92, 101, 252, 271, 313 (n. 5), 356 (n. 22)
Sharpe and Dohme, 143
Shaw, Robert, 47, 65-66
Sherwin Williams, 58, 311 (n. 61)
Shortridge, Samuel M., 200, 202
Shortridge hearings, 200-202, 215
Sibert, William, 99-100, 194
Sill, Theodore W., 181
Simon, George, 166-68, 246
Smith, Albert W., 60, 71, 106-7, 135, 272, 313 (n. 6)
Smokeless powder. *See* Nitrocellulose
Society of Chemical Industry, 177
Solvay & Cie., 33
Solvay Process Company, 33, 252
South America, 244-45
Squire, Andrew, 276
Stainless steel, 175, 229
Stamm, Christian, 48-49
Standard Aniline Products, 57
Standardization, 71-72, 75-76, 208-9
Standard Oil, 24, 249, 264
Staudinger, Herman, 265
Sterling Products, Inc.: Bayer purchase of, 165, 174; German relations, 244-45, 248-49
Stettinius, Edward R., Sr., 83, 86
Stieglitz, Julius, 155, 223, 348 (n. 66)
Stine, Charles M. A., 260-61, 265
Stone, I. F., 66, 70, 144, 147
Stone, Lauson, 131
Student Army Training Corps (SATC), 129
Sutherland, Leslie, 105

Swarthmore College, 226
Switzerland, 23, 44, 125, 182, 214, 250, 256, 279; immigrant chemists, 38, 57, 59, 257; importers, 61, 73, 194-95, 198, 269-70, 311 (n. 62), 359 (n. 53); Swiss chemical firms, 60-61, 142, 269, 355 (n. 18)
Sykes, Walter F., 198
Synthetic Organic Chemicals Manufacturers' Association (SOCMA), 206-14, 237, 246; customs administration, 207, 209-10; Department of Commerce, 212-13; organization of, 206-7
Synthetic Patents Company, 160, 163-64. *See also* Bayer Company
Syphilis, 21, 23, 152, 154, 221

Taggesell, R. C., 181
Tariffs, 2, 21-22, 31, 138-39; and ACS proposal (1914), 121-22, 148; and American selling price, 207, 209; and American valuation, 199, 202, 207-8, 210-11, 293; definitions, 147-48, 207-8; Democratic party positions on, 138-40, 144, 148; and dumping, 141; flexible provision for, 205, 207, 210-12; and foreign valuation, 207; and infant industry, 4, 146-47, 236, 291; and licensing proposal, 192-93, 196-99; and reparations, 179; Republican party positions on, 140, 143, 148, 198-99; and U.S. valuation, 208
—specific legislation: Emergency Tariff Act (1921), 192, 198; Fordney-McCumber tariff (1922), 199, 202-3, 205, 210; Hill bill (1915-16), 143-48; Longworth bill (1919), 191-98; Revenue Act of 1916, 147-49, 190-92; Smoot-Hawley tariff (1930), 210, 292; Tariff Act of 1883, 31, 121; Underwood-Simmons tariff (1913), 140-41
Taussig, Frank W., 205-6, 210, 342 (n. 1)
Teagle, Walter, 249
Technical exchanges, 30, 142, 291; with British firms, 37, 88, 267; and Du Pont, 37, 68-69, 82, 243, 256-57, 266-67; with French firms, 82; with German firms, 37, 52, 167, 241, 243, 245-46, 248-49, 267,

287; and military, 15, 82, 86–92, 97–99, 111; and National Aniline and Chemical Company, 243, 252; with Swiss firms, 60–61, 142, 269–70
Tetraethyllead, 264, 273, 281
Tetryl, 80, 313 (n. 3)
Textile Alliance, 186–90, 201
Textile Foundation, 190
Textile industry (U.S.): and dyes shortages, 44–46, 70; and German dyes, 27–28; and market information, 124, 211–12; and tariffs on dyes, 30–31, 121–22, 126, 144–46, 192–93, 196–98, 288; and reparations dyes, 180, 182–90
Tillman, Benjamin, 158
TNA (tetranitroaniline), 90, 94, 193, 316 (n. 23)
TNT (trinitrotoluene) 3, 36–37, 80–82, 87–89, 91–93. *See also* Du Pont Company; Explosives
Toluene, 34, 59, 67, 82, 85, 92–94, 111, 145, 193–94, 313 (n. 4), 317 (nn. 32–33)
Townes, Leigh R., 117
Trade associations, 43, 71–76, 178, 206–7, 210, 254. *See also* American Dyes Institute; Synthetic Organic Chemicals Manufacturers' Association
Trademarks, 32, 50, 150, 153, 165, 174, 176, 221, 245, 283
Trading with the Enemy Act (TWEA), 5, 139, 149–52, 156, 166; amendments, 157–58, 164–65, 166; and patents, 155; provisions, 149–53; Section 10(f), 152, 155, 216, 234–35; and World War II, 293. *See also* Alien Property Custodian; Garvan, Francis P.; Palmer, A. Mitchell; *USA v. The Chemical Foundation, Inc.*
Treaty of Versailles, 9–10, 178–90, 230–31; chemicals provisions, 185, 187–88
Trench Warfare Section, 98–99

Underwood, Oscar, 148
Union Carbide & Chemical Corporation: acetylene, 274–79, 281; and "American" chemicals, 274; engineers, 281–82; and ethyl alcohol, 281; and ethylene, 274, 278, 280–81; and Mellon Institute, 277–78; merger, 274; and Prestone, 280; raw materials, 274, 279
Union Carbide Company (1898), 278
Union Dye and Chemical Company (Federal Dyestuff & Chemical Corporation), 62–64
United Piece Dye Works, 68
USA v. The Chemical Foundation, Inc., 214–30, 234; charges, 217; and cinchophen experiment, 225–27; expert witnesses, 220–27; lawyers, 217; verdict and appeals, 228
U.S. Army. *See* U.S. Department of War
U.S. Customs Service, 208; customs administration, 21, 199, 202, 205–12; market information, 14, 142–43, 213, 291, 328 (n. 8)
U.S. Department of Agriculture. *See* Bureau of Mines, war gas research by; Color Lab
U.S. Department of Commerce, 126, 139–43, 150, 204; Bureau of Foreign and Domestic Commerce, 212–13; census of dyestuffs (1916), 142–43; Chemical Advisory Committee, 212; Chemical Division, 212; foreign attachés, 213; market information studies, 141–43, 212–14; and Treaty of Versailles, 179–84. *See also* Hoover, Herbert; Redfield, William C.
U.S. Department of Justice, 150, 161, 215; Du Pont antitrust case, 36–37, 80–81. *See also USA v. The Chemical Foundation, Inc.*
U.S. Department of Navy, 84–85, 233
U.S. Department of State, 139, 150; technical advisors, 180–82, 185–87; Treaty of Versailles, 179–90. *See also* War Trade Board; Treaty of Versailles
U.S. Department of Treasury, 142, 150, 165, 189–90. *See also* U.S. Customs Service
U.S. Department of War, 15, 84–87, 98, 223. *See also* Ordnance Department
U.S. Dyestuff and Chemical Importers' Association, 73, 198
U.S. Navy. *See* U.S. Department of Navy
U.S. Tariff Commission, 204; customs

valuation, 209; flexible provision, 210-11; market information, 206, 238. *See also* U.S. Customs Service
U.S. valuation, 207
Universities, 38-39; academic consultants, 134-36; German, 16; and mobilization, 114-15, 126-33; and supply of chemists, 123, 127-37
University of Chicago, 39, 118, 122, 136-37, 155, 223, 277
University of Georgia, 125
University of Illinois, 56, 58, 129, 132-33, 135-37, 222, 261-62
University of Michigan, 60, 98, 108, 118, 134, 136, 272
University of Missouri, 131
University of North Carolina, 125

Van Dyk & Company, 58
Van Winckel, William, 61
Vat dyes, 17, 22-23, 183, 195, 224, 242, 253-54, 257-58, 299 (n. 8)
Veronal, 27, 246, 300 (n. 19)
Volwiler, Ernest, 58, 135, 222

Walker, George, 154
Walker, William, 100-101, 107, 110, 124
Wallace, E. S., 105
Warburg, Paul, 249
War claims, 205, 230-35; Mixed Claims Commission, 231-32; Section 10(f) lawsuits, 234-35
War Claims Act of 1928, 231-32
War Claims Arbiter, 232
War gases, 95-112, 181, 194; accidents involving, 107-9; and mobilization, 96-112, 361 (n. 71). *See also* Chemical Warfare Service; Mustard gas
War Industries Board, 84-86, 130, 183
Warren, Charles, 151, 155-56
War Trade Board, 130, 150; licensing, 196-97, 199; and reparations, 182-86
Watkins, Willard, 56
W. Beckers Aniline & Chemical Works. *See* Beckers, William
Weber, Orlando F., 207, 251-54, 355-56 (n. 20), 356 (n. 22)

Weidlein, E. R., 281, 361 (n. 70)
Weiss, William, 244-45, 247-49
Werner, Alfred, 125
Westvaco, 279
Whitaker, Milton C., 120, 130
Whitmore, Frank, 286
Widener, Peter A. B., 152
Wien, Wilhelm, 117
Wigglesworth, Henry, 182, 184, 253, 337 (n. 18)
Wilhelm II (German kaiser), 1, 24
Williams & Crowell Color Company, 160-63. *See also* Bayer Company
Willstätter, Richard, 117
Wilmot, Harrison, 209
Wilson, Rufus, 196
Wilson, Woodrow: mobilization, 79, 83-85, 101, 128; and tariffs, 140-41, 143; and Trading with the Enemy Act, 150, 156-57, 161, 174, 177; and Treaty of Versailles, 182; and *USA v. The Chemical Foundation, Inc.*, 215, 217-19
Winslow Act, 230. *See also* War claims
Winthrop Chemical Company, 244, 247-49. *See also* Sterling Products, Inc.
Witt, Otto, 25, 125
Wöhler, Friedrich, 117
Women: as chemists, 133, 326 (n. 37), 333 (n. 40); organizations of, and chemicals industry, 199
Woodford, Lois, 206
Worden, Edward C., 62, 64
World War I, 1-3; atrocities, 115-16; beginning of, 115; impact, 287-90; as industrial war, 3, 18
World War II, 192, 234, 248, 266-67, 289, 291-93, 308 (n. 28), 350 (n. 90)

Xenophobia. *See* Anti-German sentiment
Xylene, 34

Yale University, 136-37, 225, 316 (n. 27), 333 (n. 40)
Yerkes, Leonard, 184, 338 (n. 24)
Ypres, Belgium, 95

Zinsser & Company, 102, 111, 130, 135